Vacuolar Compartments

Annual Plant Reviews

A series for researchers and postgraduates in the plant sciences. Each volume in this annual series will focus on a theme of topical importance and emphasis will be placed on rapid publication.

Titles in the series:

1. Arabidopsis
Edited by M. Anderson and J. Roberts.

2. Biochemistry of Plant Secondary Metabolism
Edited by M. Wink.

3. Functions of Plant Secondary Metabolites and their Exploitation in Biotechnology
Edited by M. Wink.

4. Molecular Plant Pathology
Edited by M. Dickinson and J. Beynon.

5. Vacuolar Compartments
Edited by D.G. Robinson and J.C. Rogers.

Vacuolar Compartments

Edited by
DAVID G. ROBINSON
Zellenlehre
Universität Heidelberg
Germany

and

JOHN C. ROGERS
Institute of Biological Chemistry
Washington State University
USA

Sheffield
Academic Press

CRC Press

First published 2000

Published by
Sheffield Academic Press Ltd
Mansion House, 19 Kingfield Road
Sheffield S11 9AS, England

ISBN 1-84127-112-8
ISSN 1460-1494

Published in the U.S.A. and Canada (only) by
CRC Press LLC
2000 Corporate Blvd., N.W.
Boca Raton, FL 33431, U.S.A.
Orders from the U.S.A. and Canada (only) to CRC Press LLC

U.S.A. and Canada only:
ISBN 0-8493-0500-4
ISSN 1097-7570

Printed on acid-free paper in Great Britain by
Bookcraft Ltd, Midsomer Norton, Bath

British Library Cataloguing-in-Publication Data:
A catalogue record for this book is available from the British Library

Library of Congress Cataloging-in-Publication Data:
Vacuolar compartments / edited by David G. Robinson, John C. Rogers.
p. cm. -- (Annual plant reviews)
Includes bibliographical references (p.).
ISBN 0-8493-0500-4 (alk. paper)
1. Plant vacuoles. 2. Plant cell compartmentation. I. Robinson, David G. II. Rogers, John C. III. Series.
QK725 .V33 2000
571.6'552--dc21

00-029764

Preface

Plant cell biology is in flower, and vacuoles provide the color, together with some of the mystery.

Previous concepts of 'the' plant vacuole were symbolized by an organelle that is shown in most textbooks as a large hole against which a thin rim of cytoplasm and other organelles are compressed by the cell wall. It has long been known that multiple, different functions are associated with 'the' vacuole, including some—such as that of an acidic compartment with active hydrolytic enzymes and that of a compartment where storage proteins are protected from degradation—that seem to be mutually exclusive. Only during the past five years has it become clear that this traditional view of a vacuole as one multifunctional organelle is incorrect. We now know that most plant cells contain basically two types of vacuoles that are generally classed as lytic and protein storage vacuoles. These are served by separate vesicular pathways, which defines them as distinct organelles. Furthermore, it appears that the functions of each type can be modified such that several different vacuoles, each with a somewhat different function, can exist in one cell. This complexity has led to the concept of a plant cell 'vacuolar system'. The present challenge is to understand how the separate organelles in the system are generated and maintained by the cell, and to explain at a biochemical level the functions of each vacuole type and the mechanisms by which they are achieved.

To our knowledge, this is the first volume to provide this perspective. It is directed at academic and industrial researchers in plant cell biology, biochemistry, physiology, molecular biology and developmental biology.

The majority of the chapters in the volume focus on individual functions of plant vacuoles and the tonoplast machinery that supports them. Additional chapters present up-to-date information on the identity of different vacuole types and the traffic of proteins to them. Two additional chapters provide a basis for comparing the plant system to vacuoles in other organisms. The volume begins with a review of the yeast vacuole and of the traffic of proteins to it—particularly important, as the yeast vacuole is a paradigm for the plant cell lytic vacuole. A second comparison is provided by a review of contractile vacuoles in protists, both plant and animal.

On a practical level, it is likely that many pending and future applications in plant biotechnology will encounter the scientific challenges posed by the plant vacuolar system. This is because the use of plant cells to produce many different types of complex metabolic product and protein will require those products to be accumulated stably in vacuoles. Success would then depend on providing not only the proper enzymatic machinery to make the desired product, but also a vacuole with the proper internal environment and the machinery to deliver the product to the vacuole. We hope that the information contained in this volume will serve as a guide towards this ultimate goal.

We wish to thank Sally W. Rogers of the Institute of Biological Chemistry, Washington State University for her expert editorial assistance in the preparation of this volume.

John C. Rogers
David G. Robinson

Contributors

Dr Bronwyn J. Barkla	Departmento de Biología Molecular de Plantas, Instituto de Biotecnología, Universidad Nacional Autónoma de México, A.P. 510-3, Colonia Miraval, Cuernavaca, Morelos, 62250 México
Dr Markus Geisler	Laboratoire de Physiologie Végétale, Institut de Botanique, Université de Neuchâtel, Rue Emile Argand 13, 2007 Neuchâtel, Switzerland
Dr Christine Gietl	Lehrstuhl fur Botanik, Biologikum–Weihenstephan, Technische Universität München, Am Hochanger 4, D-85354 Freising, Germany
Professor Ikuko Hara-Nishimura	Department of Botany, Graduate School of Science, Kyoto University, Sakyo-ku, Kyoto 606-8502, Japan
Dr Stefan Hillmer	Abteilung Strukturelle Zellphysiologie, Albrecht-von-Haller Institut für Pflanzenwissenschaften, Universität Göttingen, Untere Karspüle 2, D-37073 Göttingen, Germany
Dr Giselbert Hinz	Abteilung Strukturelle Zellphysiologie, Albrecht-von-Haller Institut für Pflanzenwissenschaften, Universität Göttingen, Untere Karspüle 2, D-37073 Göttingen, Germany
Dr Alejandro Iglesias	Friedrich Miescher Institute, a branch of the Novartis Research Foundation, PO Box 2543, CH-4002, Basel, Switzerland
Dr Markus Klein	Laboratoire de Physiologie Végétale, Institut de Botanique, Université de Neuchâtel, Rue Emile Argand 13, 2007 Neuchâtel, Switzerland

Professor Daniel Klionsky Department of Biology, University of Michigan, Ann Arbor, Michigan 48109, USA

Dr Olga A. Koroleva School of Biological Sciences, University of Wales, Bangor, Gwynedd, LL57 2UW, UK

Professor Roger Leigh Department of Plant Sciences, University of Cambridge, Downing Street, Cambridge, CB2 3EA, UK

Professor Peter Luykx Department of Biology, University of Miami, Coral Gables, Florida 33124-0421, USA

Professor Masayoshi Maeshima Graduate School of Bioagricultural Sciences, Nagoya University, Chikusa-ku, Nagoya 464-8601, Japan

Professor Enrico Martinoia Laboratoire de Physiologie Végétale, Institut de Botanique, Université de Neuchâtel, Rue Emile Argand 13, 2007 Neuchâtel, Switzerland

Professor Fred Meins Jr. Friedrich Miescher Institute, a branch of the Novartis Research Foundation, PO Box 2543, CH-4002, Basel, Switzerland

Dr Yuji Moriyasu School of Food and Nutritional Sciences, University of Shizuoka, 52-1 Yada, Shizuoka 422-8526, Japan

Professor Jean-Marc Neuhaus Laboratoire de Biochimie, Institut de Botanique, Université de Neuchâtel, Rue Emile Argand 9, 2007 Neuchâtel, Switzerland

Dr Takeshi Noda Department of Cell Biology, National Institute for Basic Biology, Okazaki 444-8585, Japan

Professor Yoshinori Ohsumi Department of Cell Biology, National Institute for Basic Biology, Okazaki 444-8585, Japan

Dr Omar Pantoja Departmento de Biología Molecular de Plantas, Instituto de Biotecnología, Universidad Nacional Autónoma de México, A.P. 510-3, Colonia Miraval, Cuernavaca, Morelos, 62250 México

Dr Rafael Ratajczak
Darmstadt University of Technology, Institute of Botany, Schnittspahnstraße 3-5, D-64287 Darmstadt, Germany

Professor Philip A. Rea
Plant Science Institute, Department of Biology, University of Pennsylvania, Philadelphia, PA 19104, USA

Professor David G. Robinson
Zellenlehre, Universität Heidelberg, Im Neuerheimer Feld 230, D-69120 Heidelberg, Germany

Professor John Rogers
Institute of Biological Chemistry, Washington State University, Pullman, Washington 99164-6340, USA

Dr Rocío Sánchez-Fernández
Plant Science Institute, Department of Biology, University of Pennsylvania, Philadelphia, PA 19104, USA

Dr Markus Schmid
Lehrstuhl fur Botanik, Biologikum–Weihenstephan, Technische Universität München, Am Hochanger 4, D-85354 Freising, Germany

Dr David J. Simpson
Department of Physiology, Carlsberg Laboratory, Gamle Carlsberg Vej 10, DK-2500 Copenhagen, Valby, Denmark

Professor A. Deri Tomos
School of Biological Sciences, University of Wales, Bangor, Gwynedd, LL57 2UW, UK

Dr Rosario Vera-Estrella
Departmento de Biología Molecular de Plantas, Instituto de Biotecnología, Universidad Nacional Autónoma de México, A.P. 510-3, Colonia Miraval, Cuernavaca, Morelos, 62250 México

Professor Thea Wilkins
Department of Agronomy and Range Science, University of California, Davis, CA 95616-8515, USA

Contents

1 The yeast vacuole: a paradigm for plant cell biologists?

Takeshi Noda, Yoshinori Ohsumi and Daniel J. Klionsky

1.1 Introduction

The budding yeast, *Saccharomyces cerevisiae*, has provided a simple system for the study of cell biology. Due to the existence of stable haploid life cycles and the subsequent ease of genetic techniques, yeast has proved to be invaluable in furthering our understanding of various intracellular processes. For example, tremendous advances have been made in the areas of protein targeting and organelle biogenesis by carrying out genetic and molecular genetic studies in yeast. This chapter focuses on the vacuole, an organelle that is analogous to the lysosome of animal and plant cells, to the storage vacuole of plants and even to the contractile vacuole of the slime mold. In recent years, our understanding of vacuole biogenesis and transport to this organelle has increased enormously, and many of the processes can be discussed at a molecular level. Much of current biology has a basis in information learned from the analysis of model systems, in particular bacteria and viruses. Similarly, it is likely that studies performed in yeast will prove to be applicable to multicellular organisms, including plants.

1.2 Vacuolar functions

1.2.1 The vacuole as a storage compartment

A single cell of the yeast, *Saccharomyces cerevisiae*, (4–6 μm in diameter) has one to a few vacuole(s) (1–4 μm in diameter). The lumen is quite aqueous compared to the cytosol, making it is possible to observe the vacuole using a phase contrast microscope. A yeast mutant without a morphologically detectable vacuole can still grow in suitable conditions. However, it is assumed that cells lacking a vacuole cannot live in naturally occurring severe conditions. The membrane of the vacuole is equipped with a vacuolar type proton ATPase, so that the lumen is kept at acidic pH (Stevens and Forgac, 1997). The vacuole functions as a storage compartment for such molecules as polyphosphate, metal ions and amino acids (reviewed in Klionsky *et al.*, 1990). In this capacity, the vacuole can function as a homeostatic organelle, maintaining the proper cytosolic pH and serving as a reserve for critical nutrients. These functions are required for response to environmental changes.

1.2.2 The vacuole as a degradative compartment

In addition to these storage roles, the vacuole functions as a degradative compartment. Wild-type yeast cells living in a natural environment are diploid. Upon encountering conditions of nutrient depletion, diploid cells sporulate, conferring a high level of resistance to severe conditions. However, mutants defective in autophagic vacuolar degradation (see Section 1.3.4) are not able to sporulate, leading to cell death (Tsukada and Ohsumi, 1993).

The vacuole is characterized as a degradative compartment, primarily because many kinds of hydrolases are localized within it (Klionsky *et al.*, 1990; Van Den Hazel *et al.*, 1996). Besides proteases, the vacuole is equipped with alkaline phosphatase (ALP), exopolyphosphatase, acid trehalase, α-mannosidase and RNase (Jones *et al.*, 1982; Klionsky and Emr, 1989; Yoshihisa and Anraku, 1990; Alizadeh and Klionsky, 1996; Andreeva *et al.*, 1998). So far, the following have been defined as vacuolar resident proteases: proteinase A (PrA), proteinase B (PrB), carboxypeptidase Y (CPY), aminopeptidase Y (APY), carboxypeptidase S (CPS), dipeptidyl aminopeptidase B (DPAP B) and aminopeptidase I (API) (for review, see Van Den Hazel *et al.*, 1996). All of these proteins, with the exception of API, transit to the vacuole through a portion of the secretory pathway as inactive precursors (see Section 1.3). Upon delivery to the vacuole, they are proteolytically processed, with the exception of DPAP B, and the processing event is required for exhibition of their protease activity. Processing is catalyzed by PrA and/or PrB, including the autocatalysis of both PrA and PrB. Besides this processing role, PrA and PrB have a broad substrate specificity, so that they may carry out the initial cleavage of degradation substrates, providing greater access for the exopeptidases. In addition, PrA and PrB have a related fundamental role in protein degradation, in that they are required for the breakdown of subvacuolar vesicles (see Section 1.3.4).

1.3 Complex routes to the vacuole

The vacuole is an organelle bounded by a single membrane bilayer. How is this organelle formed and how are the vacuolar hydrolases and degradation substrates transported to the vacuole? Currently, eight routes have been described for protein and/or organelle delivery to the vacuole (Figure 1.1); (Bryant and Stevens, 1998; Scott and Klionsky, 1998; Klionsky and Ohsumi, 1999; Kim and Klionsky, 2000). In addition, the vacuole itself moves dynamically, and in some cases changes its shape to carry out its role (Jones *et al.*, 1993; Sakai *et al.*, 1998).

1.3.1 Carboxypeptidase Y (CPY) pathway

CPY is a soluble vacuolar resident protease. CPY synthesized in the cytoplasm is translocated into the endoplasmic reticulum (ER) and then transits to the

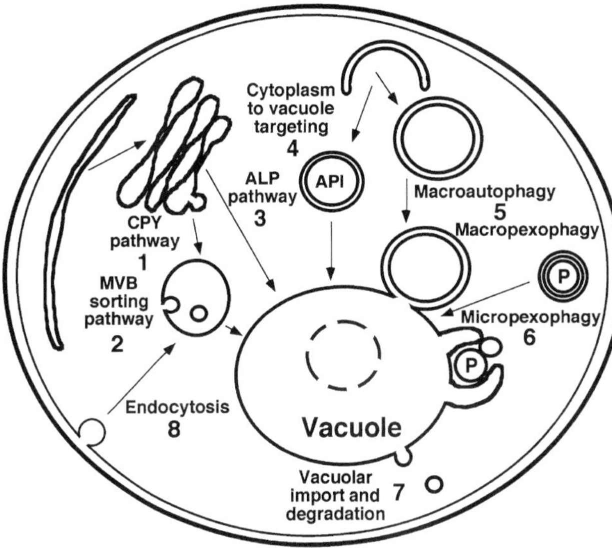

Figure 1.1 Multiple pathways are used for the delivery of resident hydrolases and substrates to the yeast vacuole (see text for details). Most resident hydrolases transit through a portion of the secretory pathway. (1) The carboxypeptidase Y (CPY) pathway comprises the 'classical' route from the endoplasmic reticulum and Golgi complex to the endosome/prevacuolar compartment and ultimately the vacuole. (2) In the multivesicular body (MVB) sorting pathway, proteins are invaginated into vesicles within the endosome lumen prior to vacuolar delivery. (3) The alkaline phosphatase (ALP) pathway delivers proteins to the vacuole from the *trans*-Golgi network and bypasses the endosome. (4) The cytoplasm-to-vacuole targeting pathway is a biosynthetic mechanism that overlaps with the degradative process of macroautophagy (5). In both processes, double-membrane vesicles sequester proteins within the cytosol. Subsequent fusion with the vacuole results in delivery to the lumen. Macropexophagy is a specific process used for peroxisome breakdown and is morphologically similar to macroautophagy. In contrast, micropexophagy (6) involves engulfment of peroxisomes directly at the vacuole surface. The vacuolar import and degradation pathway (7) is a vesicle-mediated process used for the degradation of specific cytosolic proteins under conditions where they are no longer needed. Plasma membrane and extracellular proteins are delivered to the vacuole by endocytosis (8). Abbreviations: API, aminopeptidase I; P, peroxisome.

Golgi complex through the early secretory pathway. At the *trans*-Golgi network (TGN), CPY is diverted to the vacuole. The molecular framework of this route has been revealed due to the isolation of mutants defective in localization (*vps*, vacuolar protein sorting) or activation (*pep*, peptidase) of CPY (reviewed in Bryant and Stevens, 1998; Burd *et al.*, 1998; Conibear and Stevens, 1998; Klionsky *et al.*, 1990).

At the late-Golgi, CPY is captured by its membrane protein receptor, Vps10p, and is then recruited to a transport vesicle. This post-Golgi vesicle targets to a prevacuolar compartment (PVC)/endosome. The detailed process of budding of the transport vesicle is not fully understood; however, some of the components required have been identified. Firstly, Vps1p is a GTPase homologous to dynamin of mammalian cells. Dynamin proteins form into a ring around the site where an endocytic vesicle is formed. Recruitment and activation of a lysophosphatidic acid acyl transferase (endophilin I) by dynamin allows alteration of the lipid composition, facilitating the change of membrane curvature required for generation of the vesicle (Schmidt *et al.*, 1999). In mammalian cells, budding of the endocytic vesicle is accompanied by binding of clathrin triskelions to form a protein coat. In yeast, clathrin functions together with Vps1p, as demonstrated by the similar phenotypes of *vps1* and *chc1* (clathrin heavy chain) mutants. Secondly, Vps34p is a phosphatidylinositol 3-kinase (PI3-kinase) which makes a functional complex with a protein kinase, Vps15p. Mammalian PI3-kinase is required for the formation of constitutive transport vesicles from the Golgi (Jones and Howell, 1997). Such analogies allow us to predict that these proteins function in generating the vesicle. In yeast, phosphatidylinositol (3)-phosphate (PI(3)P) plays pivotal roles in additional aspects of targeting (see Section 1.3.2).

Once the transport vesicle is formed, it docks and fuses with the PVC. The docking and fusion events are catalyzed by components that are members of protein families that function in several intracellular membrane fusion processes. These include: Rab GTPase (Vps21p); GDP/GTP exchange factor (Vps9p); t-SNARE (Pep12p); v-SNARE (Vti1p); Sec1 family protein (Vps45p); and NSF (Sec18p) (Horazdovsky *et al.*, 1995; Hama *et al.*, 1999). Detailed characterizations of these general fusion machinery components have been reviewed previously (for example, see Jahn and Sudhof, 1999). When the genes encoding these proteins are mutated, transport vesicles (~60 nm in diameter) accumulate. Vac1/Pep7p has a FYVE domain, which is sufficient to bind to the head group of PI(3)P (Burd and Emr, 1998). Vac1p also binds to the activated Rab (Vps21p) and the Sec1 family protein (Vps45p), suggesting that it coordinates these fusion machinery proteins as a downstream effector of Rab (Peterson *et al.*, 1999; Tall *et al.*, 1999). Following fusion of the vesicle with the PVC, the CPY receptor (Vps10p) releases the cargo (CPY) and returns to the Golgi via a retrograde transport vesicle. The retromer protein complex (Vps35p, Vps29p, Vps26p, Vps5p and Vps17p) is proposed to be a coat for this transport vesicle (Seaman *et al.*, 1998).

Mutation in any of a series of the class E *VPS* genes results in an exaggerated morphology of the PVC, apparently composed of multilayered, curved cisternae. Class E *VPS* gene products control not only the morphology of the PVC but also its function. Exit from the PVC to the Golgi complex and to the vacuole is severely impaired in the class E mutants. Vps4p, one of the class E Vps proteins, is an AAA type ATPase. Through its ATPase activity, Vps4p controls complex formation and functional localization of itself and the other class E Vps proteins (Babst *et al.*, 1998). Another class E protein, Vps27p, also contains a FYVE domain, which may explain its functional localization at the PVC, perhaps through binding PI(3)P (Burd and Emr, 1998).

1.3.2 Multivesicular body (MVB) sorting pathway

The PVC is now assumed to be a multivesicular body (Odorizzi *et al.*, 1998; Wurmser and Emr, 1998). Numerous small vesicles (40–50 nm in diameter) reside within its lumen. The current proposal is that these vesicles are generated by inward invagination of PVC membrane. Carboxypeptidase S (CPS) is a vacuolar protease, which exists as a membrane-spanning protein until final processing in the vacuole. CPS transits through the CPY pathway to the PVC. Prior to vacuolar delivery, it resides in the membrane of small vesicles within the PVC. However, when the *FAB1* gene is mutated, CPS remains in the outer surface of the PVC. *FAB1* encodes a phosphatidylinositol 3,5-kinase, which further phosphorylates PI(3)P generated by Vps34p (Gary *et al.*, 1998). Therefore, PI(3,5)P is required for vesicle invagination at the PVC. Consistent with certain other proteins that function at the PVC, Fab1p also has an FYVE domain that possibly leads to localization at the PVC via binding to PI(3)P (Burd and Emr, 1998).

At present, no clear picture has been obtained about how the contents of the PVC are delivered to the vacuole. One model is that the PVC fuses directly with the vacuole and another suggests the existence of transport vesicle intermediates (Odorizzi *et al.*, 1998; Vida and Gerhardt, 1999). In either case, components of the general fusion machinery family are involved at the final fusion step. These include: the t-SNARE, Vam3p; the SNAP-25 homolog, Vam7p; and the Sec1 family protein, Vps33p; together with a hetero-oligomeric complex composed of Vps18p, Vps11p and Vps16p (Darsow *et al.*, 1997; Sato *et al.*, 1998). Upon delivery to the vacuole, the small vesicles, which had resided within the PVC, are degraded in the lumen. Degradation is dependent upon the activity of PrA/PrB and acidic pH (Wurmser and Emr, 1998).

1.3.3 Alkaline phosphatase (ALP) pathway

ALP is a membrane protein containing an internal uncleaved signal sequence at its N-terminus (Klionsky and Emr, 1989). As with CPY, ALP passes through the early secretory pathway to the Golgi; however, vacuolar delivery of ALP was found to be relatively pH-independent (Klionsky and Emr, 1989; Morano and Klionsky, 1994). Similarly, targeting of ALP was unaffected by many of the *vps* mutants that block delivery of CPY (Klionsky and Emr, 1989). Further analysis confirmed that ALP transport was independent of the Vps45 and Vps27 proteins and the endosomal t-SNARE, Pep12p (Cowles *et al.*, 1997b; Piper *et al.*, 1997), components that are required for delivery to the vacuole through the CPY pathway. In contrast, sorting of ALP, but not CPY, was defective in a *vps41* mutant. This observation suggests that Vps41p acts in an alternative pathway that is needed for delivery of ALP. The current model is that ALP diverts from the CPY pathway at the TGN. The route taken by ALP skips the PVC and perhaps goes directly to the vacuole. Most of the genes involved in the ALP pathway were identified through a screen for mutants (*vam*) defective in vacuolar morphology (Wada *et al.*, 1992). Interestingly, null mutants of most of the genes involved in the CPY pathway have a vacuole with a largely normal morphology. In contrast, most mutants in the ALP pathway have a severely altered vacuole. This may indicate that the membrane supply from the ALP pathway is responsible for the shape or maintenance of the vacuole.

A novel clathrin adaptor-like complex, the AP-3 complex, but not clathrin, recruits ALP to this pathway depending on the presence of a dileucine-like motif in the cytoplasmic tail of ALP (Cowles *et al.*, 1997a; Stepp *et al.*, 1997; Vowels and Payne, 1998). The vacuolar t-SNARE, Vam3p, is also transported to the vacuole via this route, in a similar manner (Darsow *et al.*, 1998). Immunochemical methods allowed the isolation of transport intermediates of this pathway (50–130 nm in diameter) (Rehling *et al.*, 1999). Formation of the transport intermediate requires Vps41/Vam2p, which binds to Vps39/Vam6p and the AP-3 subunit (Nakamura *et al.*, 1997a; Rehling *et al.*, 1999). The transport intermediate shares with the CPY pathway the machinery for fusion with the vacuole, such as t-SNAREs (Vam3p, Vam7p), Vps18p and possibly a Rab protein (Ypt7/Vam4p) involved in docking (Darsow *et al.*, 1997; Rieder and Emr, 1997; Sato *et al.*, 1998). Vti1p, a v-SNARE involved in Golgi-to-PVC fusion, is also involved in the fusion step (Fischer von Mollard and Stevens, 1999).

1.3.4 Cytoplasm to vacuole targeting (Cvt) pathway and macroautophagy

Aminopeptidase I (API) is a soluble vacuolar resident protease, though it travels to the vacuole by a route distinct from the CPY, MVB and ALP pathways (Klionsky *et al.*, 1992). Precursor API (prAPI) contains a unique vacuolar targeting signal within its amino terminal propeptide (Oda *et al.*, 1996). Once synthesized in the cytoplasm, prAPI does not enter the ER but oligomerizes and assembles into a higher ordered complex (Kim *et al.*, 1997). This Cvt complex is subsequently enwrapped by a double-layered membrane structure, termed a Cvt vesicle (Baba *et al.*, 1997; Scott *et al.*, 1997). The outer membrane of the Cvt vesicle eventually fuses with the vacuolar membrane, and the inner membrane structure, the Cvt body, is released into the vacuolar lumen. So far, this biosynthetic Cvt pathway has been described only in *Saccharomyces cerevisiae*, and its existence in other organisms awaits determination. Likewise, α-mannosidase bypasses the secretory pathway, implying that it is transported to the vacuole by a similar/or the same route (Yoshihisa and Anraku, 1990).

The above CPY, MVB, ALP and Cvt pathways are used for biosynthesis of vacuolar enzymes. On the other hand, substrates destined for degradation must also be transported to the vacuole. Macroautophagy is a pathway responsible for bulk degradation of cytoplasmic constituents. When cells are subjected to nutrient starvation, such as limiting nitrogen, carbon, sulfur and/or phosphate, they begin to carry out macroautophagy (Takeshige *et al.*, 1992; Noda *et al.*, 1995). Ultimately, about 30% of total cytosol is transferred to the vacuole to be degraded by macroautophagy (Scott *et al.*, 1996).

Interestingly, macroautophagy occurs by a process quite similar to the Cvt pathway, in spite of their different purposes. During autophagy, a sequestering vesicle, termed an 'autophagosome', forms in the cytosol. It is a double-layered membrane structure, similar to the Cvt vesicle. However, the autophagosome enwraps cytoplasmic material nonselectively, and its diameter (300–900 nm) and corresponding volume are severalfold greater than the Cvt vesicle (~150 nm) (Baba *et al.*, 1994, 1997). Fusion with the vacuolar membrane and release of the inner membrane structure, termed an 'autophagic body', into the vacuolar lumen follow completion of vesicle synthesis. The autophagic body is then disrupted, allowing release of the contents, cytoplasmic materials, and subsequent degradation. Most notably, both the Cvt and autophagy pathways share most of the same machinery, which was revealed by a complementation test between mutants defective in the Cvt pathway (*cvt*) and in macroautophagy (*apg* and *aut*; autophagy and autophagy, respectively) (Tsukada and Ohsumi, 1993; Thumm *et al.*, 1994; Harding *et al.*, 1995, 1996; Scott *et al.*, 1996).

Analysis of API and a marker of autophagy, Pho8Δ60p (Noda *et al.*, 1995; Scott *et al.*, 1996), has led to a model for the function of both pathways. Under growing conditions the Cvt pathway proceeds, and under starvation conditions autophagy is induced. Induction of autophagy depends on a signal transmitted by inactivation of Tor kinase, which also controls the other starvation responses (Noda and Ohsumi, 1998). The other protein kinase, Apg1/Aut3p, is also involved in these pathways, perhaps together with the Apg13 protein (Funakoshi *et al.*, 1997; Matsuura *et al.*, 1997; Straub *et al.*, 1997).

The Cvt and autophagy pathways share certain unique characteristics that distinguish them from vesicle transport through the secretory pathway. In particular, the cargo of the secretory pathway is localized to a restricted site, the ER lumen, through the action of various cytosolic and membrane components, including signal recognition particle (SRP) and the ER translocon. Packaging of cargo into sites of vesicle budding ensures their enclosure within the developing vesicle. In contrast, cargo sequestration in the Cvt and autophagy pathways requires enwrapping by membrane within the cytosol. One major question concerns the origin of the sequestering membrane. Tlg2p, a t-SNARE that localizes at Golgi and endocytic structures, was shown to be required for Cvt vesicle but not autophagosome formation (Abeliovich *et al.*, 1998, 1999; Nichols *et al.*, 1998; Seron *et al.*, 1998; Coe *et al.*, 1999). Together with Tlg2p, a Sec1 family protein

(Vps45p) is required only for the Cvt pathway (Abeliovich *et al.*, 1999). As mentioned previously, Vps45p is involved in fusion of post-Golgi vesicles with the PVC; however, it is possible that distinctive pools function in these different pathways. A similar situation is seen with Vac8p, a protein that is required for the Cvt pathway but not for autophagy (Wang *et al.*, 1996, 1998; Pan and Goldfarb, 1998). Vac8p also functions in vacuolar inheritance (see Section 1.4). The two roles are probably separable, as acylated forms of Vac8p are needed for inheritance but not for import of prAPI. Another example of proteins that carry out functions in two different pathways is seen with Vps30/Apg6p. This protein is required for PVC to Golgi retrograde transport and also for the autophagy/Cvt pathways (Seaman *et al.*, 1997; Kametaka *et al.*, 1998). Apg6p forms a complex with Apg14p, but Apg14p is not required for the Vps pathway.

A mutation in *VPS4*, a class E protein containing an AAA ATPase motif, also affects macroautophagy (Shirahama *et al.*, 1997). Both Vps30/Apg6p and Vps4p function at the PVC. Their involvement in autophagy/Cvt import may reflect some aspects of the membrane source of the vesicles used in these pathways but the meaning is still the subject of investigation. Another protein that may function at the membrane of origin is Apg9p, which is an integral membrane protein that is localized adjacent to the vacuole (Noda *et al.*, 2000). The Apg9 protein does not cofractionate with most of the typical endomembrane markers but does display some overlap with the endosomal marker, Pep12p. Apg9p is not detected on autophagosomes or autophagic bodies. This means that if it functions in supplying membrane to the forming autophagosome/Cvt vesicle, it remains at the site of origin or is present in the vesicles at too low a level to allow detection.

In addition to the membrane source, another major question concerns the mechanism of vesicle formation. Cloning of the genes and characterization of the gene products involved in the Cvt/autophagy pathways has started to provide insight into this process. A reaction quite similar to ubiquitination is involved in formation of the Cvt vesicle and autophagosome. The C-terminal glycine of Apg12p is conjugated to an internal lysine residue of Apg5p via an isopeptide bond (Mizushima *et al.*, 1998). Two other proteins, Apg7/Cvt2p and Apg10p, function in this conjugation reaction as E1 and E2 enzymes, respectively (Kim *et al.*, 1999; Shintani *et al.*, 1999; Tanida *et al.*, 1999). Conjugates of Apg12p and Apg5p are further cross-linked with each other by virtue of binding to Apg16p, a coiled coil protein that forms homo-oligomers (Mizushima *et al.*, 1999). Mutants defective in the conjugation step, including a temperature sensitive *apg5* mutant, accumulate prAPI in a protease-sensitive, membrane-associated form (George *et al.*, 2000). These findings suggest that the conjugation process is involved in expanding the membrane structure of the Cvt vesicle and autophagosome.

Aut7/Apg8/Cvt5p is localized on forming autophagosomal and Cvt vesicle membranes and possible precursor structures around the forming autophagosome (Kirisako *et al.*, 1999; Huang *et al.*, 2000). The Aut7 protein is induced by starvation conditions and is degraded in the vacuole in a process dependent on the autophagic machinery. These results suggest that Aut7p transits to the vacuole inside Cvt vesicles/autophagosomes. The *aut7* mutant is impaired in formation of Cvt vesicles/autophagosomes but perhaps not as severely as the other *apg* mutants. This raises the possibility that Aut7/Apg8p functions at a rather late step of vesicle formation (Kirisako *et al.*, 1999). Aut7/Apg8p appears to function together with Aut2/Apg4p, based on the observation that the *AUT7* gene suppresses the *aut2* mutation (Lang *et al.*, 1998).

The machinery required for fusion of Cvt vesicles/autophagosomes with the vacuolar membrane is again well conserved with the MVB and ALP pathways. The Cvt body is disrupted in the vacuolar lumen in a process dependent on Cvt17p (Scott *et al.*, 1997). Disruption of the autophagic and Cvt body requires PrA and PrB and an acidic lumenal pH (Takeshige *et al.*, 1992; Nakamura *et al.*, 1997b; Scott *et al.*, 1997), conditions that are identical to those required for disruption of the small vesicles derived from the PVC. Therefore, some shared mechanism may be involved in these membrane disruption processes.

1.3.5 Microautophagy and pexophagy

In addition to macroautophagy, other degradative pathways deliver proteins and organelles to the vacuole. Microautophagy is described as a process where vacuolar membrane directly carries out the enwrapping process. This has been shown to occur for cytosol *in vitro* (Horst *et al.*, 1999), and also for degradation of peroxisomes under conditions where they are no longer required (Klionsky and Ohsumi, 1999; reviewed in Kim and Klionsky, 2000). Peroxisomes are degraded both by macro- and microautophagic mechanisms, depending on the organism and nutrient conditions. Peroxisomal uptake or pexophagy is a selective process, in that other organelles are not degraded under conditions that trigger uptake of peroxisomes. The mechanism by which peroxisomes are tagged for sequestration is not known. The morphology of the sequestration process for macropexophagy is similar to that involved in the Cvt/autophagy pathway. Double-membrane vesicles enwrap a single peroxisome and, subsequently, fuse with the vacuole. This process is the primary method seen in the methylotrophic yeast, *Hansenula polymorpha* (Veenhuis *et al.*, 1983), and is also used in *Pichia pastoris* during ethanol adaptation (Tuttle and Dunn, 1995). A micropexophagic process is used by *P. pastoris* is response to glucose (Tuttle and Dunn, 1995; Sakai *et al.*, 1998). In this case, clusters of peroxisomes are engulfed directly at the vacuolar surface. Interestingly, *GSA7*, a homolog of *APG7*, the E1 enzyme required for macroautophagy, is also required for micropexophagy in *P. pastoris* (Yuan *et al.*, 1999). In addition, selective peroxisome degradation in *S. cerevisiae* is inhibited in most of the macroautophagy mutants (Hutchins *et al.*, 1999; Kim *et al.*, 1999). These results suggest that micropexophagy also utilizes macroautophagic machinery in spite of the differences in the site and precise mechanisms of sequestration.

1.3.6 Vacuolar import and degradation

Macroautophagy is a non-selective process for the degradation of bulk cytosol under conditions of nutrient starvation. Yeast also utilizes selective processes for the turnover of specific cytosolic proteins. The best characterized cytosolic protein that is degraded in a specific manner is fructose-1,6-bisphosphatase (FBPase). FBPase is a gluconeogenic enzyme that is quite stable in the presence of non-fermentable substrates. Upon the addition of glucose, FBPase is rapidly degraded by two different mechanisms (Chiang and Schekman, 1994; Schork *et al.*, 1994). Part of the degradation occurs through a ubiquitin-dependent, proteasome-mediated process (Schork *et al.*, 1995). Recently, mutants have been isolated in this glucose-induced degradation (Gid) pathway that may provide further information about the role of the proteasome in regulating FBPase (Hammerle *et al.*, 1998). The other mechanism of turnover occurs through a vesicle-dependent process. When the vacuolar import and degradation (Vid) pathway is induced, 30 nm single-membrane vesicles form in the cytosol (Huang and Chiang, 1997; Chiang and Chiang, 1998). FBPase is not sequestered within the vesicle during its formation but appears to translocate into the vesicle following its completion. The mechanism of translocation is unknown. The cytosolic vesicles subsequently fuse with the vacuole, resulting in the release of the vesicle contents into the lumen and breakdown by vacuolar hydrolases. In either case, it is not known how a specific cytosolic protein is selectively targeted for degradation.

1.3.7 Endocytosis

Proteins within the plasma membrane, such as mating pheromone receptors, metal transporters and sugar transporters, are subjected to endocytosis under particular physiological situations (Riballo *et al.*, 1995; Gitan *et al.*, 1998; Wendland *et al.*, 1998). These proteins are internalized and delivered to a PVC through a vesicle-mediated process. Plasma membrane proteins are subsequently transported to, and degraded within, the vacuole. In addition, non-selective fluid-phase endocytosis also mediates delivery of substrates to the vacuole for degradation. Thus, the PVC serves as a convergence point between the endocytic and vacuolar delivery pathways.

Ubiquitin plays a role in endocytosis of membrane proteins, in addition to its well-known role in selective protein degradation. The cytoplasmic tail of plasma membrane proteins is conjugated with ubiquitin, which then acts as a signal for internalization (for review, see Hicke, 1997). An alternative signal,

NPFXD, present in the cytoplasmic tail of some membrane proteins also exists for internalization (Tan *et al.*, 1996).

During the past few years, a large number of proteins involved in the early stages of endocytosis, including the invagination step, have been reported. Two groups of proteins, Pan1p, Ent1p, Ent2p, Sji1p, Rsp5p, Vrp1p, Sla1p, Las17p, Arp2p, Arp3p and Rvs161p, Rvs167p, Abp1p, Srv1p, Sla2p, were found to interact physically or genetically within each group (Wendland *et al.*, 1998, 1999; Madania *et al.*, 1999 and the references therein). The detailed action of each protein will be elucidated in future studies; however, some are actin-binding proteins, and in most of these mutants actin organization is severely perturbed. In fact, normal actin function is necessary for endocytosis, together with myosin and its activator, calmodulin (Kubler and Riezman, 1993; Geli *et al.*, 1998). How actin function links to endocytosis is still to be determined.

In yeast, half of the endocytic activity still remains in a clathrin mutant, so that there is some controversy as to how clathrin functions in endocytosis (Tan *et al.*, 1993). One explanation is that some clathrin-independent endocytosis exists. For example, two subunits of the COPII coat complex, which are required for budding of ER-to-Golgi transport vesicles, are suggested to be involved in internalization (Penalver *et al.*, 1999).

In addition to these protein components, a specific lipid requirement for endocytosis has been described: the *end11* mutant defective in the internalization step was revealed to be allelic to the *erg2* mutant defective in biosynthesis of ergosterol (Munn *et al.*, 1999). Ergosterol is an abundant lipid in the plasma membrane of yeast, and specific species of ergosterol derivatives have been shown to be required for endocytosis. Such a line of study may be needed to fully understand how membrane invagination is accomplished by the combined action of the protein and lipid modification machinery.

Once the endocytic transport vesicle (30–50 nm in diameter) is formed, it transits to the early endosome (Hicke *et al.*, 1997; Prescianotto-Baschong and Riezman, 1998). Fusion of the vesicle with the early endosome is blocked by inactivation of NSF (Sec18p). The early endosome is observed as a tubular vesicular structure that extends to be 250–300 nm, mostly peripheral to the cell surface. Though the observation of this compartment is relatively rare and the characterization incomplete, the t-SNARE, Tlg2p, and input from the early secretory pathway have been postulated to be involved in the traffic or biogenesis of this transient organelle (Hicke *et al.*, 1997; Seron *et al.*, 1998). Following transport to the early endosome, endocytic flow proceeds to the late endosome, the diameter of which is approximately 200 nm. This compartment is observed most frequently peripheral to the vacuole (Prescianotto-Baschong and Riezman, 1998). A small GTPase (Arf) is reported to participate in the morphogenesis of some of these structures (Gaynor *et al.*, 1998). When class E *VPS* genes are mutated, endocytosed materials are accumulated in an exaggerated PVC, together with the cargo derived from the CPY pathway (Piper *et al.*, 1995; Rieder *et al.*, 1996). This means that endocytosis and the CPY pathway converge at the PVC. From there, they travel together to the vacuole in a manner dependent on a t-SNARE (Vam3p) and Rab (Ypt7p) protein (Odorizzi *et al.*,1998; Stefan and Blumer, 1999).

1.4 Vacuolar inheritance

As the cell cycle proceeds, the mother cell of *S. cerevisiae* buds a daughter cell. Accompanying the budding, a portion of the parental vacuole moves into the bud, and ultimately membrane scission divides the vacuole into two separate structures. This process is called vacuolar inheritance (reviewed in Bryant and Stevens, 1998). Mutants defective in vacuolar inheritance (*vac*) (Wang *et al.*, 1996) ultimately form a vacuole in the daughter cell, suggesting that the vacuole can be made by *de novo* synthesis in the absence of inheritance.

A vacuole moves toward the bud along an actin cable, in a process that depends on a myosin motor (Hill *et al.*, 1996). Vac8p is required for vacuolar inheritance. Palmitoylation and myristoylation of Vac8p cause it to localize in the vacuolar membrane, where it carries out its function (Fleckenstein *et al.*, 1998; Pan and Goldfarb, 1998; Wang *et al.*, 1998). The *vac8* mutant shows a defect in actin-dependent vacuolar movement, suggesting that Vac8p links the vacuole to a myosin motor.

Scission of a portion of the vacuole requires the action of certain proteins. Mutants defective in a PI(3,5)-kinase (Fab1p) and its possible upstream activator (Vac7p) are both defective in vacuole fission (Bonangelino *et al.*, 1997; Gary *et al.*, 1998). Interestingly, Vac7p is also required for a newly described retrograde trafficking event from the vacuole to the PVC (Bryant *et al.*, 1998). Hence, PI turnover may control both vacuole scission and retrograde transport.

As new portions of the mother vacuole reach the bud, they fuse with each other to form a novel vacuole. This process has been reconstituted *in vitro* (Ungermann *et al.*, 1999a, 1999b, and references therein). Mixing isolated vacuoles together with cytosol and ATP allows fusion to take place. This work has elucidated the requirement of fusion components and the precise kinetics of the action of fusion machinery. In summary: the process begins with a priming reaction in which NSF (Sec18p) mediates release of SNAP (Sec17p), relocation of LMA1 (a complex with thioredoxin and IB2) and disassembly of a *cis*-SNARE complex (Vam3p, Vam7p, Nyv1p, Vti1p, Ykt6p that are located on the same vacuolar membrane). Next, a Rab protein (Ypt7p) and association with a *trans*-SNARE (located on the opposite vacuolar membrane) mediate the docking of separate vacuoles. Finally, docked vacuoles fuse in a process regulated by Ca^{2+} and a calmodulin/protein phosphatase complex (Peters *et al.*, 1999).

1.5 Conclusion and future perspective

The function of the vacuole is dependent on multiple transport pathways that deliver both resident hydrolases and substrates. Moreover, these pathways share some of the molecular components with each other. This complexity has perplexed the yeast 'vacuologist' for a long time. For example, following the isolation of mutants defective in each pathway, it has been difficult to determine the direct result of the corresponding defect. The recent development of assays for each trafficking event and the use of conditional mutants that enable the rapid elimination of protein function have begun to provide tools to resolve this complexity. However, the vacuolar and endosomal system is highly dynamic, so that our current understanding of the targeting pathways and the nature of these compartments is still too rudimentary. Certainly, there are many questions that remain to be solved in this field.

Now that the whole genome sequencing of *Arabidopsis thaliana* is scheduled in the year 2000, comparison of the components involved in these pathways in a plant system will be possible. In fact, the genes involved in macroautophagy are highly conserved in *A. thaliana* (the authors' unpublished result). In that sense, it can be assumed that the identification of these genes will provide a strong foundation to resolve problems specific to a plant, which can never be answered in yeast. For example, how are the two distinct vacuoles (storage vacuole and lytic vacuole) generated and how are their functions coordinated? Are there multiple trafficking routes equivalent to the CPY, MVB and ALP pathways? Is there a Cvt pathway in plants, and what is the physiological requirement for existence of this pathway, a question that has still not been answered for yeast? How does vacuolar degradation contribute to the life cycle of the plant? What role does endocytosis play in plant cells? How is the vacuole partitioned and how is this coordinated with cell division? How does the vacuole contribute to formation of cell shape and, consequently, plant body shape? What is the homeostatic role of the plant vacuole in response to environmental change? These questions will be answered by first identifying a plant mutant corresponding to the yeast mutant, that is, by reverse genetics. Harvesting the fruit of yeast cell biology and seeding the next set of analyses in a plant field would be a hope for the researchers in the yeast field, perhaps the same as the early molecular biologists working on bacteria and phage.

References

Abeliovich, H., Grote, E., Novick, P. and Ferro-Novick, S. (1998) Tlg2p, a yeast syntaxin homolog that resides on the Golgi and endocytic structures. *J. Biol. Chem.*, **273** 11719-27.

Abeliovich, H., Darsow, T. and Emr, S.D. (1999) Cytoplasm to vacuole trafficking of aminopeptidase I requires a t-SNARE- Sec1p complex composed of Tlg2p and Vps45p. *EMBO J.*, **18** 6005-16.

Alizadeh, P. and Klionsky, D.J. (1996) Purification and biochemical characterization of the *ATH1* gene product, vacuolar acid trehalase, from *Saccharomyces cerevisiae. FEBS Lett.*, **391** 273-78.

Andreeva, N.A., Kulakovskaya, T.V. and Kulaev, I.S. (1998) Purification and properties of exopolyphosphatase isolated from *Saccharomyces cerevisiae* vacuoles. *FEBS Lett.*, **429** 194-96.

Baba, M., Takeshige, K., Baba, N. and Ohsumi, Y. (1994) Ultrastructural analysis of the autophagic process in yeast: detection of autophagosomes and their characterization. *J. Cell Biol.*, **124** 903-13.

Baba, M., Osumi, M., Scott, S.V., Klionsky, D.J. and Ohsumi, Y. (1997) Two distinct pathways for targeting proteins from the cytoplasm to the vacuole/lysosome. *J. Cell Biol.*, **139** 1687-95.

Babst, M., Wendland, B., Estepa, E.J. and Emr, S.D. (1998) The Vps4p AAA ATPase regulates membrane association of a Vps protein complex required for normal endosome function. *EMBO J.*, **17** 2982-93.

Bonangelino, C.J., Catlett, N.L. and Weisman, L.S. (1997) Vac7p, a novel vacuolar protein, is required for normal vacuole inheritance and morphology. *Mol. Cell Biol.*, **17** 6847-58.

Bryant, N.J. and Stevens, T.H. (1998) Vacuole biogenesis in *Saccharomyces cerevisiae*: protein transport pathways to the yeast vacuole. *Microbiol. Mol. Biol. Rev.*, **62** 230-47.

Bryant, N.J., Piper, R.C., Weisman, L.S. and Stevens, T.H. (1998) Retrograde traffic out of the yeast vacuole to the TGN occurs via the prevacuolar/endosomal compartment. *J. Cell Biol.*, **142** 651-63.

Burd, C.G. and Emr, S.D. (1998) Phosphatidylinositol(3)-phosphate signaling mediated by specific binding to RING FYVE domains. *Mol. Cell.*, **2** 157-62.

Burd, C.G., Babst, M. and Emr, S.D. (1998) Novel pathways, membrane coats and PI kinase regulation in yeast lysosomal trafficking. *Semin. Cell Dev. Biol.*, **9** 527-33.

Chiang, H.-L. and Schekman, R. (1994) Site of catabolite inactivation. *Nature*, **369** 284.

Chiang, M.C. and Chiang, H.L. (1998) Vid24p, a novel protein localized to the fructose-1,6-bisphosphatase-containing vesicles, regulates targeting of fructose-1,6-bisphosphatase from the vesicles to the vacuole for degradation. *J. Cell Biol.*, **140** 1347-56.

Coe, J.G., Lim, A.C., Xu, J. and Hong, W. (1999) A role for Tlg1p in the transport of proteins within the Golgi apparatus of *Saccharomyces cerevisiae. Mol. Biol. Cell*, **10** 2407-23.

Conibear, E. and Stevens, T.H. (1998) Multiple sorting pathways between the late Golgi and the vacuole in yeast. *Biochim. Biophys. Acta*, **1404** 211-30.

Cowles, C.R., Odorizzi, G., Payne, G.S. and Emr, S.D. (1997a) The AP-3 adaptor complex is essential for cargo-selective transport to the yeast vacuole. *Cell*, **91** 109-18.

Cowles, C.R., Snyder, W.B., Burd, C.G. and Emr, S.D. (1997b) Novel Golgi to vacuole delivery pathway in yeast: identification of a sorting determinant and required transport component. *EMBO J.*, **16** 2769-82.

Darsow, T., Rieder, S.E. and Emr, S.D. (1997) A multispecificity syntaxin homolog, Vam3p, essential for autophagic and biosynthetic protein transport to the vacuole. *J. Cell Biol.*, **138** 517-29.

Darsow, T., Burd, C.G. and Emr, S.D. (1998) Acidic dileucine motif essential for AP-3-dependent sorting and restriction of the functional specificity of the Vam3p vacuolar t-SNARE. *J. Cell Biol.*, **142** 913-22.

Fischer von Mollard, G. and Stevens, T.H. (1999) The *Saccharomyces cerevisiae* v-SNARE, Vti1p, is required for multiple membrane transport pathways to the vacuole. *Mol. Biol. Cell*, **10** 1719-32.

Fleckenstein, D., Rohde, M., Klionsky, D.J. and Rüdiger, M. (1998) Yel013p (Vac8p), an armadillo repeat protein related to plakoglobin and importin-α, is associated with the yeast vacuole membrane. *J. Cell Sci.*, **111** 3109-18.

Funakoshi, T., Matsuura, A., Noda, T. and Ohsumi, Y. (1997) Analyses of *APG13* gene involved in autophagy in yeast, *Saccharomyces cerevisiae. Gene*, **192** 207-13.

Gary, J.D., Wurmser, A.E., Bonangelino, C.J., Weisman, L.S. and Emr, S.D. (1998) Fab1p is essential for PtdIns(3)P 5-kinase activity and the maintenance of vacuolar size and membrane homeostasis. *J. Cell Biol.*, **143** 65-79.

Gaynor, E.C., Chen, C.Y., Emr, S.D. and Graham, T.R. (1998) ARF is required for maintenance of yeast Golgi and endosome structure and function. *Mol. Biol. Cell*, **9** 653-70.

Geli, M.I., Wesp, A. and Riezman, H. (1998) Distinct functions of calmodulin are required for the uptake step of receptor-mediated endocytosis in yeast: the type I myosin Myo5p, is one of the calmodulin targets. *EMBO J.*, **17** 635-47.

George, M.D., Baba, M., Scott, S.V., Mizushima, N., Garrison, B.S., Ohsumi, Y. and Klionsky, D.J. (2000) Apg5p functions in the sequestration step in the cytoplasm to vacuole targeting and macroautophagy pathways. *Mol. Biol. Cell*, **11** 969-82.

Gitan, R.S., Luo, H., Rodgers, J., Broderius, M. and Eide, D. (1998) Zinc-induced inactivation of the yeast *ZRT1* zinc transporter occurs through endocytosis and vacuolar degradation. *J. Biol. Chem.*, **273** 28617-24.

Hama, H., Tall, G.G. and Horazdovsky, B.F. (1999) Vps9p is a guanine nucleotide exchange factor involved in vesicle-mediated vacuolar protein transport. *J. Biol. Chem.*, **274** 15284-91.

Hammerle, M., Bauer, J., Rose, M., Szallies, A., Thumm, M., Dusterhus, S., Meeke, D., Entian, K.D. and Wolf, D.H. (1998) Proteins of newly isolated mutants and the amino-terminal proline are essential for ubiquitin-proteasome-catalyzed catabolite degradation of fructose-1,6-bisphosphatase of *Saccharomyces cerevisiae*. *J. Biol. Chem.*, **273** 25000-5005.

Harding, T.M., Morano, K.A., Scott, S.V. and Klionsky, D.J. (1995) Isolation and characterization of yeast mutants in the cytoplasm-to-vacuole protein targeting pathway. *J. Cell Biol.*, **131** 591-602.

Harding, T.M., Hefner-Gravink, A., Thumm, M. and Klionsky, D.J. (1996) Genetic and phenotypic overlap between autophagy and the cytoplasm-to-vacuole protein targeting pathway. *J. Biol. Chem.*, **271** 17621-24.

Hicke, L. (1997) Ubiquitin-dependent internalization and downregulation of plasma membrane proteins. *FASEB J.*, **11** 1215-26.

Hicke, L., Zanolari, B., Pypaert, M., Rohrer, J. and Riezman, H. (1997) Transport through the yeast endocytic pathway occurs through morphologically distinct compartments and requires an active secretory pathway and Sec18p/N-ethylmaleimide-sensitive fusion protein. *Mol. Biol. Cell*, **8** 13-31.

Hill, K.L., Catlett, N.L. and Weisman, L.S. (1996) Actin and myosin function in directed vacuole movement during cell division in *Saccharomyces cerevisiae*. *J. Cell Biol.*, **135** 1535-49.

Horazdovsky, B.F., DeWald, D.B. and Emr, S.D. (1995) Protein transport to the yeast vacuole. *Curr. Opin. Cell Biol.*, **7** 544-51.

Horst, M., Knecht, E.C. and Schu, P.V. (1999) Import into and degradation of cytosolic proteins by isolated yeast vacuoles. *Mol. Biol. Cell*, **10** 2879-89.

Huang, P.H. and Chiang, H.L. (1997) Identification of novel vesicles in the cytosol-to-vacuole protein degradation pathway. *J. Cell Biol.*, **136** 803-10.

Huang, W.-P., Scott, S.V., Kim, J. and Klionsky, D.J. (2000) The itinerary of a vesicle component, Aut7p/Cvt5p, terminates in the yeast vacuole via the autophagy/Cvt pathways. *J. Biol. Chem.*, **275** 5845-51.

Hutchins, M.U., Veenhuis, M. and Klionsky, D.J. (1999) Peroxisome degradation in *Saccharomyces cerevisiae* is dependent on machinery of macroautophagy and the Cvt pathway. *J. Cell Sci.*, **112** 4079-87.

Jahn, R. and Sudhof, T. (1999) Membrane fusion and exocytosis. *Annu. Rev. Biochem.*, **68** 863-911.

Jones, E.W., Zubenko, G.S. and Parker, R.R. (1982) *PEP4* gene function is required for expression of several vacuolar hydrolases in *Saccharomyces cerevisiae*. *Genetics*, **102** 665-77.

Jones, H.D., Schliwa, M. and Drubin, D.G. (1993) Video microscopy of organelle inheritance and motility in budding yeast. *Cell Motil. Cytosk.*, **25** 129-42.

Jones, S.M. and Howell, K.E. (1997) Phosphatidylinositol 3-kinase is required for the formation of constitutive transport vesicles from the TGN. *J. Cell Biol.*, **139** 339-49.

Kametaka, S., Okano, T., Ohsumi, M. and Ohsumi, Y. (1998) Apg14p and Apg6/Vps30p form a protein complex essential for autophagy in the yeast, *Saccharomyces cerevisiae*. *J. Biol. Chem.*, **273** 22284-91.

Kim, J., Scott, S.V., Oda, M.N. and Klionsky, D.J. (1997) Transport of a large oligomeric protein by the cytoplasm-to-vacuole protein targeting pathway. *J. Cell Biol.*, **137** 609-18.

Kim, J., Dalton, V.M., Eggerton, K.P., Scott, S.V. and Klionsky, D.J. (1999) Apg7p/Cvt2p is required for the cytoplasm-to-vacuole targeting, macroautophagy and peroxisome degradation pathways. *Mol. Biol. Cell*, **10** 1337-51.

Kim, J. and Klionsky, D.J. (2000) Autophagy, cytoplasm-to-vacuole targeting pathway and pexophagy in yeast and mammalian cells. *Annu. Rev. Biochem.*, **69** (in press).

Kirisako, T., Baba, M., Ishihara, N., Miyazawa, K., Ohsumi, M., Yoshimori, T., Noda, T. and Ohsumi, Y. (1999) Formation process of autophagosome is traced with Apg8/Aut7p in yeast. *J. Cell Biol.*, **147** 435-46.

Klionsky, D.J. and Emr, S.D. (1989) Membrane protein sorting: biosynthesis, transport and processing of yeast vacuolar alkaline phosphatase. *EMBO J.*, **8** 2241-50.

Klionsky, D.J. and Ohsumi, Y. (1999) Vacuolar import of proteins and organelles from the cytoplasm. *Annu. Rev. Cell Dev. Biol.*, **15** 1-32.

Klionsky, D.J., Herman, P.K. and Emr, S.D. (1990) The fungal vacuole: composition, function and biogenesis. *Microbiol. Rev.*, **54** 266-92.

Klionsky, D.J., Cueva, R. and Yaver, D.S. (1992) Aminopeptidase I of *Saccharomyces cerevisiae* is localized to the vacuole independent of the secretory pathway. *J. Cell Biol.*, **119** 287-99.

Kubler, E. and Riezman, H. (1993) Actin and fimbrin are required for the internalization step of endocytosis in yeast. *EMBO J.*, **12** 2855-62.

Lang, T., Schaeffeler, E., Bernreuther, D., Bredschneider, M., Wolf, D.H. and Thumm, M. (1998) Aut2p and Aut7p, two novel microtubule-associated proteins are essential for delivery of autophagic vesicles to the vacuole. *EMBO J.*, **17** 3597-607.

Madania, A., Dumoulin, P., Grava, S., Kitamoto, H., Scharer-Brodbeck, C., Soulard, A., Moreau, V. and Winsor, B. (1999) The *Saccharomyces cerevisiae* homolog of human Wiskott-Aldrich syndrome protein, las17p, interacts with the Arp2/3 complex. *Mol. Biol. Cell*, **10** 3521-38.

Matsuura, A., Tsukada, M., Wada, Y. and Ohsumi, Y. (1997) Apg1p, a novel protein kinase required for the autophagic process in *Saccharomyces cerevisiae. Gene*, **192** 245-50.

Mizushima, N., Noda, T., Yoshimori, T., Tanaka, Y., Ishii, T., George, M.D., Klionsky, D.J., Ohsumi, M. and Ohsumi, Y. (1998) A protein conjugation system essential for autophagy. *Nature*, **395** 395-98.

Mizushima, N., Noda, T. and Ohsumi, Y. (1999) Apg16p is required for the function of the Apg12p-Apg5p conjugate in the yeast autophagy pathway. *EMBO J.*, **18** 3888-96.

Morano, K.A. and Klionsky, D.J. (1994) Differential effects of compartment deacidification on the targeting of membrane and soluble proteins to the vacuole in yeast. *J. Cell Sci.*, **107** 2813-24.

Munn, A.L., Heese-Peck, A., Stevenson, B.J., Pichler, H. and Riezman, H. (1999) Specific sterols required for the internalization step of endocytosis in yeast. *Mol. Biol. Cell*, **10** 3943-57.

Nakamura, N., Hirata, A., Ohsumi, Y. and Wada, Y. (1997a) Vam2/Vps41p and Vam6/Vps39p are components of a protein complex on the vacuolar membranes and involved in the vacuolar assembly in the yeast, *Saccharomyces cerevisiae. J. Biol. Chem.*, **272** 11344-49.

Nakamura, N., Matsuura, A., Wada, Y. and Ohsumi, Y. (1997b) Acidification of vacuoles is required for autophagic degradation in the yeast, *Saccharomyces cerevisiae. J. Biochem.*, **121** 338-44.

Nichols, B.J., Holthuis, J.C. and Pelham, H.R. (1998) The Sec1p homolog, Vps45p, binds to the syntaxin, Tlg2p. *Eur. J. Cell Biol.*, **77** 263-68.

Noda, T. and Ohsumi, Y. (1998) Tor, a phosphatidylinositol kinase homolog, controls autophagy in yeast. *J. Biol. Chem.*, **273** 3963-66.

Noda, T., Matsuura, A., Wada, Y. and Ohsumi, Y. (1995) Novel system for monitoring autophagy in the yeast, *Saccharomyces cerevisiae. Biochem. Biophys. Res. Commun.*, **210** 126-32.

Noda, T., Kim, J., Huang, W.-P., Baba, M., Tokunaga, C., Ohsumi, Y. and Klionsky, D.J. (2000) Apg9p/Cvt7p is an integral membrane protein required for transport vesicle formation in the Cvt and autophagy pathways. *J. Cell Biol.*, **148** 465-80.

Oda, M.N., Scott, S.V., Hefner-Gravink, A., Caffarelli, A.D. and Klionsky, D.J. (1996) Identification of a cytoplasm-to-vacuole targeting determinant in aminopeptidase I. *J. Cell Biol.*, **132** 999-1010.

Odorizzi, G., Babst, M. and Emr, S.D. (1998) Fab1p PtdIns(3)P 5-kinase function essential for protein sorting in the multivesicular body. *Cell*, **95** 847-58.

Pan, X. and Goldfarb, D.S. (1998) YEB3/VAC8 encodes a myristylated armadillo protein of the *Saccharomyces cerevisiae* vacuolar membrane that functions in vacuole fusion and inheritance. *J. Cell Sci.*, **111** 2137-47.

Penalver, E., Lucero, P., Moreno, E. and Lagunas, R. (1999) Clathrin and two components of the COPII complex, Sec23p and Sec24p, could be involved in endocytosis of the *Saccharomyces cerevisiae* maltose transporter. *J. Bacteriol.*, **181** 2555-63.

Peters, C., Andrews, P.D., Stark, M.J., Cesare-Tadio, S., Glatz, A., Pedtelejnikov, A., Mann, M. and Mayer, A. (1999) Control of the terminal step of intracellular membrane fusion by protein phosphatase 1. *Science*, **285** 1084-87.

Peterson, M.R., Burd, C.G. and Emr, S.D. (1999) Vac1p coordinates Rab and phosphatidylinositol 3-kinase signaling in Vps45p-dependent vesicle docking/fusion at the endosome. *Curr. Biol.*, **9** 159-62.

Piper, R.C., Cooper, A.A., Yang, H. and Stevens, T.H. (1995) *VPS27* controls vacuolar and endocytic traffic through a prevacuolar compartment in *Saccharomyces cerevisiae*. *J. Cell Biol.*, **131** 603-17.

Piper, R.C., Bryant, N.J. and Stevens, T.H. (1997) The membrane protein alkaline phosphatase is delivered to the vacuole by a route that is distinct from the VPS-dependent pathway. *J. Cell Biol.*, **138** 531-45.

Prescianotto-Baschong, C. and Riezman, H. (1998) Morphology of the yeast endocytic pathway. *Mol. Biol. Cell*, **9** 173-89.

Rehling, P., Darsow, T., Katzmann, D.J. and Emr, S.D. (1999) Formation of AP-3 transport intermediates requires Vps41 function. *Nat. Cell Biol.*, **1** 346-53.

Riballo, E., Herweijer, M., Wolf, D.H. and Lagunas, R. (1995) Catabolite inactivation of the yeast maltose transporter occurs in the vacuole after internalization by endocytosis. *J. Bacteriol.*, **177** 5622-27.

Rieder, S.E. and Emr, S.D. (1997) A novel RING finger protein complex essential for a late step in protein transport to the yeast vacuole. *Mol. Biol. Cell*, **8** 2307-27.

Rieder, S.E., Banta, L.M., Kohrer, K., McCaffery, J.M. and Emr, S.D. (1996) Multilamellar endosome-like compartment accumulates in the yeast *vps28* vacuolar protein sorting mutant. *Mol. Biol. Cell*, **7** 985-99.

Sakai, Y., Koller, A., Rangell, L.K., Keller, G.A. and Subramani, S. (1998) Peroxisome degradation by microautophagy in *Pichia pastoris*: identification of specific steps and morphological intermediates. *J. Cell Biol.*, **141** 625-36.

Sato, T.K., Darsow, T. and Emr, S.D. (1998) Vam7p, a SNAP-25-like molecule, and Vam3p, a syntaxin homolog, function together in yeast vacuolar protein trafficking. *Mol. Cell. Biol.*, **18** 5308-19.

Schmidt, A., Wolde, M., Thiele, C., Fast, W., Kratzin, H., Podtelejnikov, A.V., Witke, W., Huttner, W.D. and Soling, H.D. (1999) Endophilin I mediates synaptic vesicle formation by transfer of arachidonate to lysophosphatidic acid. *Nature*, **401** 133-41.

Schork, S.M., Bee, G., Thumm, M. and Wolf, D.H. (1994) Catabolite inactivation of fructose-1,6-bisphosphatase in yeast is mediated by the proteasome. *FEBS Lett.*, **349** 270-74.

Schork, S.M., Thumm, M. and Wolf, D.H. (1995) Catabolite inactivation of fructose-1,6-bisphosphatase of *Saccharomyces cerevisiae*: degradation occurs via the ubiquitin pathway. *J. Biol. Chem.*, **270** 26446-50.

Scott, S.V. and Klionsky, D.J. (1998) Delivery of proteins and organelles to the vacuole from the cytoplasm. *Curr. Opin. Cell Biol.*, **10** 523-29.

Scott, S.V., Hefner-Gravink, A., Morano, K.A., Noda, T., Ohsumi, Y. and Klionsky, D.J. (1996) Cytoplasm-to-vacuole targeting and autophagy employ the same machinery to deliver proteins to the yeast vacuole. *Proc. Natl. Acad. Sci. USA*, **93** 12304-08.

Scott, S.V., Baba, M., Ohsumi, Y. and Klionsky, D.J. (1997) Aminopeptidase I is targeted to the vacuole by a nonclassical vesicular mechanism. *J. Cell Biol.*, **138** 37-44.

Seaman, M.N., Marcusson, E.G., Cereghino, J.L. and Emr, S.D. (1997) Endosome-to-Golgi retrieval of the vacuolar protein sorting receptor, Vps10p, requires the function of the *VPS29*, *VPS30* and *VPS35* gene products. *J. Cell Biol.*, **137** 79-92.

Seaman, M.N., McCaffery, J.M. and Emr, S.D. (1998) A membrane coat complex essential for endosome-to-Golgi retrograde transport in yeast. *J. Cell Biol.*, **142** 665-81.

Seron, K., Tieaho, V., Prescianotto-Baschong, C., Aust, T., Blondel, M.O., Guillaud, P., Devilliers, G., Rossanese, O.W., Glick, B.S., Riezman, H., Keranen, S. and Haguenauer-Tsapis, R. (1998) A yeast t-SNARE involved in endocytosis. *Mol. Biol. Cell*, **9** 2873-89.

Shintani, T., Mizushima, N., Ogawa, Y., Matsuura, A., Noda, T. and Ohsumi, Y. (1999) Apg10p, a novel protein-conjugating enzyme essential for autophagy in yeast. *EMBO J.*, **18** 5234-41.

Shirahama, K., Noda, T. and Ohsumi, Y. (1997) Mutational analysis of Csc1/Vps4p: involvement of endosome in regulation of autophagy in yeast. *Cell Struct. Funct.*, **22** 501-09.

Stefan, C.J. and Blumer, K.J. (1999) A syntaxin homolog encoded by *VAM3* mediates downregulation of a yeast G protein-coupled receptor. *J. Biol. Chem.*, **274** 1835-41.

Stepp, J.D., Huang, K. and Lemmon, S.K. (1997) The yeast adaptor protein complex, AP-3, is essential for the efficient delivery of alkaline phosphatase by the alternate pathway to the vacuole. *J. Cell Biol.*, **139** 1761-74.

Stevens, T.H. and Forgac, M. (1997) Structure, function and regulation of the vacuolar (H^+)-ATPase. *Annu. Rev. Cell Dev. Biol.*, **13** 779-808.

Straub, M., Bredschneider, M. and Thumm, M. (1997). *AUT3*, a serine/threonine kinase gene, is essential for autophagocytosis in *Saccharomyces cerevisiae*. *J. Bacteriol.*, **179** 3875-83.

Takeshige, K., Baba, M., Tsuboi, S., Noda, T. and Ohsumi, Y. (1992) Autophagy in yeast demonstrated with proteinase-deficient mutants and conditions for its induction. *J. Cell Biol.*, **119** 301-11.

Tall, G.G., Hama, H., DeWald, D.B. and Horazdovsky, B.F. (1999) The phosphatidylinositol 3-phosphate binding protein, Vac1p, interacts with a Rab GTPase and a Sec1p homolog to facilitate vesicle-mediated vacuolar protein sorting. *Mol. Biol. Cell*, **10** 1873-89.

Tan, P.K., Davis, N.G., Sprague, G.F. and Payne, G.S. (1993) Clathrin facilitates the internalization of seven transmembrane segment receptors for mating pheromones in yeast. *J. Cell Biol.*, **123** 1707-16.

Tan, P.K., Howard, J.P. and Payne, G.S. (1996) The sequence NPFXD defines a new class of endocytosis signal in *Saccharomyces cerevisiae*. *J. Cell Biol.*, **135** 1789-800.

Tanida, I., Mizushima, N., Kiyooka, M., Ohsumi, M., Ueno, T., Ohsumi, Y. and Kominami, E. (1999) Apg7p/Cvt2p: a novel protein-activating enzyme essential for autophagy. *Mol. Biol. Cell*, **10** 1367-79.

Thumm, M., Egner, R., Koch, B., Schlumpberger, M., Straub, M., Veenhuis, M. and Wolf, D.H. (1994) Isolation of autophagocytosis mutants of *Saccharomyces cerevisiae*. *FEBS Lett.*, **349** 275-80.

Tsukada, M. and Ohsumi, Y. (1993) Isolation and characterization of autophagy-defective mutants of *Saccharomyces cerevisiae*. *FEBS Lett.*, **333** 169-74.

Tuttle, D.L. and Dunn, W.A., Jr. (1995) Divergent modes of autophagy in the methylotrophic yeast, *Pichia pastoris*. *J. Cell Sci.*, **108** 25-35.

Ungermann, C., von Mollard, G.F., Jensen, O.N., Margolis, N., Stevens, T.H. and Wickner, W. (1999a) Three v-SNAREs and two t-SNAREs, present in a pentameric cis-SNARE complex on isolated vacuoles, are essential for homotypic fusion. *J. Cell Biol.*, **145** 1435-42.

Ungermann, C., Wickner, W. and Xu, Z. (1999b) Vacuole acidification is required for trans-SNARE pairing, LMA1 release and homotypic fusion. *Proc. Natl. Acad. Sci. USA*, **96** 11194-99.

Van Den Hazel, H.B., Kielland-Brandt, M.C. and Winther, J.R. (1996) Review: biosynthesis and function of yeast vacuolar proteases. *Yeast*, **12** 1-16.

Veenhuis, M., Douma, A., Harder, W. and Osumi, M. (1983) Degradation and turnover of peroxisomes in the yeast, *Hansenula polymorpha*, induced by selective inactivation of peroxisomal enzymes. *Arch. Microbiol.*, **134** 193-203.

Vida, T. and Gerhardt, B. (1999) A cell-free assay allows reconstitution of Vps33p-dependent transport to the yeast vacuole/lysosome. *J. Cell Biol.*, **146** 85-98.

Vowels, J.J. and Payne, G.S. (1998) A dileucine-like sorting signal directs transport into an AP-3-dependent, clathrin-independent pathway to the yeast vacuole. *EMBO J.*, **17** 2482-93.

Wada, Y., Ohsumi, Y. and Anraku, Y. (1992) Genes for directing vacuolar morphogenesis in *Saccharomyces cerevisiae*. I. Isolation and characterization of two classes of *vam* mutants. *J. Biol. Chem.*, **267** 18665-70.

Wang, Y.X., Zhao, H., Harding, T.M., Gomes de Mesquita, D.S., Woldringh, C.L., Klionsky, D.J., Munn, A.L. and Weisman, L.S. (1996) Multiple classes of yeast mutants are defective in vacuole partitioning yet target vacuole proteins correctly. *Mol. Biol. Cell*, **7** 1375-89.

Wang, Y.X., Catlett, N.L. and Weisman, L.S. (1998) Vac8p, a vacuolar protein with armadillo repeats, functions in both vacuole inheritance and protein targeting from the cytoplasm to vacuole. *J. Cell Biol.*, **140** 1063-74.

Wendland, B., Emr, S.D. and Riezman, H. (1998) Protein traffic in the yeast endocytic and vacuolar protein sorting pathways. *Curr. Opin. Cell Biol.*, **10** 513-22.

Wendland, B., Steece, K.E. and Emr, S.D. (1999) Yeast epsins contain an essential N-terminal ENTH domain, bind clathrin and are required for endocytosis. *EMBO J.*, **18** 4383-93.

Wurmser, A.E. and Emr, S.D. (1998) Phosphoinositide signaling and turnover: PtdIns(3)P, a regulator of membrane traffic, is transported to the vacuole and degraded by a process that requires lumenal vacuolar hydrolase activities. *EMBO J.*, **17** 4930-42.

Yoshihisa, T. and Anraku, Y. (1990) A novel pathway of import of α-mannosidase, a marker enzyme of vacuolar membrane, in *Saccharomyces cerevisiae*. *J. Biol. Chem.*, **265** 22418-25.

Yuan, W., Stromhaug, P.E. and Dunn, W.A., Jr. (1999) Glucose-induced autophagy of peroxisomes in *Pichia pastoris* requires a unique E1-like protein. *Mol. Biol. Cell*, **10** 1353-66.

2 Vacuolar processing enzymes and aquaporins

Ikuko Hara-Nishimura and Masayoshi Maeshima

2.1 Introduction

Higher plants have vacuoles of various types within their cells. Vacuoles with hydrolytic enzymes were first described in the senescing corolla of morning glory (Matile and Winkenbach, 1971). Various hydrolytic enzymes have been shown immunocytochemically and biochemically to be localized in vacuoles of several organs (Nishimura and Beevers, 1979; Holwerda *et al.*, 1992; Kinoshita *et al.*, 1999). Plant vacuoles, therefore, have been regarded as a lytic compartment similar to animal lysosomes and yeast vacuoles. In contrast to lytic vacuoles, protein-storage vacuoles (PSVs) are also found as a storage compartment in plant cells. In barley roots (Paris *et al.*, 1996) and maturing pea cotyledons (Robinson *et al.*, 1995), both types of vacuoles were reported to exist within the same cells. The PSVs develop mainly in storage organs, such as maturing seeds. It has been shown that the PSVs are converted into lytic vacuoles in cotyledon cells during seed germination (Hara-Nishimura *et al.*, 1991b). The dynamic transformation and biogenesis of vacuoles are closely related to cell differentiation and programmed cell death in plants.

Lytic vacuoles and PSVs are the most common types of vacuole in higher plant cells, and they are distinguished from each other by the presence of specific marker proteins. Some homologs of aquaporins, tonoplast intrinsic proteins (TIPs), have been used to identify the types of vacuole (Paris *et al.*, 1996). It is known that γ-TIP is localized in the membranes of lytic vacuoles and α-TIP in the membranes of PSVs. Their distinct patterns of localization and high concentrations in the vacuolar membranes have made TIPs ideal markers for plant vacuoles. Vacuolar differentiation is necessarily associated with a replacement of one type of TIP by another, as described in Section 2.3.2. The functional differentiation of plant vacuoles is also accompanied by accumulation of soluble lumenal proteins with physiological activities. Most soluble proteins are incorporated into the vacuoles in an inactive form and then converted into an active form by the action of vacuolar processing enzymes (VPEs) (Hara-Nishimura *et al.*, 1998a). The vacuolar processing system could be involved in maturation of vacuolar proteins in both lytic vacuoles and PSVs, in order for vacuoles to acquire and maintain their active functions (Kinoshita *et al.*, 1999). This chapter describes the functional differentiation of TIPs and VPEs.

2.2 Vacuolar processing systems

2.2.1 *Vacuolar processing enzymes (VPEs)*

In higher plants, proprotein precursors of most seed proteins are synthesized on the rough endoplasmic reticulum (rER) and are then transported to PSVs in maturing seed cells (Akazawa and Hara-Nishimura, 1985; Chrispeels, 1991; Okita and Rogers, 1996). Just after arriving in the vacuoles, the precursor proteins are converted into their respective mature forms by proteolytic cleavages (Hara-Nishimura and Nishimura, 1987; Hara-Nishimura *et al.*, 1993a). The post-translational cleavages occur at the carbonyl sides of asparagine residues in precursors of various seed proteins of different plants, including storage proteins, lectins and toxins, as reviewed by Hara-Nishimura and co-workers 1995. An enzyme responsible for maturation of seed proteins was first purified from maturing seeds of castor bean and designated 'vacuolar processing enzyme' (VPE) (Hara-Nishimura *et al.*, 1991a, 1993b).

Molecular characterization revealed that VPE is a novel cysteine proteinase with a specificity towards asparaginyl bonds and exhibits no homology to cysteine proteinases in the papain family (Hara-Nishimura *et al.*, 1993b). VPE homologs have been found in plants, including: soybean (Shimada *et al.*, 1994), Jack bean (Takeda *et al.*, 1994), *Arabidopsis* (Kinoshita *et al.*, 1995a,b), vetch (Becker *et al.*, 1995) and citrus (Alonso and Granell, 1995); and in animals, e.g. *Schistosoma mansoni* (Klinkert *et al.*, 1989), human (Chen *et al.*, 1997) and mouse (Chen *et al.*, 1998), as shown in Figure 2.1. The plant homologs can be separated into two subfamilies: one specific to seeds and the other specific to vegetative organs (Kinoshita *et al.*, 1995a,b). This is consistent with the fact that plant vacuoles are classified into two types, PSVs in seeds and lytic vacuoles in vegetative organs. Recent work on the vegetative VPE showed that the VPE-mediated processing system could be distributed in a wide range of plant species and plays a crucial role in maturation of various vacuolar proteins both in PSVs and lytic vacuoles in plants.

2.2.2 *Vacuolar processing system in protein-storage vacuoles (PSVs)*

It has been shown that the delivery of proprotein precursors of seed proteins to the PSVs is mediated by the vesicles, with a density of 1.24 g/cm^3 in maturing pumpkin and castor bean seeds (Hara-Nishimura *et al.*, 1985, 1993; Fukasawa *et al.*, 1988). Isolation of the vesicles from maturing pumpkin seeds revealed that they contained a large amount of proprotein precursors of various seed proteins, including 11S globulin and 2S albumin (Hara-Nishimura *et al.*, 1998b). Thus, these vesicles were designated precursor-accumulating (PAC) vesicles. *In vitro* processing assays of these proproteins have demonstrated that VPE is

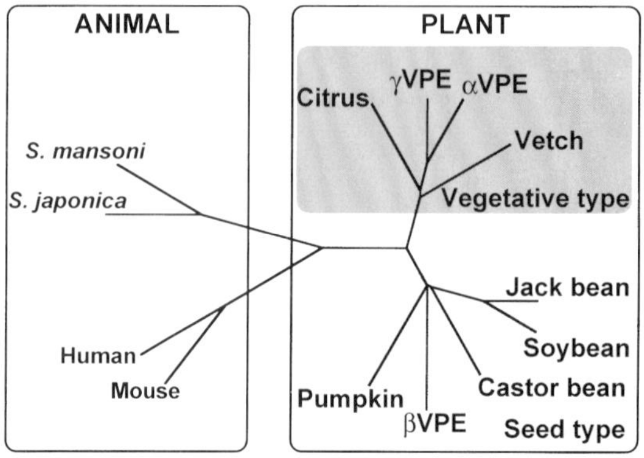

Figure 2.1 A phylogenetic tree of vacuolar processing enzyme (VPE) homologs. VPE homologs of a novel family of cysteine proteinases are distributed in higher plants and animals. Plant homologs can be separated into a seed type and a vegetative type. The computer program, CLUSTAL V (Higgins *et al.*, 1992), was used to calculate the relative lengths of branches. References to the published sequences are as follows: *Arabidopsis* αVPE and βVPE (Kinoshita *et al.*, 1995a); *Arabidopsis* γVPE (Kinoshita *et al.*, 1995b); castor bean (Hara-Nishimura *et al.*, 1993b); soybean (Shimada *et al.*, 1994); pumpkin (unpublished data); Jack bean (Takeda *et al.*, 1994); vetch (Becker *et al.*, 1995); citrus (Alonso and Granell, 1995); human (Chen *et al.*, 1997), *Schistosoma mansoni* (Klinkert *et al.*, 1989) and *Schistosoma japonicum* (accession number S31908).

the authentic processing enzyme mediating maturation of several seed proteins of pumpkin and castor bean, including 2S albumins, 11S globulins, 7S globulin, ricin and *Ricinus communis* agglutinin (Hara-Nishimura and Nishimura, 1987; Hara-Nishimura *et al.*, 1991a; Shimada *et al.*, 1994; Hiraiwa *et al.*, 1997a; Yamada *et al.*, 1999). Figure 2.2 shows seed proteins that could be converted into mature forms by the action of VPE in the PSVs. The detailed mechanism of the vacuolar processing was first investigated for 2S albumin, a major storage protein of pumpkin (Hara-Nishimura *et al.*, 1993a) VPE recognizes exposed asparagine residues on the molecular surface of the precursor proteins and then cleaves the peptide bonds at the carbonyl sides of the asparagine residues.

Recently, a unique mechanism for vacuolar processing of a 100 kDa protein component (PV100) of pumpkin PAC vesicles was reported (Yamada *et al.*, 1999). Molecular characterization of PV100 showed that its preproprotein precursor was composed of a hydrophobic signal peptide and the following three domains: an 11 kDa Cys-rich domain with four CxxxC motifs (C, Cys); a 34 kDa Arg/Glu-rich domain composed of six homologous repeats; and a 50 kDa vicilin-like domain (as shown in Figure 2.3). Both immunocytochemistry and immunoblots with anti-PV100 antibodies showed that <10 kDa proteins and the 50 kDa vicilin-like protein accumulated in the vacuoles. To identify the mature proteins derived from PV100, soluble proteins of the PSVs were

Pumpkin **2S albumin** (D16560)	-EVEE***N***RQGRE-RGIE***N***PWRRE-
Castor bean **2S albumin** (S11500)	-QTRT***N***PSQQG-RRSD***N***QERSL-
Arabidopsis **2S albumin-**1 (JA0161)	-DDAT***N***PIGPK-DDME***N***PQGQQ
Rapeseed **2S albumin** (A01329)	-DDAT***N***SAGPF-DDME***N***PQGPQ-
Brazil nut **2S albumin** (S14946)	-VEEE***N***QEECR-
Sunflower **2S albumin** HaG5 (S01062)	-VTES***N***IDIPF-
Pumpkin **11S globulin** (M36407)	-SESE***N***GLEET-
Pea **legumin** A (A22866)	-RQGD***N***GLEET-
Pea **legumin** B (S00336)	-EERK***N***GLEET-
Sunflower **11S globulin** HaG3 (JA0089)	-GGWS***N***GVEET-
Soybean **glycinin** A1aBx (A23497)	-KSRR***N***GIDET-
Soybean **glycinin** A2B1a (A91341)	-KRSR***N***GIDET-
Soybean **glycinin** A3B4 (A92524)	-CQTR***N***GVEEN-
Soybean **glycinin** A5A4B3 (A91145)	-QEQS***N***RRGSR-CETR***N***GVEEN-
Oat **globulin** (JA0160)	-DQSF***N***GLEEN-
Pea **vicilin** (A03344)	-QGKE***N***DKEEE-
Castor bean **ricin D** (A24041)	-VPNF***N***ADVCM-
Castor bean **agglutinin** (A24261)	-VPNF***N***ADVCM-
Indian licorice **abrin**-a (S32429)	-CNPP***N***ANQSP-
Jack bean **concanavalin A** (A03357)	-FPDA***N***VIRNSTTIDF***N***AAY***N***ADTIV-
	-KLKS***N***EIPDI-
Acacia **trypsin inhibitor** (JH0607)	-YCEG***N***SDDES-
Bean **α-amylase inhibitor** (A27350)	-HRQA***N***SAVGL-
Pea **lectin** (Y00440)	-LEEE***N***VTSYT-
Rice **lectin** (JQ1102)	-GRNA***N***GELCP-

Figure 2.2 Putative target proteins for seed vacuolar processing enzymes (VPEs) in protein-storage vacuoles. Proprotein processing occurs on the C-terminal side of an asparagine residue that is indicated by *N*. The processing site was determined by aligning the deduced precursor sequences with the N-terminal sequences of the respective mature proteins. The database accession numbers are indicated in parentheses.

separated and their molecular structures were determined. Mass spectrometry and peptide sequencing showed that one Cys-rich C2 peptide, three Arg/Glu-rich RE peptides and the vicilin-like protein were produced by cleaving Asn-Gln bonds of PV100, and that all had a pyroglutamate at their N-terminus.

In vitro processing of PV100 with the purified VPE revealed that VPE is responsible for cleaving Asn-Gln bonds of PV100 to produce multiple seed proteins. Figure 2.3 shows the hydrophobicity plot of PV100 and a mechanism for vacuolar processing of PV100 to produce C2 peptide, RE peptides and a vicilin-like protein. PV100 has nine Asn-Gln bonds in the polypeptide. All six Asn-Gln bonds to be cleaved are located in the hydrophilic region of the PV100 sequence, while the other three non-cleavable Asn-Gln bonds are found in the hydrophobic region of the vicilin-like domain. This result is consistent with previous data showing that VPE recognizes Asn residues that are located in the hydrophilic region and are exposed on the surface of precursor molecules (Hara-Nishimura *et al.*, 1993a). VPE cleavage of an Asn-Gln bond gives a new N-terminal Gln residue which might be converted, spontaneously, into a

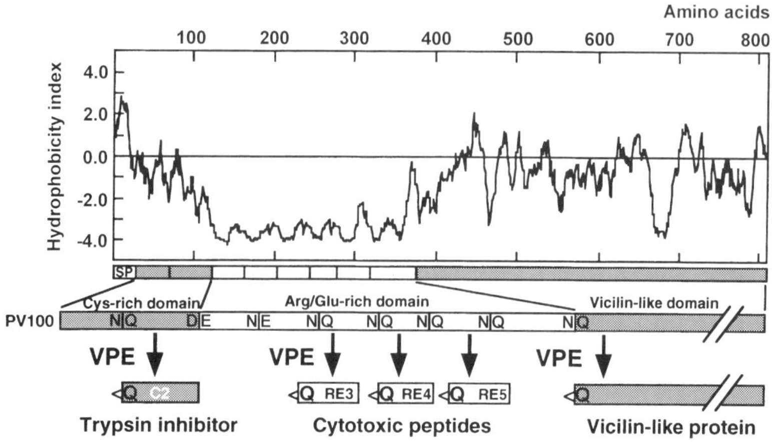

Figure 2.3 Hydropathy profile of PV100 and a mechanism for the vacuolar processing enzyme (VPE)-mediated cleavage at Asn-Gln bonds to produce multiple functional proteins. The mean hydrophobicity index was computed according to the algorithm of Kyte and Doolittle (1982) with a window of 10 residues. VPE is responsible for maturation of multiple seed proteins by cleaving Asn-Gln bonds, which are found in the hydrophilic region of the PV100. Gln at the new N-termini of the mature proteins might be spontaneously converted into pyroglutamate (< Q) under the acidic condition in the vacuoles. The cysteine-rich C2 peptide, the Arg/Glu-rich RE3, RE4 and RE5 peptides and the vicilin-like protein are produced. SP represents a signal peptide.

pyroglutamate under the acidic conditions in the vacuoles. Proteins with an N-terminus blocked by pyroglutamate are resistant to aminopeptidases that are localized in vacuoles. This suggested that the Asn-Gln cassettes in a single precursor sequence not only provide sites that can be cleaved by VPE, but also produce aminopeptidase-resistant functional proteins in the vacuoles.

The C2 peptide was shown to have trypsin inhibitory activity (Yamada *et al.*, 1999), and the RE peptides are known to be cytotoxic (Naisbitt *et al.*, 1988). These peptides might play a role in protecting the seeds from animals. PV100 itself, however, showed no trypsin inhibitory activity. This indicates that VPE-mediated maturation is coupled with the activation of functional vacuolar proteins. This finding also demonstrated that VPE converts a single precursor into multiple functional proteins, including a trypsin inhibitor, cytotoxic peptides and a vicilin-like storage protein. Thus, PV100 is thought to be not only a precursor of a vicilin storage protein but also a unique precursor for multiple seed proteins, with different biological functions. It is interesting that multiple vacuolar proteins are derived from a single precursor. Vacuolar targeting signals have been shown to be present in the N- and/or C-terminal propeptide(s) of some vacuolar proteins, including barley aleurain (Holwerda *et al.*, 1992), barley lectin (Bednarek and Raikhel, 1991), sweet potato sporamin (Matsuoka and Nakamura, 1991) and tobacco chitinase (Neuhaus *et al.*, 1991). The propeptides have been thought to be cleaved off and degraded after arrival of the proproteins in vacuoles. However, the possibility cannot be excluded that the propeptides exhibit some functions in the vacuoles after being removed from the precursor proteins, just as the C2 peptide and the RE peptides are accumulated to act as functional proteins in the vacuoles.

2.2.3 Vacuolar processing system in lytic vacuoles

Arabidopsis has three VPE homologs; βVPE is specific to seeds, and αVPE and γVPE are specific to vegetative organs (Figure 2.1). The *βVPE* promoter was found to direct the expression of the β-glucuronidase reporter gene in seeds of transgenic *Arabidopsis* plants. The decrease of *βVPE* gene expression in *Arabidopsis* cotyledons after seed germination is consistent with the observation that both the activity and protein levels of VPE rapidly decrease in castor bean endosperm during seed germination (Hiraiwa *et al.*, 1993). These results indicate that *Arabidopsis* βVPE is a counterpart to the seed VPEs of castor bean and pumpkin (Hara-Nishimura and Nishimura, 1987; Hara-Nishimura *et al.*, 1991a, 1995), and plays a role in maturation of seed proteins in PSVs of maturing *Arabidopsis* seeds.

VPE activity was also detected in vegetative tissues, including mature leaves, roots and hypocotyls of castor bean, pumpkin and mung bean, although the specific activity was much lower than that of the seeds of these plants (Hiraiwa *et al.*, 1993). The vegetative *Arabidopsis* γVPE was expressed in the yeast, *Saccharomyces cerevisiae*, to show that γVPE, like the seed VPE, has the same ability to cleave the peptide bond at the carbonyl side of asparagine residues (Kinoshita *et al.*, 1999). An immunocytochemical analysis revealed that γVPE is localized in lytic vacuoles of *Arabidopsis* leaves. These findings indicate that γVPE functions in lytic vacuoles as βVPE does in PSVs.

Both *αVPE* and *γVPE* promoters directed expression in senescent tissues but not in young intact tissues. The mRNA levels of both *αVPE* and *γVPE* were increased in primary leaves during senescence, in parallel with increased mRNA levels of a senescence-associated gene (*SAG2*). Two VPE homologs with a substrate specificity towards asparagine residues were purified from cotyledons of vetch (proteinase B) (Becker *et al.*, 1995) and *Vigna mungo* (VmPE-1) (Okamoto and Minamikawa, 1995). The level of mRNA for proteinase B in cotyledons increases after germination (Becker *et al.*, 1995). This is in good agreement with the results from gene expression experiments in cotyledons using the *αVPE* and *γVPE* gene promoters. The vegetative VPE might regulate the activation of some functional proteins in lytic vacuoles.

The temporal and spatial patterns of gene expression of the vegetative VPEs implies that target proteins of vegetative VPEs include senescence-inducible hydrolytic enzymes localized in the vacuoles, such as various cysteine proteinases that appear in tissues undergoing senescence and degeneration of different plants. These include: actinidin in ripening kiwi fruits (Lin *et al.*, 1993); C14 in ripening tomato (Drake *et al.*, 1996); Cp in senescent carnation and daylily flowers (Jones *et al.*, 1995; Valpuesta *et al.*, 1995); Tpp in senescent pea ovaries (Granell *et al.*, 1992); oryzain α and oryzain β in degenerating endosperm of rice (Watanabe *et al.*, 1991); and RD21 in drought-stressed *Arabidopsis* leaves (Koizumi *et al.*, 1993). These proteinases are members of the papain family. The inactive precursor of kiwi fruit actinidin, one of the better characterized proteinases, has been shown to be converted into an active enzyme by the removal of a C-terminal propeptide at an Asn-Gln bond (Baker, 1980; Paul *et al.*, 1995). The cleavable asparagine residue in a hydrophilic region of the precursor polypeptides is conserved among all the cysteine proteinases described above. It seems likely that the vegetative VPEs with a substrate specificity towards asparagine residues might be involved in maturation and activation of the inducible cysteine proteinases in the lytic vacuoles of senescing and degenerating tissues. This is also supported by the evidence that a VPE homolog, VmPE-1, is involved in post-translational processing of a cysteine endopeptidase, SH-EP, to activate the enzyme in germinating cotyledons of *Vigna mungo* (Okamoto and Minamikawa, 1995).

The expression of *αVPE* and *γVPE* were also upregulated by treatment with wounding, ethylene and salicylic acid. These gene expression patterns of the vegetative *VPEs* were associated with the accumulation of vacuolar proteins that are known to respond to these treatments. It has been shown that accumulation of proteinase inhibitors in vacuoles of tomato leaves (Graham *et al.*, 1985) and tobacco stigmas (Atkinson *et al.*, 1993) is induced by wound treatment. The inhibitors are synthesized as proprotein precursors and then converted into mature inhibitors by post-translational cleavages at specific asparagine residues (Graham *et al.*, 1985; Atkinson *et al.*, 1993). Vegetative *VPE* mRNAs were upregulated in response to mechanical wounding (Kinoshita *et al.*, 1999). The wound-inducible VPEs could mediate the maturation of the inhibitors by cleaving peptide bonds at the asparagine residues.

When pathogens attack plants, salicylic acid and ethylene mediate the accumulation of defensive proteins (PR proteins) in vacuoles and extracellular matrix. A basic chitinase of tobacco leaves, one of the PR proteins that accumulates in the vacuoles, has been shown to be converted from the proprotein precursor form by removal of a C-terminal propeptide at an asparagine residue (Sticher *et al.*, 1993). Treatment of leaves with salicylic acid and ethylene also upregulated vegetative *VPE* mRNAs. The vegetative VPEs that were upregulated in response to ethylene could be involved in the maturation of the chitinase to defend the tissues against pathogens. The vegetative VPEs might regulate activation of such functional proteins to give lytic vacuoles new functions, including defense of the cells.

The lower expression of the vegetative *VPE* genes in young leaves revealed that VPE function might not be associated with vacuolar enlargement. In contrast, high expression of these *VPE* genes was observed in senescent and stressed tissues, in which lytic vacuoles have been shown to accumulate newly synthesized proteins that are converted into their mature forms and degrade cellular components. The high expression of vegetative VPEs could reflect the active accumulation of functional proteins in the vacuoles and the rapid degradation of unnecessary proteins to promote cell death and to cause cell differentiation in response to stress. Thus, the VPEs could be a good marker to identify vacuoles that acquire different functions in tissues.

2.2.4 Autocatalytic activation of VPEs

The castor bean VPE precursor is composed of a signal peptide, an N-terminal propeptide, the mature VPE domain and a C-terminal propeptide (Hara-Nishimura *et al.*, 1993b). The precursor in PAC vesicles is a latent form and is converted proteolytically into an active form upon arrival in the PSVs, after fusion of the PAC vesicles and the vacuoles (Hara-Nishimura *et al.*, 1993b, 1995; Hiraiwa *et al.*, 1993). Therefore, post-translational removal of propeptides from the precursor is required for activation of the VPE.

To examine the mechanism of activation of a VPE, the precursor to castor bean VPE (proVPE) was expressed in a *pep4* strain of the yeast, *Saccharomyces cerevisiae*, under the control of the *GAL1* promoter (Hiraiwa *et al.*, 1997b). During growth of the transformant, the level of proVPE increased slightly until a rapid decrease occurred after 9 h induction with galactose. In contrast, mature VPE (mVPE) appeared simultaneously with the disappearance of proVPE. While proVPE had no activity, mVPE had a vacuolar processing activity. Both proVPE and mVPE were glycosylated in yeast cells, although VPE is not glycosylated in plant cells in spite of the presence of two *N*-linked glycosylation sites. The specific activity of the glycosylated mVPE was similar to that of VPE in plant cells. The yeast transformant accumulated both proVPE and mVPE in the vacuole. These findings suggest that an inactive VPE precursor synthesized on the rER is transported to the vacuole in the yeast cells and then processed to make an active VPE within the vacuole.

VPEs are cysteine proteinases whose activity is abolished by treatment with thiol reagents. It has been shown that His180 and Cys222 residues of castor bean VPE are essential for the proteolytic activity (Hiraiwa *et al.*, 1997b). To clarify mechanisms for conversion of proVPE into the active enzyme, point mutated proVPEs were expressed in the yeast *pep4* mutant (Hiraiwa *et al.*, 1999). A proVPE with a Gly substitution for the active site, Cys222, showed no ability to convert itself into mVPE, although a wild proVPE had that ability. This proVPE C222G mutant, however, was properly processed to mature form by the action of purified VPE from castor bean endosperm. This suggested an autocatalytic mechanism for conversion of the VPE precursor into the mature enzyme.

The cleavage of the N-terminal propeptide of the VPE homolog from Jack bean occurs on the C-terminal side of the Asp residue at position 35

(Abe *et al.*, 1993). The respective Asp residues are conserved in the VPE homologs (Hiraiwa *et al.*, 1999). Originally, VPE was characterized as an asparaginyl endopeptidase (Hara-Nishimura, 1998; Hara-Nishimura *et al.*, 1991a). *In vitro* assays with synthetic peptides, however, demonstrated that a VPE exhibited an activity towards Asp residues (Becker *et al.*, 1995; Hiraiwa *et al.*, 1997a), and that a VPE cleaved an Asp-Gln bond to remove its N-terminal propeptide (Hiraiwa *et al.*, 1999).

The molecular mass of castor bean VPE suggests that cleavage of the C-terminal propeptide might occur at Asp429-Asp430. The Asp-Asp sequence is conserved among VPE homologs (Hiraiwa *et al.*, 1999). To examine whether the di-aspartic acid motif is the processing site of the C-terminal propeptide, a mutant VPE (D429G/D430G), in which Asp429 and Asp430 were substituted with Gly residues, was constructed. The transformant with the mutant accumulated the precursor, but not the mature form. It should be noted that an intermediate VPE with the C-terminal propeptide but no N-terminal propeptide was not detected in the transformant. The VPE is autocatalytically processed to be converted into the active enzyme by removal of the C-terminal propeptide, and subsequent removal of the N-terminal propeptide. The C-terminal propeptide of VPEs, therefore, might act as an inhibitor that regulates processing activity.

Figure 2.4 shows a hypothetical model for a vacuolar processing system. Autocatalytic conversion of the VPE precursor into the mature form indicates that no other factor is necessary to activate VPE. VPE functions as a key enzyme in the system responsible for maturation and/or activation of various vacuolar proteins in the vacuoles of higher plants. The putative target vacuolar

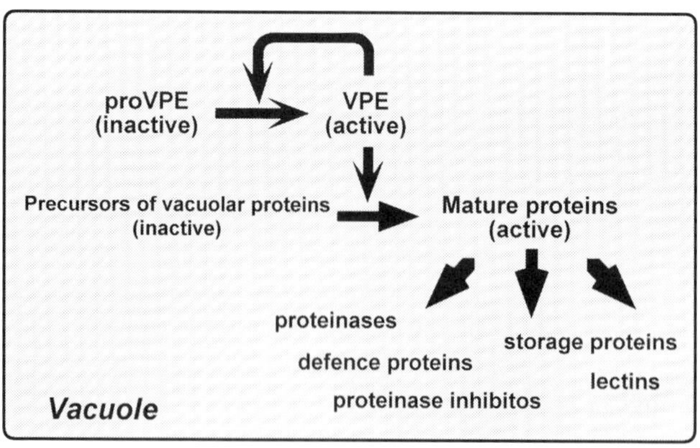

Figure 2.4 A hypothetical model for a vacuolar processing system in plant vacuoles. The inactive precursor to VPE (proVPE) is converted into the active vacuolar processing enzyme (VPE) by self-catalytic proteolysis (Hiraiwa *et al.*, 1997b). No other factor is necessary to activate VPE. VPE functions as a key enzyme of the system responsible for maturation and/or activation of various vacuolar proteins.

proteins include lytic enzymes, defense proteins, proteinase inhibitors, lectins and storage proteins. VPE might regulate activation of such functional proteins to give plant vacuoles new functions.

2.3 Aquaporins in plant vacuoles

2.3.1 Variety of aquaporins in vacuolar and plasma membranes

Aquaporins facilitate water transport across biomembranes in an osmotic pressure-dependent manner, and have been thought to play indispensable physiological roles in plants. The discovery of aquaporins has propelled studies on the transcellular and intracellular water transport at the molecular level (Maurel, 1997; Maurel *et al.*, 1997a; Chrispeels *et al.*, 1999; Kjellbom *et al.*, 1999; Tyerman *et al.*, 1999). Aquaporins belong to an ubiquitous membrane intrinsic protein (MIP) family (Agre *et al.*, 1995; Park and Saier, 1996; Maurel, 1997; Chrispeels *et al.*, 1999). The MIP family proteins have common, basic structures; namely six membrane-spanning domains and two Asn-Pro-Ala motifs. Aquaporins exist as a homotetramer of identical polypeptides having a molecular mass of 23–30 kDa (Figure 2.5). The three-dimensional structure of animal aquaporins has been determined from two-dimensional crystallography (Cheng *et al.*, 1997; Walz *et al.*, 1997). α-TIP has also been demonstrated to exist as a 6×6 nm square tetramer (Daniels *et al.*, 1999). Each subunit is formed by a heart-shaped ring comprised of six α-helices.

Plant aquaporins possess some unique characteristics. Firstly, there are two distinct types of aquaporin, plasmalemma intrinsic proteins (PIPs) and tonoplast intrinsic proteins (TIPs), localized in plasma and vacuolar membranes, respectively. This dual distribution is unique to the plant aquaporins. The presence of aquaporins in the endomembrane system is peculiar to plants. It is closely related to the presence of a huge central vacuole. TIPs have been thought to participate in the rapid osmotic equilibration between the cytosol and vacuoles. The precedent from studies of a yeast glycerol facilitator (Coury *et al.*, 1999) indicates that TIPs and PIPs may function only at their final locations, namely, in the vacuolar and plasma membranes, respectively. Thus, the proteins may not function in the ER or Golgi apparatus during biosynthesis and intracellular trafficking. There is also another type of plant aquaporin, nodulin 26, which is a nodule-specific symbiosome membrane protein (Weaver *et al.*, 1994).

Secondly, TIPs are abundant in land plants, such as radish, mung bean, *A. thaliana* and beet (Johnson *et al.*, 1989; Maeshima, 1992; Marty-Mazars *et al.*, 1995). This high abundance is similar to that for the band 3 protein (anion antiporter) in the red blood cell membrane. There are a few exceptions in plants. The TIP content is relatively low in tobacco suspension cells (Matsuoka *et al.*, 1997), and is extremely low in the vacuolar membrane of constitutive crassulacean acid metabolism (CAM) plants (Maeshima *et al.*, 1994b). A high

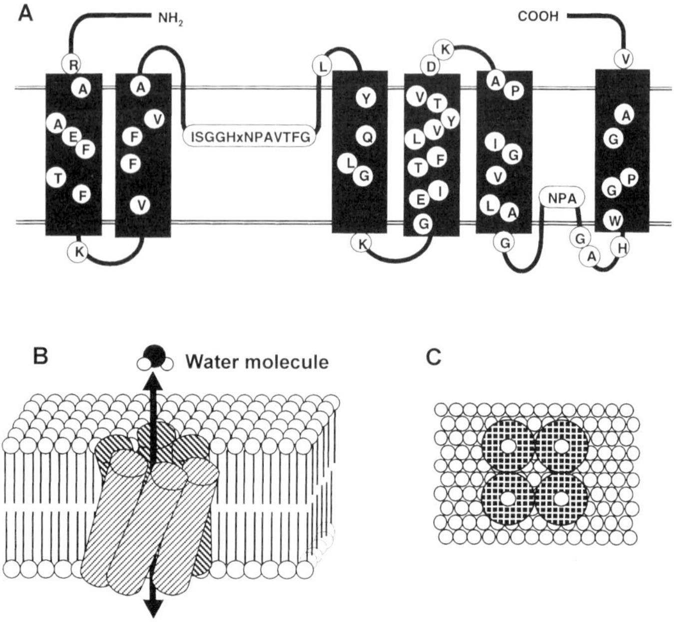

Figure 2.5 Schematic representation of aquaporins. (A) Aquaporins contain six transmembrane domains. The amino acid residues in circles are identical among six PIPs (PAQ1, 1b, 1c, 2, 2b and 2c) and two TIPs (γ-VM23 and δ-VM23) of radish (Suga and Maeshima, unpublished data, available on the DNA database). (B) Three-dimensional model of aquaporin monomer. The schema is based on the structure for mammalian aquaporin, AQP1 (Cheng *et al.*, 1997; Walz *et al.*, 1997). (C) Tetrameric form of aquaporin. Abbreviations: PIPs, plasmalemma intrinsic proteins; TIPs, tonoplast intrinsic proteins.

water permeability of vacuolar membrane containing abundant TIP is effective to prevent plasmolysis.

Thirdly, many different molecular species of aquaporin are present both in plasma and vacuolar membranes (Maurel, 1997). *A. thaliana* has been reported to contain 23 isoforms of aquaporin (Weig *et al.*, 1997). The localization of many different aquaporins in the same membrane indicates that they may work cooperatively. The identity of amino acid sequences between TIPs and PIPs is less than 40%. Such sequence diversity is characteristic of aquaporins of animals and plants. TIPs, 23–26 kDa on sodium dodecyl sulphate-polyacrylamide gel electrophoresis (SDS-PAGE), are smaller than PIPs (30 kDa); in particular, the N-terminal hydrophilic portions of TIPs are shorter than those of PIPs. There are a few regions specific to TIP or PIP. The TIP- or PIP-specific region may contribute to membrane trafficking, assembly, and regulation of aquaporin function.

Reports on the cDNA cloning of plant aquaporins have been increasing year by year, but the water channel activity for many has not been determined. In general, the water channel activity has been determined using *Xenopus* oocytes into which the mRNA for an aquaporin has been injected. The aquaporin expressed in the oocyte facilitates water transport across the cell membrane. The osmotic water permeability can be calculated from the rate of shrinking or swelling of the oocyte. Using the oocyte expression system, water transport activity has been demonstrated for PIPs (Daniels *et al.*, 1994; Kammerloher *et al.*, 1994; Kaldenhoff *et al.*, 1995; Yamada *et al.*, 1995), TIPs (Maurel *et al.*, 1993, 1995; Daniels *et al.*, 1996; Sarda *et al.*, 1997; Higuchi *et al.*, 1998) and nodulin 26 (Dean *et al.*, 1999). At present, all plant aquaporins examined have enhanced the osmotic water permeability. However, the reason why TIP expressed in *Xenopus* oocytes is localized on the oocyte plasma membrane has not been explained. γ-TIP fused with a green fluorescent protein (GFP) was clearly demonstrated to be localized in the vacuolar membrane of tobacco cells, as shown in Chapter 6 of this volume (Tsuchiya, Matsuoka and Maeshima, unpublished data).

Recently, plant aquaporins, such as tobacco TIP (Gerbeau *et al.*, 1999) and nodulin 26 (Dean *et al.*, 1999), have been reported to possess the ability to translocate not only water molecules but also small neutral solutes, such as glycerol and urea. In some plant species, the glycerol content in the tissues increases during adaptation to low temperatures in late autumn. There is a possibility that PIPs and TIPs take part in the inter- and intracellular transport of glycerol. Prasad *et al.* (1998) also demonstrated that in human red blood cells aquaporin transports CO_2, not bicarbonate, across membranes. Plant physiologists take a great interest in whether the CO_2 molecule is transported across the plasma membrane by a mechanism of simple diffusion or by an aquaporin-mediated system. In order to obtain an answer to this question, the substrate of each aquaporin should be determined using a proteoliposome reconstituted with the purified aquaporin. For the transport assay, it may be possible to use protoplasts or intact vacuoles from yeast that heterologously express a plant aquaporin.

The quantity of an aquaporin in a membrane is also related to the question of its physiological importance. TIP is abundant in the vacuolar membrane of radish taproot and accounts for about 40% by weight of the total vacuolar membrane proteins (Maeshima, 1992; Higuchi *et al.*, 1998). In that tissue, the major molecular species is γ-TIP (γ-VM23). The content of PIP has also been estimated to be about 10% of the plasma membrane protein of radish taproot (Imagawa, Ohshima, Suga and Maeshima, unpublished data). The physiological meaning of the abundance of TIP and PIP, the differences in the physiological functions of individual isoforms, and the physiological relationships between vacuolar and plasma membrane aquaporins have not yet been clarified. Hereafter, this chapter focuses on TIP, especially on the relationship between the heterogeneity of vacuolar function and the variety of TIP.

Functional and morphological differentiation of plant vacuoles is closely related to their membrane functions, and depends upon the selectivity and capacity of the membrane transport of solutes. Some vacuolar membrane transporters and channels are constitutively expressed as an essential element in most vacuoles; other membrane components may vary depending upon the types of vacuole, cell, tissue, growth stage and physiological condition. For example, transporters specific to flower pigments, amino acids, flavonoids, alkaloids, phytoalexins, ascorbate, and toxic metal ions may define differentiated, specific vacuolar functions in plant cells. Aquaporin function, dependent on osmotic gradients, is a passive element. Thus, the rate of the aquaporin-mediated water transport depends upon the other active transporters and ion channels. These membrane components should be considered when examining the physiological roles of aquaporins in living cells. More extensive research is necessary to understand the exact substrate and function of each aquaporin.

2.3.2 *Expression and turnover of α-TIP and γ-TIP in seeds and vegetative tissues*

α-TIP and γ-TIP are well-characterized plant aquaporins. γ-TIP is the major TIP in vegetative tissues (Johnson *et al.*, 1989; Maeshima, 1992; Marty-Mazars *et al.*, 1995). It has been reported that γ-TIP was expressed at a constant level throughout development in cotton fibre cells (Smart *et al.*, 1998), but was most abundant in growing pear fruits (Shiratake *et al.*, 1998) and young cells of mung bean hypocotyls (Maeshima, 1990) and radish taproots (Maeshima *et al.*, 1996).

In contrast to γ-TIP, α-TIP has been characterized more extensively. The protein-storage vacuole (PSV) is a typical example of the dynamic changes that plant vacuoles can undergo. PSVs are formed from central vacuoles during seed formation, retained as protein bodies in dry seeds, and transformed into central vacuoles after seed germination. α-TIP was the first aquaporin to be biochemically identified in plants, and to be cloned as a cDNA. It is a specific marker for the PSV membrane of seeds and vegetative tissues, and shares about 60% amino acid sequence identity with γ-TIP (Johnson *et al.*, 1990; Höfte *et al.*, 1992). α-TIP has been detected as a major integral membrane protein of PSVs in developing and dry seeds (Johnson *et al.*, 1989; Melroy and Herman, 1991). In *Arabidopsis*, α-TIP has been reported to be expressed in the middle and late stages of seed formation (Höfte *et al.*, 1992). Inoue and co-workers (1995a,b) have demonstrated that α-TIP (MP28) was accumulated in the PSV membrane of pumpkin cotyledons at the early stage of seed formation, before accumulation of storage proteins. α-TIP has been demonstrated to possess water channel activity, as assessed by the *Xenopus* oocyte swelling assay (Maurel *et al.*, 1995). α-TIP probably facilitates water transport into immature PSVs in early stages of seed formation. At the late stage of seed maturation, α-TIP may facilitate water export from PSVs for efficient desiccation.

α-TIP synthesized during seed formation is retained during seed desiccation. After imbibition of water, PSVs in seeds are transformed to lytic, central vacuoles (Figure 2.6). The α-TIP level has been reported to decrease during seed germination (Johnson *et al.*, 1989; Melroy and Herman, 1991). In the case

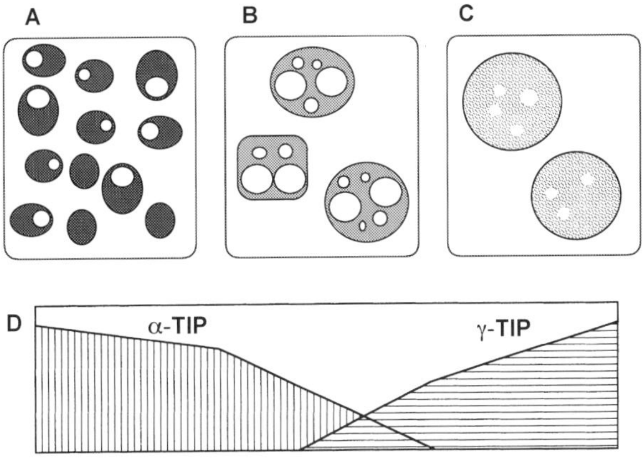

Figure 2.6 Transformation of protein-storage vacuoles (PSVs) to lytic vacuoles in pumpkin cotyledons during germination. Several PSVs exist in cotyledon cells (A). PSVs are fused to each other (B) and transformed to lytic vacuoles (C) after water imbibition. (D) Changes in the levels of α-TIP and γ-TIP during germination. Abbreviation: TIP, tonoplast intrinsic protein.

of pumpkin seed, α-TIP disappeared entirely from cotyledons after germination and was replaced with γ-TIP in a few days (Maeshima *et al.*, 1994a). The expression of γ-TIP in *Brassica napus* seeds has been reported to be enhanced by treatment with abscissic acid (ABA) (Gao *et al.*, 1999). Interestingly, vacuolar H^+-pyrophosphatase (V-PPase) that is newly synthesized during germination has been detected in PSV membranes from germinating pumpkin cotyledons (Maeshima *et al.*, 1994a). From studies on soybean PSVs using electron-microscopic immunocytochemistry, Melroy and Herman (1991) reported that internalized autophagic vesicles in PSVs were labeled with anti-α-TIP antibodies. These reports indicate that the PSV membrane should contain both the pre-existing PSV components (α-TIP) and the newly synthesized proteins (V-PPase). In germinating seeds, vesicles (prevacuolar compartment) carrying new components, such as vacuolar proton pumps, may fuse with the pre-existing PSV membrane; the old and new components would then mix freely within the membrane. The gradual turnover of membrane components may be completed through invagination and digestion. The half-life of the membrane components synthesized after germination has been estimated to be less than 2 days. The authors propose that this flow dilution mechanism is responsible for the replacement of α-TIP with γ-TIP during transformation of a PSV into a lytic vacuole.

In contrast to the transformation of PSVs to lytic vacuoles in cotyledons, barley aleurone cells contain two distinct types of vacuoles, PSVs and secondary vacuoles (Swanson *et al.*, 1998). Both were present in the same cells for several days during seed germination, although their numbers changed. Both vacuolar membranes were labeled with anti-α-TIP antibodies and the proteolytic activity in secondary vacuoles was higher than in PSVs. These studies indicated that secondary vacuoles are acidic (< pH 6), lytic organelles. Secondary vacuoles may fuse with PSVs in germinating aleurone cells. The treatment of aleurone cells with gibberellic acid decreased the lumenal pH of PSVs and secondary vacuoles (Swanson *et al.*, 1998), probably through activation of vacuolar H^+-ATPase and V-PPase (Swanson and Jones, 1996).

Rogers and co-workers (1996) have shown the presence of multiple distinct vacuolar compartments in the same cell of vegetative tissues. Using antibodies to α-TIP and γ-TIP, they distinguished PSVs carrying barley lectin, from lytic vacuoles carrying aleurain, a protease. In root tip cells, both types of vacuole are relatively small (1–3 μm). Microheterogeneity of vacuoles has been investigated in cells of root tip, cotyledon, leaf, flower petal and storage organs using additional antibodies specific to δ-TIP (Jauh *et al.*, 1998, 1999). It was found that δ-TIP was expressed in the cells of petunia flower petal and potato tuber, where neither α-TIP nor γ-TIP were detected (Jauh *et al.*, 1998). cDNAs for δ-TIPs have been cloned from *Arabidopsis* seedlings (Daniels *et al.*, 1996), germinated cotton cotyledons (Ferguson *et al.*, 1997), and radish seedlings (Higuchi *et al.*, 1998). It was also demonstrated that δ-TIP facilitated osmotic water permeability (Daniels *et al.*, 1996) and that the tissue expression patterns of δ-TIP were different from those of α-TIP and γ-TIP (Daniels *et al.*, 1996). Thus δ-TIP is expected to be a useful marker for storage vacuoles (not specific to PSVs) in vegetative tissues, although the universality of this principle remains to be examined in other plants.

2.3.3 Functional regulation and physiological roles of aquaporins

In addition to regulation of the quantity of TIP in a membrane, functional regulation of TIPs has been demonstrated. Johnson and Chrispeels (1992) have shown that a tonoplast-bound protein kinase phosphorylated bean cotyledon α-TIP in a Ca^{2+}-dependent manner. Spinach leaf PIPs were phosphorylated in response to an increased Ca^{2+} concentration and apoplastic turgor pressure (Johansson *et al.*, 1996). Water transport activity was enhanced twofold by phosphorylation of α-TIP at certain serine residues in the cytoplasmic, hydrophilic regions (Maurel *et al.*, 1995). In the oocyte expression system, cAMP-dependent protein kinase phosphorylated α-TIP but not γ-TIP (Maurel *et al.*, 1995, 1997a). Identification of an endogenous protein kinase specific to α-TIP, and detailed mechanisms of α-TIP phosphorylation remain to be resolved. It must be noted that the unphosphorylated form of α-TIP also has considerable water transport activity; additionally, γ-TIP and δ-TIP seem to constitutively express the water channel activity without regulation by phosphorylation.

Zeuthen and Klaerke (1999) reported that a mammalian aquaporin (AQP3, glycerol permeable channel) was gated by H^+. They determined the osmotic water permeability and the glycerol permeability in *Xenopus* oocytes expressing AQP3 at various conditions of external pH. The permeabilities of water and glycerol were abolished at acid pH values, with pK values of 6.4 and 6.1, respectively. Thus, this aquaporin may act as a glycerol and water channel at a physiological pH, but predominantly as a glycerol channel at around pH 6.1. It must be determined whether the water channel activity of TIPs is modulated by the pH of vacuoles or the cytoplasm.

Although the oocyte expression system has documented aquaporin function for TIPs, it has been more difficult to demonstrate this function *in planta*. Aquaporin-mediated water transport in plants or living cells has been suggested by results from experiments with sulphydryl reagents, potent inhibitors of aquaporin function (Maggio and Joly, 1995; Tazawa *et al.*, 1997). The osmotic water permeability of plasma membrane and tonoplast vesicles isolated from tobacco cells was directly assayed with a stopped-flow spectrophotometer (Maurel *et al.*, 1997b), and was reversibly inhibited by mercuric chloride in both types of membrane vesicle. In this respect, Barone *et al.* (1997) demonstrated that mercuric compounds induced conformational changes in a PIP. These results were consistent with the involvement of aquaporins in water permeability in plant cells and tissues, although it is important to note that certain plant aquaporins are insensitive to mercuric chloride (Daniels *et al.*, 1994).

Several groups have reported that TIPs are abundant in vacuolar membranes of special cells where a rapid water flux is predicted. In the mature pulvini of *Mimosa pudica*, there are motor cells that can rapidly shrink in response to mechanical stimuli, such as touch. The motor cell contains two types of vacuole, a small tannin-rich vacuole and a large central vacuole. The γ-TIP content in the central vacuole estimated on the basis of membrane area is 100-fold higher than that of the tannin-rich vacuole, and is 10-fold higher than that of vacuoles in young, nonreactive cells (Fleurat-Lessard *et al.*, 1997). This observation provides a direct correlation between TIP abundance and the volume regulation of the motor cells. Additionally, Serraj *et al.* reported that γ-TIP is abundant in the inner cortical cells of soybean nodules when compared with vascular-bundle cells and rhizobium-infected cells. They proposed that a large quantity of γ-TIP provides rapid, reversible intracellular water transport and is essential for an efficient O_2 supply in the nodule. These observations indicate that the level of γ-TIP in each tissue is related to the state of functional differentiation of vacuoles and cells. Studies in transgenic plants using a γ-TIP gene promoter fused to β-glucuronidase demonstrated high-level TIP expression in the elongating zone of young roots, the vascular bundles of stems, and the stipules and flower receptacles (Ludevid *et al.*, 1992). In cauliflower heads, expression of γ-TIP has been demonstrated in meristematic tissue in addition to the cells associated with vascular bundles (Barrieu *et al.*, 1998b). Maize γ-TIP has been shown to be predominantly expressed and accumulated in the epidermal and conducting tissues (Barrieu *et al.*, 1998a) and in the zones of cell division and elongation (Chaumont *et al.*, 1998). These results indicate that γ-TIP may be expressed at a high level in cells where there may be a demand for rapid water flow across the vacuolar membrane. The abundance of γ-TIP transcript and protein in meristematic cells seems to be related not only to the active biogenesis of vacuolar membrane but also the rapid turnover of vacuolar membrane in these cells as compared with fully differentiated cells.

For PIPs, a physiological role has been demonstrated in transgenic plants. Kaldenhoff *et al.* (1998) reported that reduced expression of a specific PIP by means of an antisense technique markedly decreased osmotic water permeability in protoplast swelling assays. Interestingly, the reduced water permeability at the cellular level was compensated for by a significant increase in root mass (Kaldenhoff *et al.*, 1998). Furthermore, circadian patterns to expression of PIP mRNAs have been suggested for tobacco cells (Yamada *et al.*, 1997) and *Lotus japonica* (Henzler *et al.*, 1999). If the amounts of PIP proteins reflect their transcriptional levels, a rapid turnover of PIP protein should occur in these plants. Diurnal changes in PIP proteins have not been determined, and the mechanism of regulation of vacuolar membrane turnover remains to be examined.

Some important questions remain to be answered concerning plant aquaporins: their actual substrates, their exact roles in water transport, biochemical regulation of the gate (open/closed), and the precise patterns of cell-specific expression for each aquaporin isoform, as discussed above. It is still unclear whether osmotic water permeability depends on the quantity of an aquaporin in the membrane, or upon the specific activity for water conductance of an aquaporin in the membrane. The osmotic water permeability of vacuolar membrane has been reported to be markedly higher than that for plasma membrane in the stopped-flow spectrophotometric assay (Maurel *et al.*, 1997b). The reason for high abundance of TIPs is also a question for plant scientists. A large quantity of water (more than 100 l/day) passes through the plasma membrane of the human kidney nephron. The aquaporin content relative to total membrane protein has been determined to be 0.88% for total membrane of rat renal cortex, 3.8% for membrane of rat renal brush border, and 2.4% for human red cell membranes (Nielsen *et al.*, 1993). In contrast, the levels of TIPs and PIPs in most plants are several times greater than those of mammalian aquaporins. An abundance of aquaporins enables plant cells and vacuoles to transport water quickly and in large quantities using a small osmotic gradient across the membrane. In addition to water channel function, other possible functions can be proposed to account for their high abundance, such as the stabilization of membranes and membrane enzymes, and the promotion or prevention of membrane fusion during biogenesis and functional differentiation of vacuoles.

References

Abe, Y., Shirane, K., Yokosawa, H., Matsushita, M., Mitta, M., Kato, I. and Ishii, S. (1993) Asparaginyl endopeptidase of Jack bean seed. *J. Biol. Chem.*, **268** 3525-29.

Agre, P., Brown, D. and Nielsen, S. (1995) Aquaporin water channels: unanswered questions and unresolved controversies. *Curr. Opin. Cell Biol.*, **7** 472-83.

Akazawa, T. and Hara-Nishimura, I. (1985) Topographic aspects of biosynthesis, extracellular secretion, and intracellular storage of proteins in plant cells. *Annu. Rev. Plant Physiol.*, **36** 441-72.

Alonso, J.M. and Granell, A. (1995) A putative vacuolar processing protease is regulated by ethylene and also during fruit ripening in citrus fruit. *Plant Physiol.*, **109** 541-47.

Atkinson, A.H., Heath, R.L., Simpson, R.J., Clarke, A.E. and Anderson, M.A. (1993) Proteinase inhibitors in *Nicotiana alata* stigmas are derived from a precursor protein, which is processed into five homologous inhibitors. *Plant Cell*, **5** 203-13.

Baker, E.N. (1980) Structure of actinidin, after refinement at 1.7 A resolution. *J. Mol. Biol.*, **141** 441-84.

Barone, L.M., Shih, C. and Wasserman, B.P. (1997) Mercury-induced conformational changes and identification of conserved surface loops in plasma membrane aquaporins from higher plants: topology of PMIP31 from *Beta vulgaris* L. *J. Biol. Chem.*, **272** 30672-77.

Barrieu, F., Chaumont, F. and Chrispeels, M.J. (1998a) High expression of the tonoplast aquaporin, ZmTIP1, in conducting tissues of maize. *Plant Physiol.*, **117** 1153-63.

Barrieu, F., Thomas, D., Marty-Mazars, D., Charbonnier, M. and Marty, F. (1998b) Tonoplast intrinsic proteins from cauliflower (*Brassica oleracea* L. var. botrytis): immunological analysis, cDNA cloning and evidence for expression in meristematic tissues. *Planta*, **204** 335-44.

Becker, C., Shutov, A.D., Nong, V.H., Senyuk, V.I., Jung, R., Horstmann, C., Fischer, J., Nielsen, N.C. and Munck, L. (1995) Purification, cDNA cloning and characterization of proteinase B, an asparagine-specific endopeptidase from germinating vetch (*Vicia sativa* L.) seeds. *Eur. J. Biochem.*, **228** 456-62.

Bednarek, S.Y. and Raikhel, N.V. (1991) The barley lectin carboxy-terminal propeptide is a vacuolar protein sorting determinant in plants. *Plant Cell*, **3** 1195-206.

Chaumont, F., Barrieu, F., Herman, E.M. and Chrispeels, M.J. (1998) Characterization of a maize tonoplast aquaporin expressed in zones of cell division and elongation. *Plant Physiol.*, **117** 1143-52.

Chen, J.M., Dando, P.M., Rawlings, N.D., Brown, M.A., Young, N.E., Stevens, R.A., Hewitt, E., Watts, C. and Barrett, A.J. (1997) Cloning, isolation and characterization of mammalian legumain and asparaginyl endopeptidase. *J. Biol. Chem.*, **272** 8090-98.

Chen, J.M., Dando, P.M., Stevens, R.A.E., Fortunato, M. and Barrett, A.J. (1998) Cloning and expression of mouse legumain, a lysosomal endopeptidase. *Biochem. J.*, **335** 111-17.

Cheng, A., van Hoek, A.N., Yeager, M., Verkman, A.S. and Mitra, A.K. (1997) Three-dimensional organization of a human water channel. *Nature*, **387** 627-30.

Chrispeels, M.J. (1991) Sorting of proteins in the secretory system. *Annu. Rev. Plant Physiol.*, **42** 21-53.

Chrispeels, M.J., Crawford, N.M. and Schroeder, J.I. (1999) Proteins for transport of water and mineral nutrients across the membranes of plant cells. *Plant Cell*, **11** 661-75.

Coury, L.A., Hiller, M., Mathai, J.C., Jones, E.W., Zeidel, M.L. and Brodsky, J.L. (1999) Water transport across yeast vacuolar and plasma membrane-targeted secretory vesicles occurs by passive diffusion. *J. Bacteriol.*, **181** 4437-40.

Daniels, M.J., Mirkov, T.E. and Chrispeels, M.J. (1994) The plasma membrane of *Arabidopsis thaliana* contains mercury-insensitive aquaporin that is a homolog of the tonoplast water channel protein, TIP. *Plant Physiol.*, **106** 1325-33.

Daniels, M.J., Chaumont, F., Mirkov, T.E. and Chrispeels, M.J. (1996) Characterization of a new vacuolar membrane aquaporin sensitive to mercury at a unique site. *Plant Cell*, **8** 587-99.

Daniels, M.J., Chrispeels, M.J. and Yeager, M. (1999) Projection structure of a plant vacuole membrane aquaporin by electron cryo-crystallography. *J. Mol. Biol.*, **294** 1337-49.

Dean, R.M., Rivers, R.L., Zeidel, M.L. and Roberts, D.M. (1999) Purification and functional reconstitution of soybean nodulin 26: an aquaporin with water and glycerol transport properties. *Biochemistry*, **38** 347-53.

Drake, R., John, I., Farrell, A., Cooper, W., Schuch, W. and Grierson, D. (1996) Isolation and analysis of cDNAs encoding tomato cysteine proteases expressed during leaf senescence. *Plant Mol. Biol.*, **30** 755-67.

Ferguson, D.L., Turley, R.B. and Kloth, R.H. (1997) Identification of a δ-TIP cDNA clone and determination of related A and D genome subfamilies in *Gossypium* species. *Plant Mol. Biol.*, **34** 111-18.

Fleurat-Lessard, P., Frangne, N., Maeshima, M., Ratajczak, R., Bonnemain, J.L. and Martinoia, E. (1997) Increased expression of vacuolar aquaporin and H^+-ATPase related to motor cell function in *Mimosa pudica* L. *Plant Physiol.*, **114** 827-34.

Fukasawa, T., Hara-Nishimura, I. and Nishimura, M. (1988) Biosynthesis, intracellular transport and *in vitro* processing of 11S globulin precursor proteins of developing castor bean endosperm. *Plant Cell Physiol.*, **29** 339-345.

Gao, Y.P., Young, L., Bonham-Smith, P. and Gusta, L.V. (1999) Characterization and expression of plasma and tonoplast membrane aquaporins in primed seed of *Brassica napus* during germination under stress conditions. *Plant Mol. Biol.*, **40** 635-44.

Gerbeau, P., Guclu, J., Ripoche, P. and Maurel, C. (1999) Aquaporin Nt-TIPa can account for the high permeability of tobacco cell vacuolar membrane to small neutral solutes. *Plant J.*, **18** 577-87.

Graham, J.S., Pearce, G., Merryweather, J., Titani, K., Ericsson, H.L. and Ryan, C.A. (1985) Wound-induced proteinase inhibitors from tomato leaves. I. The cDNA-deduced primary structure of preinhibitor I and its post-translational processing. *J. Biol. Chem.*, **260** 6555-60.

Granell, A., Harris, N., Pisabarro, A.G. and Carbonell, J. (1992) Temporal and spatial expression of a thiol protease gene during pea ovary senescence, and its regulation by gibberellin. *Plant J.*, **2** 907-15.

Hara-Nishimura, I. (1998) Asparaginyl endopeptidase, in *Handbook of Proteolytic Enzymes* (eds. A.J. Barrett, N.D. Rawlings and J.F. Woessner), Academic Press, London, UK, pp. 746-49.

Hara-Nishimura, I. and Nishimura, M. (1987) Proglobulin processing enzyme in vacuoles isolated from developing pumpkin cotyledons. *Plant Physiol.*, **85** 440-45.

Hara-Nishimura, I., Nishimura, M. and Akazawa, T. (1985) Biosynthesis and intracellular transport of 11S globulin in developing pumpkin cotyledons. *Plant Physiol.*, **77** 747-52.

Hara-Nishimura, I., Inoue, K. and Nishimura, M. (1991a) A unique vacuolar processing enzyme responsible for conversion of several proprotein precursors into the mature forms. *FEBS Lett.*, **294** 89-93.

Hara-Nishimura, I., Inoue, K. and Nishimura, M. (1991b) Reversible transformation between vacuoles and protein bodies in pumpkin cotyledons, in *Molecular Approaches to Compartmentation and Metabolic Regulation* (eds. A. Huang and L. Taiz), American Society of Plant Physiologists, Rockville, Md, pp. 245-46.

Hara-Nishimura, I., Takeuchi, Y., Inoue, K. and Nishimura, M. (1993a) Vesicle transport and processing of the precursor to 2S albumin in pumpkin. *Plant J.*, **4** 793-800.

Hara-Nishimura, I., Takeuchi, Y. and Nishimura, M. (1993b) Molecular characterization of a vacuolar processing enzyme related to a putative cysteine proteinase of *Schistosoma mansoni*. *Plant Cell*, **5** 1651-59.

Hara-Nishimura, I., Shimada, T., Hiraiwa, N. and Nishimura, M. (1995) Vacuolar processing enzyme responsible for maturation of seed proteins. *J. Plant Physiol.*, **145** 632-40.

Hara-Nishimura, I., Kinoshita, T., Hiraiwa, N. and Nishimura, M. (1998a) Vacuolar processing enzymes in protein-storage vacuoles and lytic vacuoles. *J. Plant Physiol.*, **152** 668-74.

Hara-Nishimura, I., Shimada, T., Hatano, K., Takeuchi, Y. and Nishimura, M. (1998b) Transport of storage proteins to protein-storage vacuoles is mediated by large precursor-accumulating vesicles. *Plant Cell*, **10** 825-36.

Henzler, T., Waterhouse, R.N., Smyth, A.J., Carvajal, M., Cooke, D.T., Schaffner, A.R., Steudle, E. and Clarkson, D.T. (1999) Diurnal variations in hydraulic conductivity and root pressure can be correlated with the expression of putative aquaporins in the roots of *Lotus japonicus*. *Planta*, **210** 50-60.

Higgins, D.G., Bleasby, A.J. and Fuchs, R. (1992) Clustal V: improved software for multiple sequence alignment. *Comput. Appl. Biosci.*, **8** 189-91.

Higuchi, H., Suga, S., Tsuchiya, T., Hisada, H., Morishima, S., Okada, Y. and Maeshima, M. (1998) Molecular cloning, water channel activity and tissue specific expression of two isoforms of radish vacuolar aquaporin. *Plant Cell Physiol.*, **39** 905-13.

Hiraiwa, N., Takeuchi, Y., Nishimura, M. and Hara-Nishimura, I. (1993) A vacuolar processing enzyme in maturing and germinating seeds: its distribution and associated changes during development. *Plant Cell Physiol.*, **34** 1197-204.

Hiraiwa, N., Kondo, M., Nishimura, M. and Hara-Nishimura, I. (1997a) An aspartic proteinase is involved in the maturation of storage proteins in concert with the vacuolar processing enzyme. *Eur. J. Biochem.*, **246** 133-41.

Hiraiwa, N., Nishimura, M. and Hara-Nishimura, I. (1997b) Expression and activation of the vacuolar processing enzyme in *Saccharomyces cerevisiae*. *Plant J.*, **12** 819-29.

Hiraiwa, N., Nishimura, M. and Hara-Nishimura, I. (1999) Vacuolar processing enzyme is self-catalytically activated by sequential removal of the C-terminal and N-terminal propeptides. *FEBS Lett.*, **44** 213-16.

Höfte, H., Hubberd, L., Reizer, J., Ludevid, D., Herman, E.M. and Chrispeels, M.J. (1992) Vegetative and seed-specific forms of tonoplast intrinsic protein in the vacuolar membrane of *Arabidopsis thaliana*. *Plant Physiol.*, **99** 561-70.

Holwerda, B.C., Padgett, H.S. and Rogers, J.C. (1992) Proaleurain vacuolar targeting is mediated by short contiguous peptide interactions. *Plant Cell*, **4** 307-18.

Inoue, K., Motozaki, A., Takeuchi, Y., Nishimura, M., Hara-Nishimura, I. (1995a) Molecular characterization of proteins in protein-body membrane that disappear most rapidly during transformation of protein bodies into vacuoles. *Plant J.*, **7** 235-43.

Inoue, K., Takeuchi, Y., Nishimura, M. and Hara-Nishimura, I. (1995b) Characterization of two integral membrane proteins located in the protein bodies of pumpkin seeds. *Plant Mol. Biol.*, **28** 1089-101.

Jauh, G.Y., Fischer, A.M., Grimes, H.D., Ryan, J.C.A. and Rogers, J.C. (1998) δ-Tonoplast intrinsic protein defines unique plant vacuole functions. *Proc. Natl. Acad. Sci. USA*, **95** 12995-99.

Jauh, G.Y., Phillips, T.E. and Rogers, J.C. (1999) Tonoplast intrinsic protein isoforms as markers for vacuolar functions. *Plant Cell*, **11** 1867-82.

Johansson, I., Larsson, C., Ek, B. and Kjellbom, P. (1996) The major integral proteins of spinach leaf plasma membranes are putative aquaporins and are phosphorylated in response to Ca^{2+} and apoplastic water potential. *Plant Cell*, **8** 1181-91.

Johnson, K.D. and Chrispeels, M.J. (1992) Tonoplast-bound protein kinase phosphorylates tonoplast intrinsic protein. *Plant Physiol.*, **100** 1787-95.

Johnson, K.D., Herman, E.M. and Chrispeels, M.J. (1989) An abundant, highly conserved tonoplast protein in seeds. *Plant Physiol.*, **91** 1006-13.

Johnson, K.D., Höfte, H. and Chrispeels, M.J. (1990) An intrinsic tonoplast protein of protein storage vacuoles in seeds is structurally related to a bacterial solute transporter (GlpF). *Plant Cell*, **2** 525-32.

Jones, M.L., Larsen, P.B. and Woodson, W.R. (1995) Ethylene-regulated expression of a carnation cysteine proteinase during flower petal senescence. *Plant Mol. Biol.*, **28** 505-12.

Kaldenhoff, R., Klling, A., Meyers, J., Karmann, U., Ruppel, G. and Richter, G. (1995) The blue light-responsive AthH2 gene of *Arabidopsis thaliana* is primarily expressed in expanding as well as in differentiating cells and encodes a putative channel protein of the plasmalemma. *Plant J.*, **7** 87-95.

Kaldenhoff, R., Grote, K., Zhu, J.J. and Zimmerman, U. (1998) Significance of plasmalemma aquaporins for water-transport in *Arabidopsis thaliana*. *Plant J.*, **14** 121-28.

Kammerloher, W., Fischer, U., Piechottka, G.P. and Schäffner, A.R. (1994) Water channels in the plant plasma membrane cloned by immunoselection from a mammalian expression system. *Plant J.*, **6** 187-99.

Kinoshita, T., Nishimura, M. and Hara-Nishimura, I. (1995a) Homologues of a vacuolar processing enzyme that are expressed in different organs in *Arabidopsis thaliana. Plant Mol. Biol.*, **29** 81-89.

Kinoshita, T., Nishimura, M. and Hara-Nishimura, I. (1995b) The sequence and expression of the γ-*VPE* gene, one member of a family of three genes for vacuolar processing enzymes in *Arabidopsis thaliana. Plant Cell Physiol.*, **36** 1555-62.

Kinoshita, T., Yamada, K., Hiraiwa, N., Nishimura, M. and Hara-Nishimura, I. (1999) Vacuolar processing enzyme is upregulated in the lytic vacuoles of vegetative tissues during senescence and under various stressed conditions. *Plant J.*, **19** 43-53.

Kjellbom, P., Larsson, C., Johansson, I., Karlsson, M. and Johanson, U. (1999) Aquaporins and water homeostasis in plants. *Trends Plant Sci.*, **4** 308-14.

Klinkert, M.-Q., Felleisen, R., Link, G., Ruppel, A. and Beck, E. (1989) Primary structure of Sm31/32 diagnostic proteins of *Schistosoma mansoni* and their identification as proteases. *Mol. Biochem. Parasitol.*, **33** 113-22.

Koizumi, M., Yamaguchi-Shinozaki, K., Tsuji, H. and Shinozaki, K. (1993) Structure and expression of two genes that encode distinct drought-inducible cysteine proteinases in *Arabidopsis thaliana. Gene*, **129** 175-82.

Kyte, J. and Doolittle, R.F. (1982) A simple method for displaying the hydropathic character of a protein. *J. Mol. Biol.*, **157** 105-32.

Lin, E., Burns, D.J.W. and Gardner, R.C. (1993) Fruit developmental regulation of the kiwi fruit actinidin promoter is conserved in transgenic petunia plants. *Plant Mol. Biol.*, **23** 489-99.

Ludevid, D., Höfte, H., Himelblau, E. and Chrispeels, M.J. (1992) The expression pattern of the tonoplast intrinsic protein γ-TIP in *Arabidopsis thaliana* is correlated with cell enlargement. *Plant Physiol.*, **100** 1633-39.

Maeshima, M. (1990) Development of vacuolar membranes during elongation of cells in mung bean hypocotyls. *Plant Cell Physiol.*, **31** 311-17.

Maeshima, M. (1992) Characterization of the major integral protein of vacuolar membrane. *Plant Physiol.*, **98** 1248-54.

Maeshima, M., Hara-Nishimura, I., Takeuchi, Y. and Nishimura, M. (1994a) Accumulation of vacuolar H^+-pyrophosphatase and H^+-ATPase during reformation of the central vacuole in germinating pumpkin seeds. *Plant Physiol.*, **106** 61-69.

Maeshima, M., Mimura, T. and Sato, T. (1994b) Distribution of vacuolar H^+-pyrophosphatase and a membrane integral protein in a variety of green plants. *Plant Cell Physiol.*, **35** 323-28.

Maeshima, M., Nakanishi, Y., Matsuura-Endo, C. and Tanaka, Y. (1996) Proton pumps of vacuolar membrane in growing plant cell. *J. Plant Res.*, **109** 119-25.

Maggio, A. and Joly, R.J. (1995) Effects of mercuric chloride on the hydraulic conductivity of tomato root system. *Plant Physiol.*, **109** 331-35.

Marty-Mazars, D., Clémencet, M.C., Dozolme, P. and Marty, F. (1995) Antibodies to the tonoplast from the storage parenchyma cells of beet root recognize a major intrinsic protein related to TIPs. *Eur. J. Cell Biol.*, **66** 106-18.

Matile, P. and Winkenbach, F. (1971) Function of lysozomes and lysozomal enzymes in the senescing corolla of morning glory. *J. Exp. Bot.*, **22** 759-71.

Matsuoka, K. and Nakamura, K. (1991) Propeptide of a precursor to a plant vacuolar protein required for vacuolar targeting. *Proc. Natl. Acad. Sci. USA*, **88** 834-38.

Matsuoka, K., Higuchi, T., Maeshima, M. and Nakamura, K. (1997) A vacuolar type H^+-ATPase in a nonvacuolar organelle is required for the sorting of soluble vacuolar protein precursors in tobacco cells. *Plant Cell*, **9** 533-46.

Maurel, C. (1997) Aquaporins and water permeability of plant membranes. *Annu. Rev. Plant Physiol. Plant Mol. Biol.*, **48** 399-429.

Maurel, C., Reizer, J., Schroeder, J.I. and Chrispeels, M.J. (1993) The vacuolar membrane protein γ-TIP creates water specific channels in *Xenopus* oocytes. *EMBO J.*, **12** 2241-47.

Maurel, C., Kado, R.T., Guern, J. and Chrispeels, M.J. (1995) Phosphorylation regulates the water channel activity of the seed-specific aquaporin α-TIP. *EMBO J.*, **14** 3028-35.

Maurel, C., Chrispeels, M., Lurin, C., Tacnet, F., Geelen, D., Ripoche, P. and Guern, J. (1997a) Function and regulation of seed aquaporins. *J. Exp. Bot.*, **48** 421-30.

Maurel, C., Tacnet, F., Guclu, J., Guern, J. and Ripoche, P. (1997b) Purified vesicles of tobacco cell vacuolar and plasma membranes exhibit dramatically different water permeability and water channel activity. *Proc. Natl. Acad. Sci. USA*, **94** 7103-08.

Melroy, D.L. and Herman, E.M. (1991) TIP, an integral membrane protein of the protein-storage vacuoles of the soybean cotyledon undergoes developmentally-regulated membrane accumulation and removal. *Planta*, **184** 113-22.

Naisbitt, G.H., Lu, M., Gray, W.R. and Vernon, L.P. (1988) Properties of a small basic peptide from pumpkin seeds. *Plant Physiol.*, **88** 770-73.

Neuhaus, J.-M., Sticher, L., Meins, F.J. and Boller, T. (1991) A short C-terminal sequence is necessary and sufficient for the targeting of chitinase to the plant vacuole. *Proc. Natl. Acad. Sci. USA*, **88** 10362-66.

Nielsen, S., Smith, B.L., Christensen, E.I., Knepper, M.A. and Agre, P. (1993) CHIP28 water channels are localized in constitutively water-permeable segments of the nephron. *J. Cell Biol.*, **120** 371-83.

Nishimura, M. and Beevers, H. (1979) Hydrolysis of protein in vacuoles isolated from higher plant tissues. *Nature*, **277** 412-13.

Okamoto, T. and Minamikawa, T. (1995) Purification of a processing enzyme (VmPE-1) that is involved in post-translational processing of a plant cysteine endopeptidase (SH-EP). *Eur. J. Biochem.*, **231** 300-05.

Okita, T.W. and Rogers, J.C. (1996) Compartmentation of proteins in the endomembrane system of plant cells. *Annu. Rev. Plant Physiol.*, **47** 327-50.

Paris, N., Stanley, C.M., Jones, R.L. and Rogers, J.C. (1996) Plant cells contain two functionally distinct vacuolar compartments. *Cell*, **85** 563-72.

Park, J.H. and Saier, J.M.H. (1996) Phylogenetic characterization of the MIP family of transmembrane channel proteins. *J. Membr. Biol.*, **153** 171-80.

Paul, W., Amiss, J., Try, R., Praekelt, U., Scott, R. and Smith, H. (1995) Correct processing of the kiwi fruit protease, actinidin, in transgenic tobacco requires the presence of the C-terminal propeptide. *Plant Physiol.*, **108** 261-68.

Prasad, G.V., Coury, L.A., Finn, F. and Zeidel, M.L. (1998) Reconstituted aquaporin 1 water channels transport CO_2 across membranes. *J. Biol. Chem.*, **273** 33123-26.

Robinson, D.G., Hoh, B., Hinz, G. and Jeong, B.-K. (1995) One vacuole or two vacuoles: do protein storage vacuoles arise *de novo* during pea cotyledon development? *J. Plant Physiol.*, **145** 654-64.

Sarda, X., Tousch, D., Ferrare, K., Legrand, E., Dupuis, J.M., Casse-Delbart, F. and Lamaze, T. (1997) Two TIP-like genes encoding aquaporins are expressed in sunflower guard cells. *Plant J.*, **12** 1103-11.

Serraj, R., Frangne, N., Maeshima, M., Fleurat-Lessard, P. and Drevon, J.J. (1998) A γ-TIP cross-reacting protein is abundant in the cortex of soybean N_2-fixing nodules. *Planta*, **206** 681-84.

Shimada, T., Hiraiwa, N., Nishimura, M. and Hara-Nishimura, I. (1994) Vacuolar processing enzyme of soybean that converts proprotein to the corresponding mature forms. *Plant Cell Physiol.*, **35** 713-18.

Shiratake, K., Kanayama, Y., Maeshima, M. and Yamaki, S. (1998) Changes in tonoplast protein and density with the development of pear fruit. *Physiol. Plant.*, **103** 312-19.

Smart, L.B., Vojdani, F., Maeshima, M. and Wilkins, T.A. (1998) Genes involved in osmoregulation during turgor-driven cell expansion of developing cotton fibers are differentially regulated. *Plant Physiol.*, **116** 1539-49.

Sticher, L., Hofsteenge, J., Neuhaus, J.-M., Boller, T. and Meins, F.J. (1993) Post-translational processing of a new class of hydroxyproline-containing proteins: prolyl hydroxylation and C-terminal cleavage of tobacco (*Nicotiana tabacum*) vacuolar chitinase. *Plant Physiol.*, **101** 1239-47.

Swanson, S.J. and Jones, R.L. (1996) Gibberellic acid induces vacuolar acidification in barley aleurone. *Plant Cell*, **8** 2211-21.

Swanson, S.J., Bethke, P.C. and Jones, R.L. (1998) Barley aleurone cells contain two types of vacuoles: characterization of lytic organelles by use of fluorescent probes. *Plant Cell*, **10** 685-98.

Takeda, O., Miura, Y., Mitta, M., Matsushita, H., Kato, I., Abe, Y., Yokosawa, H. and Ishii, S. (1994) Isolation and analysis of cDNA encoding a precursor of *Canavalia ensiformis* asparaginyl endopeptidase (legumain). *J. Biochem.*, **116** 541-46.

Tazawa, M., Ohkuma, E., Shibasaka, M. and Nakashima, S. (1997) Mercurial-sensitive water transport in barley roots. *J. Plant Res.*, **110** 435-42.

Tyerman, S.D., Bohnert, H.J., Maurel, C., Steudle, E. and Smith, J.A.C. (1999) Plant aquaporins: their molecular biology, biophysics and significance for plant water relations. *J. Exp. Bot.*, **50** 1055-71.

Valpuesta, V., Lange, N.E., Guerrero, C. and Reid, M.S. (1995) Upregulation of a cysteine protease accompanies the ethylene-insensitive senescence of daylily (*Hemerocallis*) flowers. *Plant Mol. Biol.*, **28** 575-82.

Walz, T., Hirai, T., Murata, K., Heymann, J.B., Mitsuoka, K., Fujiyoshi, Y., Smith, B.L., Agre, P. and Engel, A. (1997) The three-dimensional structure of aquaporin-1. *Nature*, **387** 624-27.

Watanabe, H., Abe, K., Emori, Y., Hosoyama, H. and Arai, S. (1991) Molecular cloning and gibberellin-induced expression of multiple cysteine proteinases of rice seeds (oryzains). *J. Biol. Chem.*, **266** 16897-902.

Weaver, C.D., Shower, N.H., Louis, C.F. and Roberts, D.M. (1994) Nodulin-26, a nodule-specific symbiosome membrane protein from soybean, is an ion channel. *J. Biol. Chem.*, **269** 17858-62.

Weig, A., Deswarte, C. and Chrispeels, M.J. (1997) The major intrinsic protein family of *Arabidopsis* has 23 members that form three distinct groups with functional aquaporins in each group. *Plant Physiol.*, **114** 1347-57.

Yamada, S., Katsuhara, M., Kelly, W.B., Michalowski, C.B. and Bohnert, H.J. (1995) A family of transcripts encoding water channel proteins: tissue-specific expression in the common ice plant. *Plant Cell*, **7** 1129-42.

Yamada, S., Komori, T., Myers, P.N., Kuwata, S., Kubo, T. and Imaseki, H. (1997) Expression of plasma membrane water channel genes under water stress in *Nicotiana excelsior. Plant Cell Physiol.*, **38** 1226-31.

Yamada, K., Shimada, T., Kondo, M., Nishimura, M. and Hara-Nishimura, I. (1999) Multiple functional proteins are produced by cleaving Asn-Gln bonds of a single precursor by vacuolar processing enzyme. *J. Biol. Chem.*, **274** 2563-70.

Zeuthen, T. and Klaerke, D.A. (1999) Transport of water and glycerol in aquaporin 3 is gated by H^+. *J. Biol. Chem.*, **274** 21631-36.

3 Contractile vacuoles

Peter Luykx

3.1 Introduction

Since cell membranes and their external surface coats and walls cannot be made completely impermeable to water, all metabolizing cells and organisms that live in dilute aqueous environments are subject to osmotic water influx, potentially diluting internal cellular concentrations of molecules and ions. The osmotic uptake of water is opposed by: 1) a porous but mechanically strong cell wall that counteracts osmotic swelling with internal hydrostatic pressure, for example, in bacteria and yeast cells, freshwater multicellular fungi, algae, and freshwater plants in general; 2) the existence, in most multicellular freshwater animals, of an extracellular space, into which is secreted a fluid isosmotic with tissues and cells in general and out of which is then pumped molecules and ions, leaving behind (in the extracellular space) a more dilute fluid, which is then transported back out to the aqueous environment, for example, in the kidneys of freshwater vertebrates and in analogous organs in freshwater invertebrates; and 3) the formation of a membrane-bound intracellular space—a contractile vacuole (CV)—into which water is moved from the cell cytoplasm and from which it is expelled at an average rate equivalent to the osmotic influx.

Contractile vacuoles are found in single cells—mostly plant and animal protists—living in fresh water, whose life styles and life cycles include cell flexibility and cell fusion, precluding their making use of rigid cell walls to counteract osmotic water influx. CVs also occur in some freshwater multicellular organisms: in freshwater sponges (Brauer and McKanna, 1978), and in the single-celled zoospores released by freshwater fungi and filamentous green algae (e.g. Marchant and Pickett-Heaps, 1970; Cain *et al.*, 1973; Marchant *et al.*, 1973).

There is little doubt that the primary function of the CV, occurring as it does in cells without rigid cell walls in dilute media, is osmoregulation. Firstly, the activity of the CV in such cells has often been shown to be inversely related to the osmotic strength of the external medium, for example, in amoebae (Hopkins, 1946); general reviews (Kitching, 1956, 1967); in *Crithidia* (trypanosome) (Cosgrove and Kessel, 1958); in *Paramecium* (ciliated protozoan) (Ishida *et al.*, 1996); and in *Chlamydomonas* (single-celled green alga) (Hellebust *et al.*, 1989; Luykx *et al.*, 1997a). Secondly, cells in which CV function is altered by specific mutations are unable to grow or survive in dilute media (Guillard, 1960; Denning and Fulton, 1989a; Luykx *et al.*, 1997b).

In addition to the contractile vacuole, mechanisms of general osmoregulation include: under conditions of high external osmolarity, the uptake of ions, the synthesis of osmolytes, and the degradation of intracellular macromolecules; and under conditions of low external osmolarity, diverse mechanisms leading to the release of ions and amino acids, with the reduction of total intracellular osmolarity (Burg, 1995; Raven, 1995; Steck *et al.*, 1997; Waldegger *et al.*, 1997; Lang *et al.*, 1998b); changes in the water permeability of the plasma membrane may also be involved. The CV and associated membranes may be regarded as a special adaptation to dilute environments, in which osmotic influx of water occurs at rates too great for other osmoregulatory mechanisms to handle.

Osmoregulation is not the only function of the contractile vacuole. In some cells, the CV may be secondarily appropriated for: transport to the cell surface of components of the plasma membrane, cell surface coats and cell wall (Weiss, 1983; Grief and Shaw, 1987; Domozych and Dairman, 1993; Sesaki *et al.*, 1997); endocytosis (de Chastellier *et al.*, 1978; Domozych and Nimmons, 1992); calcium homeostasis (Zhu *et al.*, 1993; Rooney *et al.*, 1994; Xie *et al.*, 1996; Moniakis *et al.*, 1999); and, possibly, the elimination of excess bicarbonate (Heuser *et al.*, 1993). The contractile vacuole may also participate in the regulation of intracellular pH (Nelson and Nelson, 1990).

Some osmoregulation-related characteristics of the activity of CV membranes—membrane fusion, rapid membrane cycling between different configurations, and responsiveness to external conditions—have their counterparts in other well-studied membrane phenomena: exocytosis in secretory cells, synaptic vesicle recycling in nerve terminals, and cell signal transduction. The most interesting function of the CV membrane, however, is segregation of water from the rest of the cytoplasm, apparently against an osmotic gradient. Water movement against osmotic gradients also occurs across the plasma membranes of vertebrate epithelial cells in some circumstances (Zeuthen, 1992). The question of reverse-osmotic movement of water across cell membranes is more easily studied in microorganisms, where biochemical and genetic approaches are more feasible (e.g. O'Halloran and Anderson, 1992; Bush *et al.*, 1996; Luykx *et al.*, 1997b; Becker *et al.*, 1999). The study of CV function in microorganisms may shed light on questions of water movement across cell membranes in general.

The close regulation of intracellular water content—often considered under the heading of 'cell volume regulation' (Parker, 1993; Strange, 1994; Hoffmann and Dunham, 1995; Lang *et al.*, 1998a)—appears to be important in maintaining normal metabolism. The reasons for this are not entirely clear. Although many metabolic processes have water as a reaction substrate or reaction product, the water content of most cells is about 70%. Therefore, it seems unlikely that variations in water concentration (even somewhat outside the limits set by normal osmoregulation) would significantly affect the rates of these reactions by simple mass action, or the degree of hydration of the enzymes catalyzing the reactions. There are, however, theoretical and experimental reasons for believing that the associations of proteins and other macromolecules, and their interactions with cell membranes, are markedly affected by slight changes in water content (see articles on 'macromolecular crowding' by Zimmerman and Minton, 1993 and Garner and Burg, 1994; see also the recent article on the effect of cell volume changes on internal membrane recycling, Lee and Linstedt, 1999).

Aside from sporadic (and occasionally illuminating) observations on the contractile vacuoles of particular protist species, most detailed studies of CV structure and function at the cellular and molecular levels have been carried out on the amoebae of the slime mold, *Dictyostelium discoideum*, and the ciliated protozoan, *Paramecium multimicronucleatum*. There are more observations on the CV of the single-celled green alga, *Chlamydomonas reinhardtii*, than any other plant cell, although here the information lags behind that available for the slime mold amoebae and the ciliated protozoan mentioned above. Earlier reviews of aspects of CV function were provided by Kitching (1938, 1956, 1967), Patterson (1980), Hausmann and Patterson (1984) and Zeuthen (1992).

3.2 The contractile vacuole complex and the CV cycle: an overview

What can be seen by light microscopy in living cells is the gradual expansion of a vacuole (sometimes involving visible fusion of adjacent vacuoles) to a certain size, and then a sudden 'contraction' or collapse, followed by a brief pause and then a repeat of the cycle. Figure 3.1 illustrates the time- and physical-scale of the process in two cells near opposite extremes of size: a small cell (*Chlamydomonas*) with a maximum CV diameter of about 2 μm and a CV cycle period of about 12s; and a large cell (*Paramecium*) with a maximum CV diameter of about 27 μm and a CV cycle period of about 40s.

For descriptive purposes, the CV cycle may conveniently be divided into: an early stage, just after the completion of fluid expulsion; a middle stage, during which the CV is increasing in size ('diastole'); and a late stage, signaled by the rapid contraction or collapse of the CV membrane and expulsion of fluid ('systole').

In functional terms, the following processes can be identified as essential to the contractile vacuole cycle and its initiation: 1) osmosensing and signal transduction; 2) fluid transport from the cytosol; 3) CV membrane expansion, by membrane fusion or by change in membrane configuration; 4) docking of the CV with the plasma membrane and pore formation; and 5) 'contraction'.

3.2.1 *Morphological aspects*

Electron microscopy has revealed a variety of membrane arrangements associated with the CV in different cells (Figure 3.2). It is useful to distinguish,

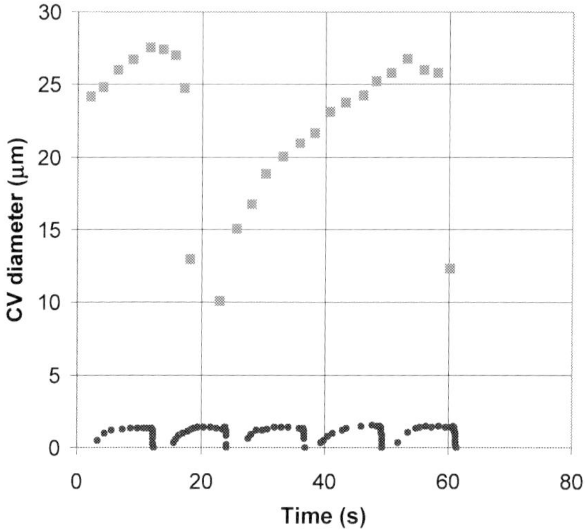

Figure 3.1 Sixty seconds of contractile vacuole (CV) activity in a small cell, *Chlamydomonas* (●) (Luykx, unpublished) and a large cell, *Paramecium* (■) (from Tominaga *et al*., 1998).

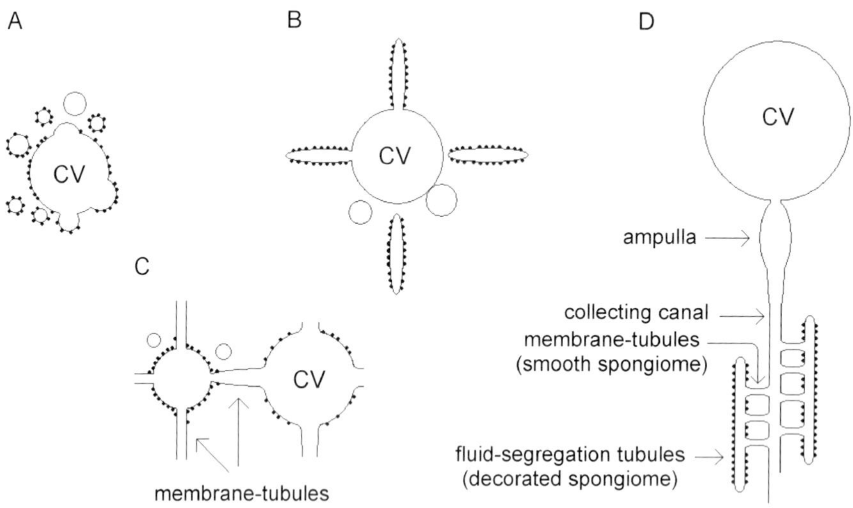

Figure 3.2 Diagrammatic representation of membrane arrangements in contractile vacuole (CV) complexes of different cell types. Water-segregation membranes are marked with black dots. A) *Chlamydomonas* and *Amoeba*; B) *Poterioochromonas*, *Pseudomicrothorax*, *Leptomonas*; C) *Dictyostelium*; D) *Paramecium*.

morphologically, the CV proper from other membranes (the 'spongiome') that are often associated with it, depending on the cell type. The contractile vacuole together with the spongiome is known as the 'contractile vacuole complex' (CVC). The morphological distinction apparent in most cells, although not obvious in *Amoeba* or *Chlamydomonas*, evidently corresponds to a functional difference, the spongiome being responsible for the segregation of water and its transport and the CV being responsible for water storage and expulsion from the cell (Heuser *et al.*, 1993; Ishida *et al.*, 1993, 1996; Nolta and Steck, 1994). In *Chlamydomonas*, there is no obvious physical division of the CVC into segregation and storage portions (Luykx *et al.*, 1997a), but even in this cell there is some evidence for two kinds of membrane in the CVC, in that the density of integral membrane particles (characteristic of water-segregating membranes; see next paragraph, and Section 3.3 below) in the fully formed CV is only about half that in the initial fusing vesicles (Robinson *et al.*, 1998), suggesting that other, non-particle-bearing membranes may also contribute to the CV membrane during expansion. In the discussions that follow, the term 'contractile vacuole complex' (CVC) rather than 'contractile vacuole' (CV) will be used to refer to the complete membrane system.

The membrane elements responsible for segregation of water from the cytoplasm have been identified experimentally in *Paramecium* as the parallel arrays of straight membranous tubules, about 50 nm in diameter, bearing 15 nm knob-like projections on their cytosolic surfaces (identified as proton pumps; see Section 3.3) (Ishida *et al.*, 1993; Fok *et al.*, 1995). Despite the lack of direct experimental evidence in other cells, the occurrence of similar membranous elements with knob-like particles in all other cell types that have been adequately studied, and the way these membranes are associated with the CV, leaves little doubt that these 'decorated membranes' constitute the fluid-segregation elements of the CVC in cells in general (McKanna, 1971, 1974, 1976).

Different arrangements of CVC membrane are found in different cell types (Figure 3.2). In the simplest arrangement, exemplified by *Amoeba* and *Chlamydomonas* (McKanna, 1973a; Luykx *et al.*, 1997a; Robinson *et al.*, 1998), small 70–120 nm diameter vesicles composed of water-segregation membrane fuse to form the CV (Figure 3.2A). In the chrysophycean flagellate, *Poterioochromonas* (Wessel and Robinson, 1979), the trypanosome, *Leptomonas* (Linder and Staehelin, 1979), and the ciliate, *Pseudomicrothorax* (Hausmann, 1983), specialized fluid-accumulating, disc-shaped, flattened or tubular vesicles apparently fuse with (although not necessarily involving complete incorporation into) contractile vacuoles that store the fluid before expelling it (Figure 3.2B). In *Dictyostelium* and some other amoebae (de Chastellier *et al.*, 1978; Baines *et al.*, 1992, 1995; Heuser *et al.*, 1993; Bush *et al.*, 1994), small vacuoles composed of the water-segregation membrane are interconnected by a network of tubular membranes; the vacuoles accumulate increasing amounts of fluid, apparently expanding by the incorporation of adjacent segments of the interconnecting tubular membranes (Figure 3.2C). It may be that in this cell there is, in addition, a population of small vesicles that contribute to the expansion of the CVs (de Chastellier *et al.*, 1978). The most complex membrane systems occur in *Paramecium* and some other ciliated protozoa (McKanna, 1974; Hausmann and Allen, 1977; Allen and Fok, 1988), where the water-segregation tubular membranes, the 'decorated spongiome', are separated from water-storage vacuoles by interconnecting membranous tubules and canals; the storage (contractile) vacuoles themselves expand by unfolding and incorporating the membranes of adjacent ampullae and other membranous elements (Figure 3.2D).

These various configurations of CVC membranes in different cells undoubtedly reflect dissimilar protein and lipid composition, and would be expected to be related to distinct functions. The significance, or the necessity, for these differences is not obvious.

The early stage of the CV cycle can be defined as that existing just after the expulsion of fluid in systole. In *Chlamydomonas*, it is characterized by the presence of numerous small (70–120 nm) 'CV vesicles' and the beginning of membrane fusion between these vesicles (Luykx *et al.*, 1997a). In *Dictyostelium*, the early stages of the CV cycle are identified simply by means of small CV size. In the earliest stages of the CV cycle in *Paramecium*, immediately after systole (at a time when the ampullae are already filling with fluid), the CV appears to consist of interconnected tubular membranes and flattened cisternae (Allen and Fok, 1988). Presumably, the fluid-segregation membranous tubules at the distal ends of the CVC membrane system remain unchanged, and probably operate continuously throughout the whole CV cycle.

In middle stages of the CV cycle, two things are happening: i) fluid is being transferred from the cytosol across the fluid-segregation membrane—the prime function of the CVC; and ii) the CV is expanding in size, either by a change in membrane configuration, by fusion with adjacent vesicles and membranous tubules, or both. In *Chlamydomonas*, it seems that the expansion of the CV is accomplished entirely by vesicle fusion, since the combined surface area of the cloud of 350–400 small (95–nm [average] diameter) 'CV vesicles' at the beginning of the CV cycle is about the same (10–11 μm^2) as the surface area calculated for the single, expanded CV (1.8–2.0 μm in diameter) at the end of the cycle (10–13 μm^2) (Hoppenrath, 1995; see also Figure 6a in Luykx *et al.*, 1997a). In *Paramecium* and other protozoa, the difference in the volume enclosed by the collapsed as compared to the expanded CV is accounted for by changes in the configuration of CV-specific membrane and its incorporation into the membrane of the expanding CV (Leedale, 1967; Linder and Staehelin, 1979; Allen and Fok, 1988; Tominaga *et al.*, 1999). In *Dictyostelium* also, CV expansion appears to occur by means of a change in membrane configuration, the CV membrane incorporating segments of the contiguous radiating tubule membrane (Heuser *et al.*, 1993).

In late stages of the cycle, the CV makes contact with the inner surface of the plasma membrane, some sort of pore is formed connecting the interior of the CV with the outside of the cell, and the CV 'contracts' or collapses to expel the fluid (systole). Electron micrographs suggest that in a few cell types the CV may be permanently docked, perhaps at a fixed site, on the plasma membrane (Leedale, 1967; McKanna, 1973b; Linder and Staehelin, 1979); however, evidence suggests that this is not typical (see Section 3.6).

3.2.2 Functional aspects

Table 3.1 presents a comparison of two cell types, *Paramecium* and *Chlamydomonas* (with CVs near the extremes in size; see Figure 3.1), with respect to CVC water transport rates, membrane surface areas and other membrane characteristics, including the membrane density of the 15 nm 'pegs' or 'knobs' (V-H^+-ATPase; see Section 3.3). Assuming that the rate of water movement from the cytosol to the interior of the CVC equals the overall rate of fluid expulsion from the CVs, it is an easy matter to calculate the water flux across a unit area of water-segregation membrane surface. The value is approximately the same in different cell types, 1–2 $\mu m^3/min/\mu m^2$ ($\mu m/min$) of membrane (Table 3.1). It is interesting to note that both the density of V-H^+-ATPase pegs and the curvature of the primary water-segregation membrane in *Chlamydomonas* and *Paramecium* are about the same, suggesting that, in spite of the differences in cell size and overall morphology of the CVCs in the two

Table 3.1 Comparison of morphological and functional parameters of the contractile vacuole (CV) complex of two cells types of very different size

	Chlamydomonas	*Paramecium*
a. Maximum CV diameter (μm)	2.2	27
b. Maximum CV volume (μm^3)	5.6	10300
c. Maximum CV surface area (μm^2)	15	2300
d. CV cycle period (s)	12	40
e. Apparent rate of CV surface expansion ($\mu m^2/s$) (c/d)	1.2	60
f. Water-segregation surface area (μm^2)	15	10000*
g. Rate of fluid expulsion ($\mu m^3/min$) (b/d)	30	15000
h. Water flux across water-segregation membrane ($\mu m^3/min/\mu m^2$) (g/f)**	2.0	1.5
i. Membrane particle density (number/μm^2)	2500***	3000
j. Water flux per membrane particle (molecules/s) (h/i)	4.5×10^5	2.8×10^5

*: The relevant membrane surface in *Paramecium* is not that of the CV itself (which in this cell serves only for fluid storage) but of the 'decorated tubule' membranes associated with the CV complex (Fok *et al.*, 1995). **: For comparison, the values calculated for *Pelomyxa* from data in Riddick (1968) and for *Amoeba* from data in Ahmad and Couillard (1974) are 0.9 and 1.0 $\mu m^3/min/\mu m^2$, respectively. ***: calculated for the small CV vesicles, which have approximately twice the number per unit surface area as the fully expanded CV (Luykx *et al.*, 1997a; Robinson *et al.*, 1998).

cell types, the basic process of water segregation is the same and operates at about the same efficiency.

Other features of CV function differ between the cell types (Figure 3.1; Table 3.1). Rates of CV expansion and contraction differ considerably, reflecting either different amounts of membrane involved or different mechanisms, for example, membrane expansion (unfolding) and membrane collapse in *Paramecium* (Allen and Fok, 1988), vesicle fusion and membrane vesiculation in *Chlamydomonas* (Luykx *et al.*, 1997a).

3.3 Biochemistry

Proteins found to be associated with the membranes of the CVC in various cell types are listed in Table 3.2, classified by the general roles they may play in CVC function. As expected from the morphological observations, a variety of proteins involved in the regulation of membrane trafficking and in the processes of membrane fusion and membrane transport have been found to be associated with the CVC. Some of the proteins, such as cell wall glycoproteins, may not be involved directly in the primary function of CVCs (segregation of water from the cytosol) but in some of the secondary functions (see Section 3.1).

The identification of the prominent CVC membrane particle in *Dictyostelium* and *Paramecium* as a vacuolar proton pump, vacuolar-type hydrogen-transporting adenosine triphosphatase (V-H^+-ATPase), is based on light- and electron-microscopic staining with polyclonal antibody against the 57 kDa (B subunit) of mammalian and yeast V-H^+-ATPases (Heuser *et al.*, 1993; Fok *et al.*, 1995), in accordance with the evolutionarily conserved nature of these pumps. In *Chlamydomonas*, however, in spite of the presence of biochemically demonstrable bafilomycin-sensitive V-H^+-ATPase activity in membrane fractions, antibodies against two higher plant V-H^+-ATPases failed to react with membrane fractions and could not be demonstrated by electron-microscopic immunocytochemistry (Robinson *et al.*, 1998). This was possibly because of antigenic differences between V-H^+-ATPases of *Chlamydomonas* and those of higher plants. The presence of a V-H^+-ATPase in the CV membranes of *Chlamydomonas*, therefore, rests on the morphological similarity of *Chlamydomonas* CV membrane particles to those of *Dictyostelium* and *Paramecium* in freeze-fractured preparations, but remains without direct support. Using electron-microscopical immunocytochemical methods, Robinson *et al.* (1998) were, however, able to demonstrate a distinct pyrophosphatase (PPase) activity associated with the CV membranes of *Chlamydomonas*. This cell type, therefore, may have two proton pumps (V-H^+-ATPase and V-PPase) in its CV membrane, as has been documented for the tonoplast membrane of other plants (Davies, 1997).

V-H^+-ATPases are widely distributed in both plant and animals cells, serving among other things to acidify cytoplasmic vesicles and vacuoles (Harvey and

Table 3.2 Proteins associated with the contractile vacuole complex in various cell types, classified according to general function

Protein	Cell	CVC location	Method	References
Hydrogen ion transport				
V-H^+-ATPase	*Chlamydomonas*		ic	Robinson *et al.*, 1998
V-H^+-ATPase	*Dictyostelium*	sp	ic	Nolta *et al.*, 1993
V-H^+-ATPase	*Dictyostelium*		im, ic	Heuser *et al.*, 1993; Bush *et al.*, 1994
V-H^+-ATPase	*Dictyostelium*	sp	bc, ic	Fok *et al.*, 1993; Nolta and Steck, 1994
V-H^+-ATPase	*Dictyostelium*		in	Temesvari *et al.*, 1996
V-H^+-ATPase	*Paramecium*	sp	ic, in	Fok *et al.*, 1995
Pyrophosphatase	*Chlamydomonas*		ic	Robinson *et al.*, 1998
Vesicle formation and traffic, exocytosis and endocytosis, membrane fusion, membrane transport				
GTPase, Rab4-like	*Dictyostelium*	sp	im	Bush *et al.*, 1994
Ypt1p (Rab)	*Chlamydomonas*		ic	Huber *et al.*, 1996
Drainin	*Dictyostelium*	cv	gk	Becker *et al.*, 1999
Clathrin	*Dictyostelium*		gk	O'Halloran and Anderson, 1992
Clathrin	*Chlamydomonas*		bc	Denning and Fulton, 1989b
Calmodulin	*Tetrahymena*	cv pore	in	Suzuki *et al.*, 1982
Calmodulin	*Tetrahymena*		in	Bergquist, 1989
Calmodulin	*Dictyostelium*	cv	bc, ic	Zhu and Clarke, 1992; Zhu *et al.*, 1993; Heuser *et al.*, 1993
Cytoplasmic contractile processes, vesicle movement				
Actin	*Dictyostelium*	cv	ic	Baines *et al.*, 1995
Actin-binding protein	*Tetrahymena*	cv pore	ic	Watanabe *et al.*, 1998
α-Actinin	*Dictyostelium*	cv	ic	Furukawa and Fechheimer, 1994
Unconventional myosin	*Dictyostelium*	cv	ic	Zhu and Clarke, 1992
Myosin IC	*Acanthamoeba*		in	Doberstein *et al.*, 1993
Myosin IC	*Acanthamoeba*	cv	ic	Baines *et al.*, 1992, 1995
Other				
Ca^{++}-ATPase (PAT1)	*Dictyostelium*	cv	ic	Moniakis *et al.*, 1999
Alkaline phosphatase	*Dictyostelium*	cv	cc	Quiviger *et al.*, 1978
Alkaline phosphatase	*Dictyostelium*		bc	Zhu and Clarke, 1992
Alkaline phosphatase	*Dictyostelium*	cv	bc, cc	Nolta and Steck, 1994
Phosphodiesterase	*Dictyostelium*	cv	bc	Nolta and Steck, 1994
Nucleotide phosphatases	*Gloeomonas*		cc	Domozych, 1989
Cell wall glycoprotein	*Chlamydomonas*		ic	Grief and Shaw, 1987

Abbreviations: cv, contractile vacuole; sp, spongiome; Methods: bc, biochemical; cc, cytochemical; ic, immunocytochemical; im, immunological; in, inhibitor studies; gk, genetic knockout.

Nelson, 1992). Their presence in contractile vacuole membranes thus brings the CVs into the group of acidified membrane-bounded compartments, which includes Golgi vesicles, endosomes, lysosomes, synaptic vesicles, secretion vacuoles and the central vacuoles of plant cells. V-PPases have also been found to be localized at plant vacuolar membranes; the hydrolysis of PP_i may also energize the acidification of the vacuole, and in addition may lead to the accumulation of K^+ in the vacuole by coupled H^+/K^+ symport (Davies, 1997).

3.4 Osmosensing and signal transduction

The 'ground state' of the CVC can be considered that shown by the cell when it is in an isosmotic or hyperosmotic medium; that is, under conditions when there is no osmotic influx of water. Under these conditions, cells display no visible CV activity. In *Chlamydomonas*, the CV membranes appear to be in a state similar to that of the early stage in the CV cycle, a collection of small spherical vesicles about 7–120 nm in diameter (Luykx *et al*., 1997a). In *Paramecium* under hyperosmotic conditions, the water-segregation membranes of the CVC are present as numerous vesicles 120–200 nm in diameter (Ishida *et al*., 1996). On transfer to hypo-osmotic media, CV activity begins after a short delay, suggesting that CVC membrane function is responding to some intracellular or surface signal initiated by the change in external conditions. The first visible response is a fusion of the vesicles, in *Chlamydomonas* to initiate the expansion of the CV, and in *Paramecium* to re-form the tubular membranes of the 'decorated spongiome' and re-connect them with the fluid-transport membrane system.

What changes occur at the cell surface or in the cell interior when the cell is transferred to a hypo-osmotic medium? What are the critical changes that can be considered the osmotic sensor system, and what are the intracellular signaling pathways that lead from the sensor to the activation of the CVC?

An osmolarity sensor could be located in any one of several systems in the cell, responding to subtle changes attendant on the initial swelling of the cell with the uptake of water from the dilute external medium. The main possible locations would seem to be: i) in the plasma membrane, which might undergo a change in solvent interactions or experience an altered membrane tension influencing mechanosensitive ion channels (Sackin, 1995; Wood, 1999); ii) in the submembrane cytoskeleton, which might undergo reorganization with cell swelling (Mills *et al*., 1994); or iii) in the cytosol, where activity of some macromolecule or class of macromolecules might be sensitive to slight changes in intracellular water content (Minton, 1994).

There are no critical observations on the nature of the putative osmosensor in any eukaryotic cells with contractile vacuoles, although Yoshimura (1998) did find evidence of a mechanosensitive Ca^{++} channel in the plasma membrane of *Chlamydomonas*, which might serve as an osmosensor upon swelling of the cell.

Two proteins in the membrane of *Escherichia coli*, MscL and EnvZ, are involved in the cell's responses to changes in the osmolarity of the medium. MscL is a mechanosensitive ion channel (Sukharev *et al.*, 1997; Sukharev, 1999), and EnvZ is the first element in a two-element signal pathway controlling the activation of genes for other membrane proteins synthesized in response to osmolarity changes (Mizuno, 1991). Their 3-dimensional structures are being elucidated (Chang *et al.*, 1998; Saint *et al.*, 1998; Tanaka *et al.*, 1998; Tomomori *et al.*, 1999); this research could soon provide molecular details of how membrane proteins respond to altered osmolarity, and these proteins could serve as general models for membrane osmosensors in other cell types.

The transduction of signals to the machinery of the CVC from the osmosensor, whatever it may be, is likely to involve signal pathways that are similar to some of those uncovered in other cell processes, especially those in which inositol-lipid signals and a rise in cytosolic calcium ion are involved (consistent with the detection of calmodulin and a Ca^{++}-ATPase associated with the CV in *Dictyostelium* and *Tetrahymena*; see Table 3.2). Vertebrate cells as well as plant cells respond to osmotic stress by: activation of phosphoinositol- and G-protein-signaling pathways (e.g. O'Neil and Leng, 1997; Tinel *et al.*, 1997; Fujii *et al.*, 1999), an increase in cytoplasmic calcium concentration, activation of ion channels and transporters, and changes in the cytoskeleton (Mills *et al.*, 1994; Hoffman and Dunham, 1995; Cessna *et al.*, 1998; Knight, 2000), but the molecular details remain to be clarified. Two common early responses of the CVC to lowered extracellular osmolarity—vesicle fusion and activation of CV membrane pumps—are reminiscent of the responses of many mammalian secretory cells and neurons to the binding of ligands by cell surface receptors. This activates phosphoinositol signal transduction pathways and results in a release of calcium from internal membrane-bounded stores and exocytotic membrane fusion (Burgoyne and Morgan, 1998; Hirose *et al.*, 1999; Lang, 1999). It is likely that similar signaling pathways will be found to be involved in the activation of CVC machinery. Meijer *et al.* (1999) observed that the levels of phosphatidyl-D-inositol-3,5-bisphosphate increased after stressing *Chlamydomonas* with hyperosmotic medium; however, no changes in this phosphoinositide were observed after transfer to low osmotic strength (Meijer, personal communication). Buczynski *et al.* (1997) also found that *Dictyostelium* cells deficient in two phosphatidylinositide-3-kinases had normal contractile vacuoles; therefore, CVC activation does not depend on these particular phosphoinositide pathways.

3.5 Water transport

The main function of the CVC is the movement of water from the cytosol to a space enclosed by a specialized membrane. How the water is moved across this membrane is the primary question concerning CVC function. Although active transport of pure water does not, in itself, seem to be ruled out by any obvious theoretical consideration, it has not yet been described in any biological system. Therefore, water transport probably occurs either passively or accompanying the transport of some other molecule or ion.

Models of how the CVC might transfer water from the cytosol can be divided into four classes, according to the mechanism by which water enters the CVC from the cytosol (Figure 3.3): i) an osmotic model, in which the contents of the CVC are made hyperosmotic to the cytosol and water moves into the CVC passively (Figure 3.3A,B); ii) a channel-diffusion model, in which the CV fluid is first formed isosmotically by diffusion of ions and water through channels in the CVC membrane, and then the ions are pumped back into the cytosol from the CVC lumen (Figure 3.3C); iii) an ion-water co-transport model, in which water accompanies ion transfer into the CVC lumen (Figure 3.3D); and iv) physical models, in which membrane tension and changes in membrane

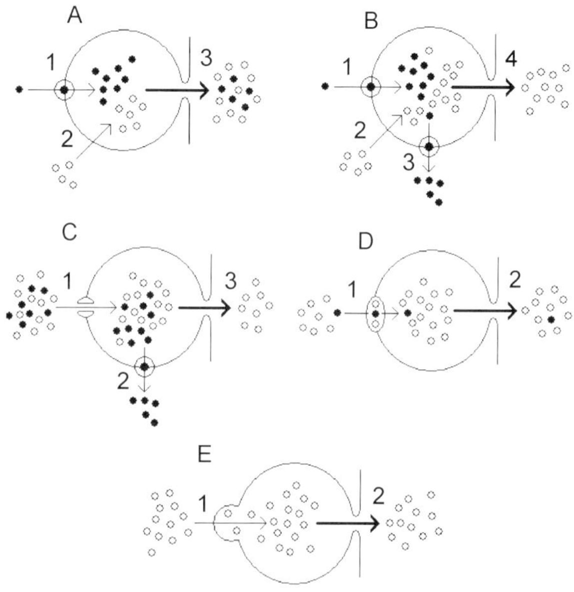

Figure 3.3 Hypotheses of water transport into the contractile vacuole. Key: Open circles, water; filled circles, ions (or other osmolytes). (A, B) Osmotic hypotheses, in which ions are pumped into the vacuole, creating hyperosmotic contents, with water following passively: A, hyperosmotic fluid expelled; B, hypo-osmotic fluid expelled after reverse ion pumping. (C) Channel-diffusion hypothesis, in which vacuolar fluid is formed isosmotically, and then made hypo-osmotic by reverse ion pumping. (D) Ion-water co-transport hypothesis, in which ion transport is accompanied by water flow into the vacuole against the osmotic gradient. (E) Physical hypothesis, in which membrane fusion drives the increase in the volume enclosed, with water moving across the membrane hydraulically.

curvature contribute to the conditions for hydraulic transfer of water into the CVC (Figure 3.3E).

3.5.1 Osmotic models

In osmotic models of water segregation in the CVC, ion pumps in the CV membrane (or, conceivably, active transport of some other osmotically active substance) raise the contents of the CVC to an osmolarity higher than that of the cytosol. Water then moves passively (osmotically) from the cytosol across the CV membrane into the lumen of the CV (Figure 3.3A). This mechanism was envisioned in earlier reviews by Kitching (1938, 1956, 1967) and Patterson (1980), and more recently by Heuser *et al.* (1993), Giglione and Gross (1995) and Robinson *et al.* (1998).

There have been three attempts to measure the composition and osmotic strength of CV fluid, by means of micropuncture and withdrawal of small fluid samples from amoebae (Schmidt-Nielsen and Schrauger, 1963; Mayer and Iverson, 1967; Riddick, 1968). These studies found the CV contents to be significantly hypo-osmotic, with osmotic strengths ranging from 24–80 mosM, only about 1/5 to 1/2 that of the cytoplasm. Although perhaps not very likely, a technical error in this kind of study is conceivable; samples withdrawn from the CV might require a deeper micropuncture, and so could systematically permit a greater dilution with the bathing medium than for the control samples drawn from the cytoplasm. Nevertheless, at face value, these studies do not support models requiring the contents of the CV to be hyperosmotic to the cytoplasm.

In a variant of the osmotic model, CV fluid may initially be formed by the movement of water from the cytosol down an osmotic gradient across a CVC membrane, but, the fluid is then modified, at a later time or in a different location, by the pumping of ions (or other osmolytes) back out into the cytosol, leaving behind a hypo-osmotic fluid in the CV proper (Figure 3.3B). This model would require two ion-specific pumps with opposite orientations in the CV membrane. In cells like *Paramecium* and *Dictyostelium*, where the water-segregation membranes are partially or completely physically separate from the CV membranes (see Section 3.2.1), the fluid may initially be formed under the influence of a decorated spongiome with hyperosmotic contents, and then becomes hypo-osmotic as it moves into the CV. In other cell types, such as *Amoeba* and *Chlamydomonas*, where there appears to be no separation of the two membranes, the early CV vesicles and vacuoles might be hyperosmotic to the cytosol, with the CV contents later becoming hypo-osmotic as ions (or other osmolytes) are pumped back into the cytosol. In a study by Riddick (1968), the smaller CVs tended to have osmolarities higher than that of the larger CVs, suggesting that in the earliest stages of CV expansion the smallest CV vesicles might have higher osmolarities still, perhaps even greater than that of the cytosol. In this variant of the model, it might also be useful to have a CV membrane that is more permeable to water in the early stages of CV expansion, when water is moving down its osmotic gradient into the CV, and less permeable in later stages of expansion, when the CV contents are becoming hypo-osmotic to the cytosol. Permeability changes in the CV membrane might be promoted by the changes in membrane curvature that occur during CV expansion.

The functional significance of differences between tubular and planar membrane configurations was discussed by Allen (1995) in connection with the *Paramecium* CV. Membranes with a high degree of curvature enclose small spaces, and this may be important for the concentration of protons or other ions in the CVC lumen as part of the mechanism of water uptake from the cytosol.

The route by which water crosses the water-segregation membrane in osmotic models is usually not specified. The rate at which water is transferred into the CV (averaged over the whole CV cycle), 1–2 $\mu m^3/min/\mu m^2$ (Table 3.1), is about the rate at which water moves across artificial lipid bilayers under an osmotic gradient of 0.2 OsM (Cass and Finkelstein, 1967; Stein, 1986). If water is transferred into the CVC by this means over, say, only the first half of the cycle, then the osmotic gradient would have to be correspondingly higher (0.4 OsM) to accomplish the observed amount of fluid transfer from the cytosol. Since such high internal osmotic gradients may be more than cells can achieve, it is likely that water follows other routes into the CVC than only across the lipid bilayer. The lipid bilayer of the CV membrane is indeed probably highly impermeable to water, compared to other cellular membranes—a membrane state that would be necessary in any model in which an osmotic gradient (CV contents hypo-osmotic) must be maintained for a significant part of the CV cycle. Riddick (1968) found that in some conditions the isolated expanded *Pelomyxa* CV lost water only slowly, consistent with a relatively impermeable CV membrane.

In view of the ubiquity of the membrane particles identified as proton pumps (see Section 3.3) on water-segregation membranes of the CVC, the proton seems to be a good candidate for an ion that could be involved in raising the osmotic strength inside the CVC. One possibility, proposed by Heuser *et al.* (1993) and Giglione and Gross (1995) for *Dictyostelium* and adopted by Robinson *et al.* (1998) for *Chlamydomonas*, is that protons are pumped from the cytosol into the CVC lumen by the V-H^+-ATPase (and/or by PPase in *Chlamydomonas*). The electrogenic potential draws a counterion into the CVC lumen through specific ion channels, and thus the osmolarity inside the CVC is raised to a level above that in the cytosol. The HCO_3^- ion is a likely candidate for an ion that could be expelled in the CV fluid because of its dispensability as a waste product of animal cell metabolism (from CO_2). In *Chlamydomonas*, CO_2 is used in photosynthesis but might still be used to contribute to CV fluid hyperosmolarity and be expelled, given that this cell has an efficient CO_2-uptake mechanism

(Somanchi *et al.*, 1998; Sültemeyer, 1998). With special water channels (e.g. aquaporins) increasing the rate of passive water flow, the osmotic strength of the fluid inside the CVC need only be slightly higher than that of the cytosol.

3.5.2 Channel-diffusion model

There are other ways, besides moving across a membrane down its activity gradient, that water could be segregated by the CVC. The initial CV fluid could be formed isosmotically, with water diffusing through membrane channels (aquaporins), along with ions, into the CVC. At mid- or late stages of the CV cycle, the channels might close, the CV membrane might become less water-permeable, and membrane pumps could then remove the ions from the CV lumen and return them to the cytosol, leaving behind a hypo-osmotic CV fluid (Figure 3.3C). This idea would be consistent with findings of hypo-osmolarity of the CV fluid in late stages of the cycle (Schmidt-Nielsen and Schrauger, 1963; Riddick, 1968), but has the disadvantage that it relies entirely on cytosol-directed ion pumps and leaves no obvious role for the apparently ubiquitous V-H^+-ATPase in CVC membranes (see Section 3.3).

3.5.3 Ion-water co-transport model

In the 1980s and early 1990s, Thomas Zeuthen, using combined electrical and volume measurements on single cells of amphibians, demonstrated that the transport of ions across the plasma membrane can be accompanied by water flow in a direction opposite to that expected from purely passive osmotic water movement. He provided a number of observations indicating that the ion and water movements were coupled (see summaries by Zeuthen and Stein, 1994; and Zeuthen, 1995). For each K^+ and Cl^- ion transported, it was estimated that about 500 molecules of water were co-transported. Zeuthen called attention to the fact that conformational changes in a number of different proteins involve the displacement of anything from a few tens to several hundreds of water molecules, and suggested that the conformational changes involved in ion-transport membrane proteins could simultaneously transport the estimated number of water molecules.

Water movements across the plasma membrane can occur against osmotic gradients of up to 100 mosM (Zeuthen, 1995), a figure larger than the 60–70 mosM gradients calculated from the measurements of cytosolic and CV fluid osmolarities in amoebae (Schmidt-Nielsen and Schrauger, 1963; Riddick, 1968). The mechanism moving water against osmotic gradients in vertebrate epithelia may, therefore, be sufficient to account for the movement of water from cytosol to a hypo-osmotic contractile vacuole lumen (Zeuthen, 1992). This mechanism (Figure 3.3D) provides a possible alternative to osmotic-flow schemes that require a hyperosmotic CV fluid, minimizes the number of ions lost to the exterior of the cell with the CV contraction, and obviates the necessity of explaining away the finding of hypo-osmolarity in the CV lumen (see Section 3.5.1).

Could the CV membrane particles identified as V-H^+-ATPase (see Section 3.3) move protons through the CVC membrane, with an accompanying 500 molecules/proton, at a rate sufficient to account for the rate of fluid uptake by the CVC (Table 3.1)? With an ATPase molecular weight of 650 kDa, a maximum turnover rate of about 4.1 μmol ADP/mg protein/min (Sze *et al.*, 1992), and a maximum of three protons transferred for each ATP hydrolyzed (Davies, 1997), it can be calculated that, for each V-H^+-ATPase, 133 protons and 66,500 water molecules are transferred into the CVC every second. This is fast enough to account for the rate of fluid uptake into contractile vacuoles (Table 3.1).

If such a mechanism of water transport exists at the plasma membrane (where ion-water co-transport was first described), should not such water-pumps at the plasma membrane be sufficient to expel all the water entering the cell by osmosis, without the necessity of a special vacuolar system? There are several possibilities. The density of ion-transport particles necessary to achieve the observed rate of water transport may be too high to be accommodated at the cell surface along with all the other transport and regulatory proteins that exist there, so that a specialized internal membrane system densely populated with efficient ion-water co-transport particles, devoted primarily to this one function, might be required. In addition, the functioning of an ion-water co-transport particle may be sensitive to ionic conditions on one side of the membrane (e.g. outside the cell), so that an internal ion-water co-transport membrane system isolated from the external medium is advantageous. Finally, the maximum rates calculated in the previous paragraph may not ordinarily obtain *in vivo*, so that for efficient CVC function what is required is a combination of ion-water co-transport and osmotic water flow into the lumen of the CVC, the latter requiring an internal membrane-bounded space where the osmolarity can be controlled and maintained above that of the cytosol.

3.5.4 Physical models

Electron-microscopic observations on the expanding contractile vacuole suggest that the expansion is essentially a change in CV membrane conformation without much change in total membrane surface area (see Section 3.2.1). In *Paramecium*, the CV membrane changes from a collapsed network of membranous tubules to an expanded spherical vacuole; in *Chlamydomonas*, from numerous small vesicles (via membrane fusion) to one large vacuole. In the latter case, a simple calculation shows that as the membrane enclosing 350–400 small vesicles is transformed into a single large vacuole (see Section 3.2.1 and Table 3.1), the volume enclosed by this membrane increases about 25-fold. In addition to an increase in the volume enclosed by the membrane, there is in both *Paramecium* and *Chlamydomonas* a substantial decrease in membrane curvature.

Changes in membrane tension associated with changes in membrane curvature might also play a more important role in CV membrane function, contributing directly to transfer of water from the cytosol (Figure 3.3E). If CV membranes cycle rapidly (within periods of less than one minute; see Figure 3.1) back and forth between states of high and low membrane curvature (vesicular or tubular membranes versus nearly planar membranes in the expanded CV, respectively), one might expect substantial changes in membrane tension, since lipids do not migrate rapidly from one leaflet of the membrane lipid bilayer to the other (e.g. see Victorov *et al.*, 1994). If, for example, membrane vesicles and tubules represent a state of high membrane tension, then relaxation into the expanded state could contribute to a hydraulic force drawing water, against the osmotic gradient, from the cytosol into the space enclosed by the CV membrane. Alternatively, the membranes of small early vesicles might be in a state of low tension, with an energy-requiring membrane fusion process driving them to a more planar configuration and a state of higher tension. Assessing the feasibility of such models would require quantitative evaluation of the forces involved, and a determination of which membrane state represents one of membrane tension.

Even if potential changes in membrane tension are ameliorated by changes in the configuration of integral membrane proteins or by alterations in the distribution of different lipids within the plane of a single lipid leaflet (e.g. Brumm *et al.*, 1996), changes in membrane curvature could contribute to changes in water permeability of the membrane during the CV cycle, as well as to alterations in the activity of ion pumps or in the rates of ion movement through integral membrane protein channels (Ajouz *et al.*, 2000), and so could contribute to the regulation of ion and water movement through the CV membrane at different stages of the cycle.

3.6 Membrane fusion and recycling

There are two kinds of specific membrane fusion associated with the CV cycle. One is the early fusion of vesicles that contribute to the expansion of the CV itself, as in *Amoeba* and *Chlamydomonas*, or the fusion of ampulla membrane to CV membrane, as in *Paramecium* (see Section 3.2.1). The other is the fusion of the membrane of the expanded CV with the plasma membrane. Not much is known about the details of either membrane fusion process but they are reminiscent of membrane-fusion and -recycling processes in other cells. Bush and co-workers (1994, 1996) in *Dictyostelium* and Huber *et al.* (1996) in *Chlamydomonas* have discovered proteins associated with CVs that are related to endosome-lysosome-secretion vesicular traffic in other cell types

(Table 3.2). The regulation of membrane fusion events in the CV cycle is likely to be complex, if the lessons from the synaptic vesicle are any guide; here the whole cycle of vesicle membrane fusion with, and retrieval from, the neuronal plasma membrane involves the participation of several dozen different proteins (Südhof, 1995).

Fusion of the CV membrane with the plasma membrane at the end of the CV cycle is probably limited to what is required to form a pore connecting the contents of the CV with the outside of the cell. In *Paramecium*, extensive fusion of the CV membrane with the plasma membrane seems unlikely, because of the highly structured nature of the cell surface. In *Chlamydomonas*, complete fusion of the CV membrane with the plasma membrane would add a plasma-membrane's equivalent of new membrane to the cell surface approximately every 2 or 3 minutes, but no extensive compensatory membrane retrieval by endocytosis has been observed (Luykx, Hoppenrath, and Robinson, unpublished observations). In *Amoeba*, the CV membrane appears to vesiculate during systole rather than undergo large-scale fusion with the plasma membrane (McKanna, 1973a). In *Dictyostelium*, immunofluorescence microscopy showed that the protein, drainin, associated with the CV membrane, remains concentrated at the site of CV discharge rather than spreading through the plasma membrane (Becker *et al.*, 1999). All these observations indicate that the extent of fusion of the CV membrane with the plasma membrane during CV discharge is limited, and does not resemble the wholesale incorporation of membrane that occurs during the synaptic vesicle cycle in neurons, or during secretion-exocytosis in mammalian cells.

These observations do not preclude the possibility of a local and restricted fusion of CV membrane with the plasma membrane. That there is some fusion between the two membranes is supported by the finding that in *Dictyostelium* a CV-associated Ca^{++}-ATPase also occurs on the plasma membrane (Moniakis *et al.*, 1999). Some fusion of CV membrane with the plasma membrane would also make sense of the frequent observation of (clathrin-) coated vesicles (see also Table 3.2) in the vicinity of collapsing CVs in a variety of cell types (Manton, 1964; Leedale, 1967; Eyden and Vickerman, 1975; Weiss, 1983; Hausmann and Patterson, 1984; Luykx *et al.*, 1997a). These vesicles could be part of a CV membrane retrieval process. O'Halloran and Anderson (1992) found that *Dictyostelium* amoebae deficient in clathrin heavy chains appeared to be unable to form contractile vacuoles, an effect that might be due to a failure of CV-specific membrane retrieval from the cell surface.

In most cell types, there appears to be no fixed site for CV docking at the plasma membrane prior to fluid expulsion. In *Dictyostelium*, for example, the few larger CVs normally present appear to be dispersed during mitosis into more numerous smaller CVs docked and functioning at many sites on the plasma membrane (Zhu *et al.*, 1993). The *osm3* mutant of *Chlamydomonas* often has several small CVs instead of the usual two; in many cells, most or all the small

CVs seem to be able to dock at different sites on the plasma membrane and discharge their contents (Luykx *et al.*, 1997b, and unpublished observations). Thus, the whole plasma membrane, or at least extensive regions of it, appear(s) to be competent to receive CVs and form functional pores for fluid discharge.

By analogy with regulated exocytosis in secretion (e.g. Valentijn *et al.*, 1999), the control of fusion of the CV membrane with the plasma membrane is likely to involve modifications of the cortical actin cytoskeleton, in order to permit the close approximation of the two membranes. In support of this analogy, Becker *et al.* (1999) found, in an immunocytochemical study in *Dictyostelium*, evidence for the presence of actin filaments in the space between the CV membrane and the plasma membrane.

The exact nature of the pore connecting the interior of the CV to the outside of the cell at the end of the CV cycle is somewhat uncertain; it may differ from one cell type to another. Given the rapidity of systole, electron micrographs of fixed, random cell samples happening to catch the CV at an instant when a pore is open would be expected to be rare. Nevertheless, a few micrographs illustrating an open pore connecting the CV membrane with the plasma membrane have been published, from *Colacium* (Leedale, 1967), *Paramecium* (McKanna, 1973b), *Acanthamoeba* (de Chastellier *et al.*, 1978); *Dictyostelium* (Baines *et al.*, 1995) and *Chlamydomonas* (Luykx *et al.*, 1997a). If these images are not the result of preparation artifacts, they suggest that in some cell types, at least, there is a real fusion (even if local and restricted in extent) between the CV membrane and the plasma membrane. In some cases, the pore is small, no more than about 15 nm in diameter; in other cases, for example, *Paramecium* (McKanna, 1973b) and *Spongilla* (sponge) (Brauer and McKanna, 1978), the fusion of the CV membrane and plasma membrane appears to be more extensive, giving an opening of as much as 400–500 nm.

Weiss *et al.* (1977) observed circular arrays of particles in freeze-fractured preparations of *Chlamydomonas* in the plasma membrane overlying the CVs, suggesting a change in plasma membrane structure accompanying contact of the CV membrane with the plasma membrane just prior to fluid expulsion. They did not, however, detect any real pores connecting the CV lumen to the exterior of the cell, and thought that the two membranes might not actually fuse; instead, their close apposition might, they suggested, be accompanied by the formation of hydrophilic channels in both membranes, comparable perhaps to the gap junctions between the tightly apposed plasma membranes of adjacent epithelial cells in animal tissues. Linder and Staehelin (1979) were, likewise, unable to find any evidence for fusion of CV membrane and plasma membrane in *Leptomonas*. They interpreted their observations on thin-sectioned and freeze-fractured preparations in terms of an 'adhesion plaque', the reorganization of which might open submicroscopic water-channels in both membranes.

Becker *et al.* (1999) recently discovered in *Dictyostelium* a protein they call 'drainin', which is apparently involved in some step in the effective fusion of the CV with the plasma membrane. Cells without functional drainin form CVs as in wild-type cells, but the CVs are apparently unable to interact appropriately with the plasma membrane and are unable to form channels for the discharge of the CV fluid.

3.7 'Contraction'

The main issues concerning the 'contraction' of the contractile vacuole are: i) whether the sudden change in the CV at systole is best regarded as a contraction or a collapse; ii) whether the change can be explained entirely by rearrangements and reconfigurations of the lipids and proteins of the CV membrane itself, or whether other proteins (e.g. cytoskeletal proteins) associated with the CV membrane are required for the contraction; and iii) whether the energy for the membrane change in systole is expended during the actual contraction or prior to it.

An alternative to the view that the systole of the CV is a true contraction is that it is a collapse of a membrane under tension. Early light-microscopic observations of the CV of *Amoeba proteus* (Wigg *et al.*, 1967) suggested that the CV membrane collapses (i.e. without an obvious decrease in membrane surface area) rather than contracts (with a decrease in surface area). The electron-microscopic observations of McKanna (1973a) on this cell indicated that the membrane is in fact breaking up into small (35 nm diameter) vesicles as it is collapsing, a view consistent with the interpretation given by Luykx *et al.* (1997a,b) for the events of systole in *Chlamydomonas*. From their observations on *Paramecium*, Naitoh *et al.* (1997) suggested that the CV membrane might store energy as it expands, and then collapse into an interconnected system of flattened cisternae and membranous tubules as water is expelled. In this cell, the CV membrane appears merely to change conformation, converting near-planar membrane into tubular membrane but retaining its integrity as a single membrane (Allen and Fok, 1988). In other cells also, the CV may simply flatten out or otherwise change configuration at systole (see Leedale, 1967; Brauer and McKanna, 1978; Linder and Staehelin, 1979).

In individual vesicles derived from the CVC of mechanically ruptured *Paramecium* cells, Tominaga *et al.* (1998) observed a periodic rounding and slackening (with a cycle duration approximately the same as that of the normal CV cycle in intact cells, but without fluid expulsion), and reasonably interpreted them as cycles of membrane tension and relaxation. In a majority of the cycles recorded, tension developed rapidly (in less than a second), and was followed by a slower relaxation over a period of several seconds. If these time periods correspond to the rapid systolic contraction and slower diastolic expansion in intact cells, then the contraction (or the fraction of a second immediately preceding it) would indeed correspond to a state of high membrane tension, and the 'contraction' is better regarded as a 'collapse'. The source of the sudden development of tension, however, remains unclear.

Prusch and Dunham (1970) isolated individual CVs by hand from *Amoeba proteus* and demonstrated that the addition of 5 mM $MgCl_2$ and 2–10 mM ATP (but not ITP, GTP or AMP) could induce a sudden shrinkage of the CV to about 1/2–1/3 its original volume. The straightforward interpretation would be that the contraction of the CV is an active process, requiring ATP. It is not clear, however, whether this phenomenon actually mimics what happens in intact cells, nor to what extent cytoskeletal or other proteins might have been present in their preparations. 'Partially isolated' CVs of mechanically-ruptured *Paramecium* cells also occasionally diminish in size during an apparently normal systole, but 2–3 times more slowly than in intact cells (Naitoh *et al.*, 1997). It may be that the hydrostatic pressure in the intact cell or the concentration of some diffusible cytosolic molecules in the vicinity of the CV contributes to the rate of CV collapse.

No submicroscopic filaments have been detected closely associated with the CV membrane, as there are for example in the subplasmalemmal actomyosin contractile ring in the equatorial region of animal cells during cytokinesis (see references in Field *et al.*, 1999). But there is some evidence from immunofluorescence microscopy for the presence of actin, actin-binding proteins and myosin isoforms around the CV (see Table 3.2). Doberstein *et al.* (1993) also found in *Acanthamoeba* that microinjection of antibodies against myosin isoforms increased the length of the CV cycles, although it is not clear whether systole was specifically affected. On the other hand, Naitoh *et al.* (1997) failed to find any effect of cytochalasin B (an inhibitor of the formation of labile actin microfilaments) on the rate of CV contraction in *Paramecium*.

These observations suggest that the change in the CV membrane at systole may best be regarded as a collapse rather than a 'contraction', that membrane tension is involved in CV contraction, and that the source of the tension may reside in CV-membrane-associated actin- and myosin-like molecules, but the details remain obscure.

3.8 Summary

If this chapter is long on speculation and short on definitive answers, it is because we do not yet really know how the CVC works—how water is segregated from the cytosol, and what the nature of the 'contraction' is. However, the questions are more sharply defined and the CVC is slightly less mysterious than it was a few years ago. It appears to be an internal membrane system with many features in common with other endomembrane systems in the cell, including membrane fusion and ion transport. In spite of the considerable variation in the morphology and behavior of CVCs in different cells, similar features have nevertheless emerged: the differentiation of CVC membranes, in most cell types, into water-segregation and water-storage compartments; internal membrane fusion; the presence of a lumen-directed V-H^+-ATPase in the membranes; and docking, fusion and pore-formation of the CV membrane with the plasma membrane, perhaps like exocytosis in cells in general. The extent to which the CV may participate in other processes, such as the transport of materials to the cell surface and the regulation of cytosolic calcium and pH levels, and under what circumstances, remains unclear.

One feature of membrane behavior during the CV cycle is striking: the rapidity of the transformation of membranes from a near-planar to a highly curved configuration and back again. How rapid changes in membrane curvature and in associated membrane tension might be related to CVC function and to the 'contraction' at the end of the CV cycle (probably better described as a 'collapse'), is unknown.

There are no critical observations that rule out any of the four classes of explanation of water segregation described in Section 3.5. What remains uncertain about the uptake of water by the CVC is whether the initial transport of water is accomplished with or against an osmotic gradient, what the exact role of the V-H^+-ATPase is, and what other ion pumps or channels might be involved in the segregation of water.

Several approaches are now available to study these questions. Membranes of the CVC have been isolated (Nolta and Steck, 1994), so that we may be able to learn more detail about their lipid and protein composition. Genetic approaches—mutant isolation and gene knockout experiments—in favorable organisms, such as *Chlamydomonas* and *Dictyostelium*, have started to define general processes and specific proteins involved in various facets of CVC function (O'Halloran and Anderson, 1992; Bush *et al.*, 1996; Luykx *et al.*, 1997b; Becker *et al.*, 1999). The use of these methods to identify the essential molecular machinery should shed considerable light on how contractile vacuoles work.

Acknowledgments

The author is grateful to Bruce Grayson and Leo Sternberg for useful discussions. The preparation of this review, and part of the work on *Chlamydomonas* described in it, were supported by the US National Science Foundation, research grant no. MCB-9603675.

References

Ahmad, M. and Couillard, P. (1974) The contractile vacuole in *Amoeba proteus*: temperature effects. *J. Protozool.*, **21** 330-36.

Ajouz, B., Berrier, C., Besnard, M., Martinac, B. and Ghazi, A. (2000) Contributions of the different extramembranous domains of the mechanosensitive ion channel, MscL, to its response to membrane tension. *J. Biol. Chem.*, **275** 1015-22.

Allen, R.D. (1995) Membrane tubulation and proton pumps. *Protoplasma*, **189** 1-8.

Allen, R.D. and Fok, A.K. (1988) Membrane dynamics of the contractile vacuole complex of *Paramecium*. *J. Protozool.*, **35** 63-71.

Baines, I.C., Brzeska, H. and Korn, E.D. (1992) Differential localization of *Acanthamoeba* myosin I isoforms. *J. Cell Biol.*, **119** 1193-203.

Baines, I.C., Corigliano-Murphy, A. and Korn, E.D. (1995) Quantification and localization of phosphorylated myosin I isoforms in *Acanthamoeba castellanii*. *J. Cell Biol.*, **130** 591-603.

Becker, M., Matzner, M. and Gerisch, G. (1999) Drainin required for membrane fusion of the contractile vacuole in *Dictyostelium* is the prototype of a protein family also represented in man. *EMBO J.*, **18** 3305-16.

Bergquist, B.L. (1989) Modification of contractile vacuole activity by calmodulin inhibitors. *Trans. Am. Microsc. Soc.*, **108** 369-79.

Brauer, E.B. and McKanna, J.A. (1978) Contractile vacuoles in cells of a fresh water sponge, *Spongilla lacustris*. *Cell Tissue Res.*, **192** 309-17.

Brumm, T., Jorgensen, K., Mouritsen, O.G. and Bayerl, T.M. (1996) The effect of increasing membrane curvature on the phase transition and mixing behavior of a dimyristoyl-sn-glycero-3-phosphatidylcholine/distearoyl-sn-glycero-3-phosphatidylcholine lipid mixture as studied by Fourier transform infrared spectroscopy and differential scanning calorimetry. *Biophys. J.*, **70** 1373-79.

Buczynski, G., Grove, B., Nomura, A., Kleve, M., Bush, J., Firtel, R.A. and Cardelli, J. (1997) Inactivation of two *Dictyostelium discoideum* genes, *DdPIK1* and *DdPIK2*, encoding proteins related to mammalian phosphatidylinositide 3-kinases, results in defects in endocytosis, lysosome to post-lysosome transport, and actin cytoskeleton organization. *J. Cell Biol.*, **136** 1271-86.

Burg, M.B. (1995) Molecular basis of osmotic regulation. *Am. J. Physiol.*, **268** F983-96.

Burgoyne, R.D. and Morgan, A. (1998) Calcium sensors in regulated exocytosis. *Cell Calcium*, **24** 367-76.

Bush, J., Nolta, K., Rodriguez-Paris, J., Kaufmann, N., O'Halloran, T., Ruscetti, T., Temesvari, L., Steck, T. and Cardelli, J. (1994) A Rab4-like GTPase in *Dictyostelium discoideum* co-localizes with V-H^+-ATPases in reticular membranes of the contractile vacuole complex and in lysosomes. *J. Cell Sci.*, **107** 2801-12.

Bush, J., Temesvari, L., Rodriguez-Paris, J., Buczynski, G. and Cardelli, J. (1996) A role for a Rab4-like GTPase in endocytosis and in regulation of contractile vacuole structure and function in *Dictyostelium discoideum*. *Mol. Biol. Cell*, **7** 1623-38.

Cain, J.R., Mattox, K.R. and Stewart, K.D. (1973) The cytology of zoosporogenesis in the filamentous green alga genus, *Klebsormidium*. *Trans. Am. Microsc. Soc.*, **92** 398-404.

Cass, A. and Finkelstein, A. (1967) Water permeability of thin lipid membranes. *J. Gen. Physiol.*, **50** 1765-84.

Cessna, S.G., Chandra, S. and Low, P.S. (1998) Hypo-osmotic shock of tobacco cells stimulates Ca^{2+} fluxes deriving first from external and then internal Ca^{2+} stores. *J. Biol. Chem.*, **273** 27286-91.

Chang, G., Spencer, R.H., Lee, A.T., Barclay, M.T. and Rees, D.C. (1998) Structure of the MscL homolog from *Mycobacterium tuberculosis*: a gated mechanosensitive ion channel. *Science*, **282** 2220-26.

Cosgrove, W.B. and Kessel, R.G. (1958) The activity of the contractile vacuole of *Crithidia fasciculata*. *J. Protozool.*, **5** 296-98.

Davies, J.M. (1997) Vacuolar energization: pumps, shunts and stress. *J. Exp. Bot.*, **48** 633-41.

de Chastellier, C., Quiviger, B. and Ryter, A. (1978) Observations on the functioning of the contractile vacuole of *Dictyostelium discoideum* with the electron microscope. *J. Ultrastr. Res.*, **62** 220-27.

Denning, G.M. and Fulton, A.B. (1989a) Electron microscopy of a contractile vacuole mutant of *Chlamydomonas moewusii* (Chlorophyta) defective in the late stages of diastole. *J. Phycol.*, **25** 667-72.

Denning, G.M. and Fulton, A.B. (1989b) Purification and characterization of clathrin-coated vesicles from *Chlamydomonas*. *J. Protozool.*, **36** 334-40.

Doberstein, S.K., Baines, I.C., Wiegand, G., Korn, E.D. and Pollard, T.D. (1993) Inhibition of contractile vacuole function *in vivo* by antibodies against myosin I. *Nature*, **365** 841-43.

Domozych, D.S. (1989) The endomembrane system and mechanism of membrane flow in the green alga, *Gloeomonas kupfferi* (Volvocales, Chlorophyta). II. A cytochemical analysis. *Protoplasma*, **149** 108-19.

Domozych, D.S. and Nimmons, T.T. (1992) The contractile vacuole as an endocytotic organelle of the chlamydomonad flagellate, *Gloeomonas kupfferi* (Volvocales, Chlorophyta). *J. Phycol.*, **28** 809-16.

Domozych, D.S. and Dairman, M. (1993) Synthesis of the inner cell wall layer of the chlamydomonad flagellate, *Gloeomonas kupfferi. Protoplasma*, **176** 1-13.

Eyden, B.P. and Vickerman, K. (1975) Ultrastructure and vacuolar movements in the free-living diplomonad, *Trepomonas agilis* Krebs. *J. Protozool.*, **22** 54-66.

Field, C., Li, R. and Oegema, K. (1999) Cytokinesis in eukaryotes: a mechanistic comparison. *Curr. Opin. Cell Biol.*, **11** 68-80.

Fok, A.K., Clarke, M., Ma, L. and Allen, R.D. (1993) Vacuolar H^+-ATPase of *Dictyostelium discoideum*: a monoclonal antibody study. *J. Cell Sci.*, **106** 1103-13.

Fok, A.K., Aihara, M.S., Ishida, M., Nolta, K.V., Steck, T.L. and Allen, R.D. (1995) The pegs on the decorated tubules of the contractile vacuole complex of *Paramecium* are proton pumps. *J. Cell Sci.*, **108** 3163-70.

Fujii, M., Ohtsubo, M., Ogawa, T., Kamata, H., Hirata, H. and Yagisawa, H. (1999) Real-time visualization of pH domain-dependent translocation of phospholipase C-δ_1 in renal epithelial cells (MDCK): response to hypo-osmotic stress. *Biochem. Biophys. Res. Commun.*, **254** 284-91.

Furukawa, R. and Fechheimer, M. (1994) Differential localization of α-actinin and the 30 kD actin-binding protein in the cleavage furrow, phagocytic cup and contractile vacuole of *Dictyostelium discoideum. Cell Motil. Cytoskeleton*, **29** 46-56.

Garner, M.M. and Burg, M.B. (1994) Macromolecular crowding and confinement in cells exposed to hypertonicity. *Am. J. Physiol.*, **266** C877-92.

Giglione, C. and Gross, J.D. (1995) Anion effects on vesicle acidification in *Dictyostelium. Biochem. Mol. Biol. Int.*, **36** 1057-65.

Grief, C. and Shaw, P.J. (1987) Assembly of cell-wall glycoproteins of *Chlamydomonas reinhardii*: oligosaccharides are added in medial and trans-Golgi compartments. *Planta*, **171** 302-12.

Guillard, R.R.L. (1960) A mutant of *Chlamydomonas moewusii* lacking contractile vacuoles. *J. Protozool.*, **7** 262-68.

Harvey, W.R. and Nelson, N. (1992) *V-ATPases*. The Company of Biologists, Ltd, Cambridge. *J. Exp. Biol.*, **172** 1-485.

Hausmann, K. (1983) Zur Organisation der kontraktilen Vacuole von *Pseudomicrothorax dubius* (Ciliophora, Kinetophragminophorea, Hypostomatia). *Arch. f. Protistenk.*, **127** 319-25.

Hausmann, K. and Allen, R.D. (1977) Membranes and microtubules of the excretory apparatus of *Paramecium caudatum. Cytobiol. Eur. J. Cell Biol.*, **15** 303-20.

Hausmann, K. and Patterson, D.J. (1984) Contractile vacuole complexes in algae, in *Compartments in Algal Cells and Their Interaction*. (eds. W. Wiessner, D. Robinson and R.C. Starr) Springer-Verlag, Berlin Heidelberg, pp. 139-46.

Hellebust, J.A., Mérida, T. and Ahmad, I. (1989) Operation of contractile vacuoles in the euryhaline green flagellate, *Chlamydomonas pulsatilla* (Chlorophyceae), as a function of salinity. *Mar. Biol.*, **100** 373-79.

Heuser, J., Zhu, Q. and Clarke, M. (1993) Proton pumps populate the contractile vacuoles of *Dictyostelium* amoebae. *J. Cell Biol.*, **121** 1311-27.

Hirose, K., Kadowaki, S., Tanabe, M., Takeshima, H. and Iino, M. (1999) Spatiotemporal dynamics of inositol 1,4,5-triphosphate that underlies complex Ca^{2+} mobilization patterns. *Science*, **284** 1527-30.

Hoffmann, E.K. and Dunham, P.B. (1995) Membrane mechanisms and intracellular signalling in cell volume regulation. *Int. Rev. Cytol.*, **161** 173-262.

Hopkins, D.L. (1946) The contractile vacuole and the adjustment to changing concentrations in fresh water amoebae. *Biol. Bull.*, **90** 158-76.

Hoppenrath, M. (1995) Ultrastrukturelle Untersuchungen zur Funktion der Kontraktilen Vakuolen von *Chlamydomonas reinhardtii*. Unpublished thesis, Cytologischen Abteilung der Pflanzenphysiologischen Instituts der Georg-August-Universität zu Göttingen, Germany.

Huber, H., Beyser, K. and Fabry, S. (1996) Small G proteins of two green algae are localized to exocytic compartments and to flagella. *Plant Mol. Biol.*, **31** 279-93.

Ishida, M., Aihara, M.S., Allen, R.D. and Fok, A.K. (1993) Osmoregulation in *Paramecium*: the locus of fluid segregation in the contractile vacuole complex. *J. Cell Sci.*, **106** 693-702.

Ishida, M., Fok, A.K., Aihara, M.S. and Allen, R.D. (1996) Hyperosmotic stress leads to reversible dissociation of the proton pump-bearing tubules from the contractile vacuole complex in *Paramecium*. *J. Cell Sci.*, **109** 229-37.

Kitching, J.A. (1938) Contractile vacuoles. *Biol. Rev.*, **13** 403-44.

Kitching, J.A. (1956) Contractile vacuoles of protozoa. *Protoplasmatogia*, **III, D3a** 1-45.

Kitching, J.A. (1967) Contractile vacuoles, ionic regulation and excretion, in *Research in Protozoology*, (ed. T.-T. Chen) Pergamon Press, Oxford, Vol. 1, pp. 307-36.

Knight, H. (2000) Calcium signaling during abiotic stress in plants. *Int. Rev. Cytol.*, **195** 269-324.

Lang, J. (1999) Molecular mechanisms and regulation of insulin exocytosis as a paradigm of endocrine secretion. *Eur. J. Biochem.*, **259** 3-17.

Lang, F., Busch, G.L., Ritter, M., Volkl, H., Waldegger, S., Gulbins, E. and Haussinger, D. (1998a) Functional significance of cell volume regulatory mechanisms. *Physiol. Rev.*, **78** 247-306.

Lang, F., Busch, G.L. and Volkl, H. (1998b) The diversity of volume regulatory mechanisms. *Cell Physiol. Biochem.*, **8** 1-45.

Lee, T.H. and Linstedt, A.D. (1999) Osmotically induced cell volume changes alter anterograde and retrograde transport, Golgi structure and COPI dissociation. *Mol. Biol. Cell*, **10** 1445-62.

Leedale, G.F. (1967) *Euglenoid Flagellates*. Chp. 14. Prentice-Hall, Inc., Englewood Cliffs, New Jersey.

Linder, J.C. and Staehelin, L.A. (1979) A novel model for fluid secretion by the trypanosomatid contractile vacuole apparatus. *J. Cell Biol.*, **83** 371-82.

Luykx, P., Hoppenrath, M. and Robinson, D.G. (1997a) Structure and behavior of contractile vacuoles in *Chlamydomonas reinhardtii. Protoplasma*, **198** 73-84.

Luykx, P., Hoppenrath, M. and Robinson, D.G. (1997b) Osmoregulatory mutants that affect the function of the contractile vacuole in *Chlamydomonas reinhardtii. Protoplasma*, **200** 99-111.

Manton, I. (1964). Observations on the fine structure of the zoospore and young germling of *Stigeoclonium*. *J. Exp. Bot.*, **15** 399-411.

Marchant, H.J. and Pickett-Heaps, J.D. (1970) Ultrastructure and differentiation of *Hydrodictyon reticulatum*. II. Formation of zooids within the coenobium. *Austr. J. Biol. Sci.*, **24** 471-86.

Marchant, H.J., Pickett-Heaps, J.D. and Jacobs, K. (1973) An ultrastructural study of zoosporogenesis and the mature zoospore of *Klebsormidium flaccidum. Cytobios*, **8** 95-107.

Mayer, L.M. and Iverson, R.M. (1967) Osmotic concentration of the contractile vacuole of *Amoeba proteus* following UV light irradiation. *Experientia*, **23** 120.

McKanna, J.A. (1971) Homologous structure of coated membranes composing the fluid segregating organelles of contractile vacuoles. *J. Protozool.*, **18** (Suppl.) 27-28 (Abstract).

McKanna, J.A. (1973a) Membrane recycling: vesiculation of the amoeba contractile vacuole at systole. *Science*, **179** 88-90.

McKanna, J.A. (1973b) Fine structure of the contractile vacuole pore in *Paramecium*. *J. Protozool.*, **20** 631-38.

McKanna, J.A. (1974) Permeability modulating membrane coats. I. Fine structure of fluid segregation organelles of peritrich contractile vacuoles. *J. Cell Biol.*, **63** 317-22.

McKanna, J.A. (1976) Fine structure of fluid segregation organelles of *Paramecium* contractile vacuoles. *J. Ultrastr. Res.*, **54** 1-10.

Meijer, H.J.G., Divecha, N., van den Ende, H., Musgrave, A. and Munnik, T. (1999) Hyperosmotic stress induces rapid synthesis of phosphatidyl-D-inositol 3,5-bisphosphate in plant cells. *Planta*, **208** 294-98.

Mills, J.W., Schwiebert, E.M. and Stanton, B.A. (1994) The cytoskeleton and cell volume regulation, in *Cellular and Molecular Physiology of Cell Volume Regulation*. (ed. K. Strange) CRC Press, Boca Raton, pp. 241-58.

Minton, A.P. (1994) Influence of macromolecular crowding on intracellular association reactions: possible role in volume regulation, in *Cellular and Molecular Physiology of Cell Volume Regulation*. (ed. K. Strange) CRC Press, Boca Raton, FL, pp. 181-91.

Mizuno, T. (1991) Control of envelope protein synthesis in *Escherichia coli*, in *Control of Cell Growth and Division* (eds. A. Ishihama and H. Yoshikawa) Japan Scientific Societies Press, Springer-Verlag, Tokyo, Berlin, pp. 141-59.

Moniakis, J., Coukell, M.B. and Janiec, A. (1999) Involvement of the Ca^{2+}-ATPase PAT1 and the contractile vacuole in calcium regulation in *Dictyostelium discoideum*. *J. Cell Sci.*, **112** 405-14 .

Naitoh, Y., Tominaga, T., Ishida, M., Fok, A.K., Aihara, M.S. and Allen, R.D. (1997) How does the contractile vacuole of *Paramecium multimicronucleatum* expel fluid? Modelling the expulsion mechanism. *J. Exp. Biol.*, **200** 713-21.

Nelson, H. and Nelson, N. (1990) Disruption of genes encoding subunits of yeast vacuolar H^+-ATPase causes conditional lethality. *Proc. Natl. Acad. Sci. USA*, **87** 3503-07.

Nolta, K.V. and Steck, T.L. (1994) Isolation and initial characterization of the bipartite contractile vacuole complex from *Dictyostelium discoideum*. *J. Biol. Chem.*, **269** 2225-33.

Nolta, K.V., Padh, H. and Steck, T.L. (1993) An immunocytochemical analysis of the vacuolar proton pump in *Dictyostelium discoideum*. *J. Cell Sci.*, **105** 849-59.

O'Halloran, T.J. and Anderson, R.G.W. (1992) Clathrin heavy chain is required for pinocytosis, the presence of large vacuoles, and development in *Dictyostelium*. *J. Cell Biol.*, **118** 1371-77.

O'Neil, R.G. and Leng, L. (1997) Osmo-mechanically sensitive phosphatidylinositol signaling regulates a Ca^{2+} influx channel in renal epithelial cells. *Am. J. Physiol.*, **273** F120-28.

Parker, J.C. (1993) In defense of cell volume? *Am. J. Physiol.*, **265** C1191-200.

Patterson, D.J. (1980) Contractile vacuoles and associated structures: their organization and function. *Biol. Rev.*, **55** 1-46.

Prusch, R.D. and Dunham, P.B. (1970) Contraction of isolated contractile vacuoles from *Amoeba proteus*. *J. Cell Biol.*, **46** 431-34.

Quiviger, B., de Chastellier, C. and Ryter, A. (1978) Cytochemical demonstration of alkaline phosphatase in the contractile vacuole of *Dictyostelium discoideum*. *J. Ultrastr. Res.*, **62** 228-36.

Raven, J.A. (1995) Costs and benefits of low intracellular osmolarity in cells of freshwater algae. *Funct. Ecol.*, **9** 701-07.

Riddick, D.H. (1968) Contractile vacuole in the amoeba, *Pelomyxa carolinensis*. *Am. J. Physiol.*, **215** 736-40.

Robinson, D.G., Hoppenrath, M., Oberbeck, K., Luykx, P. and Ratajczak, R. (1998) Localization of pyrophosphatase and V-ATPase in *Chlamydomonas reinhardtii*. *Bot. Acta*, **111** 108-22.

Rooney, E.K., Gross, J.D. and Satre, M. (1994) Characterisation of an intracellular Ca^{2+} pump in *Dictyostelium*. *Cell Calcium*, **16** 509-22.

Sackin, H. (1995) Mechanosensitive channels. *Annu. Rev. Physiol.*, **57** 333-53.

Saint, N., Lacapere, J.J., Gu, L.Q., Ghazi, A., Martinac, B. and Rigaud, J.L. (1998) A hexameric transmembrane pore revealed by two-dimensional crystallization of the large mechanosensitive ion channel (MscL) of *Escherichia coli*. *J. Biol. Chem.*, **273** 14667-70.

Schmidt-Nielsen, B. and Schrauger, C.R. (1963) *Amoeba proteus*: studying the contractile vacuole by micropuncture. *Science*, **139** 606-07.

Sesaki, H., Wong, E.F. and Siu, C. (1997) The cell adhesion molecule, DdCAD-1, in *Dictyostelium* is targeted to the cell surface by a nonclassical transport pathway involving contractile vacuoles. *J. Cell Biol.*, **138** 939-51.

Somanchi, A., Handley, E.R. and Moroney, J.V. (1998) *Chlamydomonas reinhardtii* cDNAs upregulated in low-CO_2 conditions: expression and analysis. *Can. J. Bot.*, **76** 1003-09.

Steck, T.L., Chiaraviglio, L. and Meredith, S. (1997) Osmotic homeostasis in *Dictyostelium discoideum*: excretion of amino acids and ingested solutes. *J. Eukar. Microbiol.*, **44** 503-10.

Stein, W.D. (1986) *Transport and Diffusion Across Cell Membranes*. Academic Press, Inc., San Diego, New York.

Strange, K., ed. (1994) *Cellular and Molecular Physiology of Cell Volume Regulation*. CRC Press, Boca Raton, FL.

Südhof, T. (1995) The synaptic vesicle cycle: a cascade of protein-protein interactions. *Nature*, **375** 645-53.

Sukharev, S. (1999) Mechanosensitive channels in bacteria as membrane tension reporters. *FASEB J.*, **13** (Suppl.) S55-61.

Sukharev, S.I., Blount, P., Martinac, B. and Kung, C. (1997) Mechanosensitive channels of *Escherichia coli*: the MscL gene, protein and activities. *Annu. Rev. Physiol.*, **59** 633-57.

Sültemeyer, D. (1998) Carbonic anhydrase in eukaryotic algae: characterization, regulation, and possible function during photosynthesis. *Can. J. Bot.*, **76** 962-72.

Suzuki, Y., Ohnishi, K., Hirabayashi, T. and Watanabe, T. (1982) *Tetrahymena* calmodulin: characterization of an anti-tetrahymena calmodulin and the immunofluorescent localization in *Tetrahymena*. *Exp. Cell Res.*, **137** 1-14.

Sze, H., Ward, J.M., Lai, S. and Perera, I. (1992) Vacuolar-type H^+-translocating ATPases in plant endomembranes: subunit organization and multigene families. *J. Exp. Biol.*, **172** 123-35.

Tanaka, T., Saha, S.K., Tomomori, C., Ishima, R., Liu, D., Tong, K.I., Park, H., Dutta, R., Qin, L., Swindells, M.B., Yamazaki, T., Ono., A.M., Kainosho, M., Inouye, M. and Ikura, M. (1998) NMR structure of the histidine kinase domain of the *E. coli* osmosensor, EnvZ. *Nature*, **396** 88-92.

Temesvari, L.A., Rodriguez-Paris, J.M., Bush, J.M., Zhang, L. and Cardelli, J.A. (1996) Involvement of the vacuolar proton-translocating ATPase in multiple steps of the endo-lysosomal system and in the contractile vacuole system of *Dictyostelium discoideum*. *J. Cell Sci.*, **109** 1479-95.

Tinel, H., Wehner, F. and Kinne, R.K. (1997) Arachidonic acid as a second messenger for hypotonicity-induced calcium transients in rat IMCD cells. *Pflugers Arch.*, **433** 245-53.

Tominaga, T., Allen, R.D. and Naitoh, Y. (1998) Cyclic changes in the tension of the contractile vacuole complex membrane control its exocytotic cycle. *J. Exp. Biol.*, **201** 2647-58.

Tominaga, T., Naitoh, Y. and Allen, R.D. (1999) A key function of non-planar membranes and their associated microtubular ribbons in contractile vacuole membrane dynamics is revealed by electrophysiologically controlled fixation of *Paramecium*. *J. Cell Sci.*, **112** 3733-45.

Tomomori, C., Tanaka, T., Dutta, R., Park, H., Saha, S.K., Zhu, Y., Ishima, R., Liu, D., Tong, K.I., Kurokawa, H., Qian, H, Inouye, M. and Ikura, M. (1999) Solution structure of the homodimeric core domain of *Escherichia coli* histidine kinase, EnvZ. *Nature Struct. Biol.*, **6** 729-34.

Valentijn, K., Valentijn, J.A. and Jamieson, J.D. (1999) Role of actin in regulated exocytosis and compensatory membrane retrieval: insights from an old acquaintance. *Biochem. Biophys. Res. Commun.*, **266** 652-61.

Victorov, A.V., Janes, N., Moehren, G., Rubin, E., Taraschi, T. and Hoek, J.B. (1994) Rapid transbilayer movement of phosphatidylethanol in unilamellar phosphatidylcholine vesicles. *J. Am. Chem. Soc.*, **116** 4050-52.

Waldegger, S., Busch, G.L., Kaba, N.K., Zempel, G., Ling, H., Heidland, A., Haussinger, D. and Lang, F. (1997) Effect of cellular hydration on protein metabolism. *Mineral Electr. Metab.*, **23** 201-05.

Watanabe, A., Kurasawa, Y., Watanabe, Y. and Numata, O. (1998) A new *Tetrahymena* actin-binding protein is localized in the division furrow. *J. Biochem. (Tokyo)*, **123** 607-13.

Weiss, R.L. (1983) Coated vesicles in the contractile vacuole/mating structure region of *Chlamydomonas*. *J. Ultrastr. Res.*, **85** 33-44.

Weiss, R.L., Goodenough, D.A. and Goodenough, U.W. (1977) Membrane particle arrays associated with the basal body and with contractile vacuole secretion in *Chlamydomonas*. *J. Cell Biol.*, **72** 133-43.

Wessel, D. and Robinson, D.G. (1979) Studies on the contractile vacuole of *Poterioochromonas malhamensis* Peterfi. I. The structure of the alveolate vesicles. *Eur. J. Cell Biol.*, **19** 60-66.

Wigg, D., Bovee, E.C. and Jahn, T.L. (1967) The evacuation mechanism of the water expulsion vesicle ('contractile vacuole') of *Amoeba proteus*. *J. Protozool.*, **14** 104-08.

Wood, J.M. (1999) Osmosensing by bacteria: signals and membrane-based sensors. *Microbiol. Mol. Biol. Rev.*, **63** 230-62.

Xie, Y., Coukell, M.B. and Gombos, Z. (1996) Antisense RNA inhibition of the putative vacuolar H(+)-ATPase proteolipid of *Dictyostelium* reduces intracellular Ca^{2+} transport and cell viability. *J. Cell Sci.*, **109** 489-97.

Yoshimura, K. (1998) Mechanosensitive channels in the cell body of *Chlamydomonas*. *J. Membr. Biol.*, **166** 149-55.

Zeuthen, T. (1992) From contractile vacuole to leaky epithelia: coupling between salt and water fluxes in biological membranes. *Biochim. Biophys. Acta*, **1113** 229-58.

Zeuthen, T. (1995) Molecular mechanisms for passive and active transport of water. *Int. Rev. Cytol.*, **160** 99-161.

Zeuthen, T. and Stein, W.D. (1994) Co-transport of salt and water in membrane proteins: membrane proteins as osmotic engines. *J. Membr. Biol.*, **137** 179-95.

Zhu, Q. and Clarke, M. (1992) Association of calmodulin and an unconventional myosin with the contractile vacuole complex of *Dictyostelium discoideum*. *J. Cell Biol.*, **118** 347-58.

Zhu, Q., Liu, T. and Clarke, M. (1993) Calmodulin and the contractile vacuole complex in mitotic cells of *Dictyostelium discoideum*. *J. Cell Sci.*, **104** 1119-27.

Zimmerman, S.B. and Minton, A.P. (1993) Macromolecular crowding: biochemical, biophysical and physiological consequences. *Annu. Rev. Biophys. Biomol. Struc.*, **22** 27-65.

4 Autophagy and vacuole formation

Yuji Moriyasu and Stefan Hillmer

4.1 Introduction

Autophagy is defined as a process in which a cell sequesters its own constituents in membrane-bounded compartments, where they are subsequently degraded. Little is known about the molecular mechanisms underlying autophagy in plant cells, but apparent autophagic processes have been described at various developmental stages. They range from meristematic cells, where the formation of the vacuole occurs, to senescent tissues, where programmed cell death takes place. Apart from its role in development, autophagy will probably play an essential role in intracellular protein turnover in plant cells. In mammalian cells, protein degradation occurs via autophagic/lysosomal and non-lysosomal pathways. The rate of lysosomal protein degradation, as determined by the use of inhibitors of this pathway, has been estimated as varying from 0 to 4% cell mass/h, depending on the cell type, nutritional and hormonal conditions (Seglen and Bohley, 1992). Since autophagy is a major contributor to the lysosomal pathway, it constitutes an important cellular process, especially in stressed cells, where protein degradation is enhanced (Blommaart *et al.*, 1997).

4.2 Morphological evidence for autophagy in plant cells

Autophagic processes were originally classified into macro- and micro-autophagy, according to the size of the material sequestered within lysosomes (de Duve, 1984). Over the years, this definition has been refined. Now the two types of autophagy can be differentiated in terms of morphological parameters. In macroautophagy, special membrane structures called 'isolation membranes' or 'phagophores' develop and segregate portions of cytoplasm to form autophagosomes. These then fuse with endosomes and/or lysosomes to generate autolysosomes, where their cytoplasmic contents are finally degraded. In contrast, microautophagy is a process where the lysosome itself takes up small portions of cytoplasm by pinching off a vesicle. The vesicle with the enclosed cytoplasm is then degraded inside the lysosome.

Processes analogous to macroautophagy and microautophagy have also been described in plant cells (Matile, 1975; Marty, 1997, 1999; Robinson *et al.*, 1998). The involvement of plant vacuoles in microautophagy is suggested by numerous ultrastructural observations showing the presence of cytoplasmic material, including partially digested organelles, such as mitochondria and plastids, other undefined membranes, and ribosomes within vacuoles. Poux (1963) interpreted these images as representing a degradative process taking place within vacuoles and noticed, even then, that many of the inclusions of cytoplasm in vacuoles were surrounded by a membrane. Subsequent studies indicated that the membrane wrapping the intravacuolar inclusions was tonoplastic in origin. For example, in freeze-etched specimens of root tips, the membrane which surrounds vacuolar inclusions has the same fine structure as the vacuolar membrane (Fineran, 1970). Furthermore, the fine structure of the freeze-etched tonoplast was recognized to be different at sites where invagination occurs. In particular, the number of intramembrane particles was seen to be larger at the site where invagination occurs, suggesting their active involvement in the pinocytotic process.

A process analogous to macroautophagy has also been described by plant cell biologists. Villiers (1967) first reported the presence of double membrane-bounded vacuoles enclosing partially digested cytoplasm. By definition, these vacuoles should be regarded as autolysosomes. Another example of macroautophagy has been described in endosperm cells of germinating castor beans, where double-membrane structures are seen sequestering microbodies (Vigil, 1970). In these cells, the number of microbodies (glyoxysomes) increases transiently, and thereafter decreases. At this time, some of these microbodies are sequestered by double membrane-bounded autophagic vacuoles and degraded. These autophagic vacuoles appear to contain only microbodies, segments of endoplasmic reticulum (ER) and free ribosomes.

4.3 Biochemical characterization of autophagy in plant cells

Little is known about the contribution of plant vacuoles to intracellular protein degradation (Canut *et al.*, 1986; Moriyasu, 1995; see also Chapter 2 in this volume). However, as in yeasts (see Chapter 1 in this volume), and mammalian cells (for a review, see Dunn, 1994), autophagy can be induced in suspension-cultured cells through depletion of the carbon source. The changes in several biochemical parameters induced by starvation treatment in suspension-cultured sycamore (*Acer pseudoplatanus*) have been monitored by Douce and co-workers (1991, 1992). They showed that net protein degradation occurs upon sucrose starvation and proposed that this protein degradation is due to autophagy. Net protein degradation begins when intracellular carbohydrates, including sucrose and starch, become limiting, and as a result the total amount of free amino acids increases (Genix *et al.*, 1990). This roughly accounts for the total protein that disappears. Some of these amino acids become deaminated and oxidized in place of sugar, because the concentrations of ammonia and asparagine, a scavenger of ammonia, increase (Genix *et al.*, 1990). When whole maize plants are kept in the dark for more than 2 days, the total amount of cellular protein decreases, presumably due to autophagic processes, and ammonia and asparagine in the root and leaves increase accordingly (Brouquisse *et al.*, 1998).

Sugar starvation also leads to the hydrolysis of polar lipids, especially phosphatidylcholine and phosphatidylethanolamine (Dorne *et al.*, 1987). As a result, phosphorylcholine and phosphorylethanolamine accumulate during starvation treatment (Journet *et al.*, 1986; Dorne *et al.*, 1987; Aubert *et al.*, 1994, 1996). Interestingly, there is a decrease in plastid envelope galactolipids, whereas sterols, which are mainly associated with the plasmalemma and tonoplast, remain constant, suggesting that these membranes are not directly affected by autophagy (Journet *et al.*, 1986; Douce *et al.*, 1991; Aubert *et al.*, 1996).

It has also been shown that the rate of uncoupled respiration decreases (Journet *et al.*, 1986). This has been attributed to a decrease in the number of mitochondria, since the amount of the two specific mitochondrial markers, cardiolipin and cytochrome oxidase, also decrease (Journet *et al.*, 1986). The total amount of hexose-phosphates, including glucose-P, mannose-P and fructose-P, and uridine diphosphoglucose (UDP-glucose) decreases, whereas cytoplasmic and vacuolar Pi concentrations increase (Rébeillé *et al.*, 1985; Journet *et al.*, 1986; Roby *et al.*, 1987; Aubert *et al.*, 1996). Vacuolar pH shifts from 5.7 to 6.1, partly as a result of the accumulation of ammonium in the vacuole (Aubert *et al.*, 1996) but also through loss of vacuolar proton-translocating adenosine triphosphatase (V-ATPase) and vacuolar proton-translocating pyrophosphatase (V-PPase), which disappear from the tonoplast during sucrose starvation in suspension-cultured tobacco cells (Sikora *et al.*, 1998).

The accumulation of phosphorylcholine, a degradation product of phosphatidylcholine, in the cells during sucrose starvation is a useful marker of autophagy, because phosphorylcholine appears to be metabolically very stable. In contrast, phosphorylethanolamine, which results from the degradation of phosphatidylethanolamine, only transiently accumulates in the cells. Using the accumulation of phosphorylcholine in the cells as a marker of autophagy, Aubert and co-workers (1996) have further investigated the factors controlling autophagy. When the cells are transferred to a culture medium where sucrose is replaced by glycerol or pyruvate, there is no accumulation of phosphorylcholine and no formation of autophagic vacuoles (Aubert *et al.*, 1994, 1996). Even in the presence of glycerol or pyruvate, the intracellular concentration of hexose phosphates remains low (Aubert *et al.*, 1994). In contrast, when pyruvate or glycerol is removed from the culture medium, autophagy starts, and, as a result, phosphorylcholine accumulates and autophagic vacuoles are formed. These data suggest that it is the supply of respiratory substrates to mitochondria, and not the decrease in hexose phosphates, which controls the induction of autophagy in plant cells under conditions of starvation.

In addition, there are links between changes in enzymatic activities and autophagy. Both cysteine and aspartic protease activities increase upon sucrose starvation in tobacco cultured cells (Moriyasu and Ohsumi, 1996). A vacuolar serine protease is induced by glucose starvation in excised maize root tips (James *et al.*, 1993, 1996) and four intracellular RNases have been shown to be induced in cultured tomato cells upon phosphate starvation (Löffler *et al.*, 1992). Three of these four enzymes are localized in the vacuole (Abel and Glund, 1986; Löffler *et al.*, 1992).

4.4 Autophagy in various developmental processes in plant cells (Figure 4.1)

4.4.1 Autophagy and genesis of vegetative vacuoles

Many meristematic cells of higher plants appear to lack vacuoles. Vacuoles are formed as these cells begin to divide and differentiate. Up to now, there is no unified concept for vacuole biogenesis in plants, nevertheless autophagy seems to be essential for this process. Marty and co-workers have studied the formation of the vacuoles in meristematic root cells of *Euphorbia* by high-voltage and conventional electron microscopy (Marty, 1978; Marty *et al.*, 1980). In the youngest cells, where vacuolation is supposed to be initiated, 0.1 μm diameter vesicles are detected in the vicinity of the Golgi apparatus. These vesicles appear to bud-off from the *trans*-Golgi network (TGN) but not from Golgi or from ER cisternae. Histochemistry of acid phosphatase shows that this enzyme exists only in the TGN and associated vesicles. Marty designated these vesicles as 'provacuoles'. In slightly more differentiated cells, the provacuoles elongate into tubes of 0.1 μm in diameter. Their lumen becomes filled with vesicles that are presumably derived from the invagination of the surface membrane (Marty, 1999). High-voltage electron microscopy shows that these provacuolar tubes are part of a branched structure, which forms continuous channels throughout the cytoplasm. In the next stage, several provacuolar tubes are seen to envelop portions of cytoplasm. A transverse section of this structure gives the appearance of several membrane vesicles arranged in a circular fashion. Subsequently, the tubes completely enclose the cytoplasm and form a double membrane-bounded structure. Initially, only the cavities between the two membranes show acid phosphatase activity, but in more differentiated cells, the intactness of the inner membrane is gradually lost and hydrolases in the cavities permeate into the sequestered cytoplasm. When the degradation of the segregated cytoplasm and the inner membrane is complete, a typical young vacuole is the result. The outer membrane (tonoplast) remains to confine hydrolytic enzymes within the vacuole. Finally, the young vacuoles appear to fuse together and produce a large vegetative vacuole. According to Marty, this vacuole can engulf bits of cytoplasm by microautophagy. Moreover, he claims that in mature, differentiated cells the provacuoles fuse directly with the large vacuole without performing autophagy.

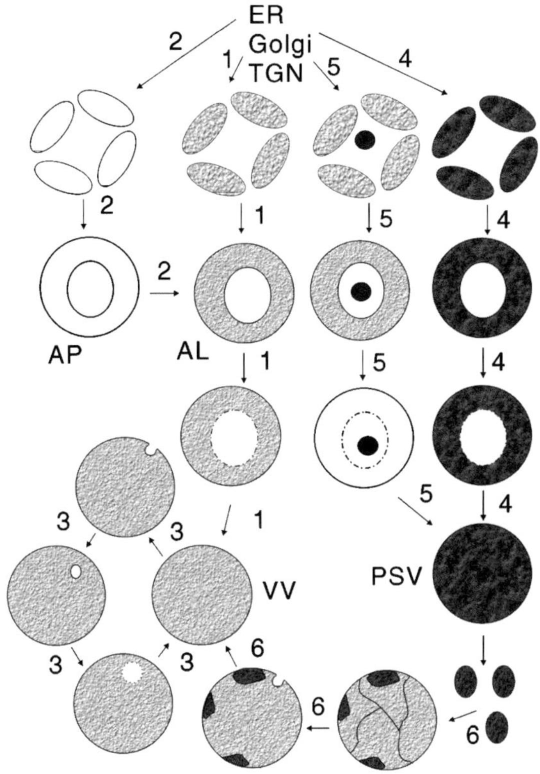

Figure 4.1 Autophagic processes in plants: an overview. When vegetative vacuoles (VVs) are formed *do novo* in meristematic cells, which are devoid of vacuoles, the autophagic process 1 appears to take place. A similar kind of autophagic process also seems to occur when mature plant cells, which already have large VVs, are placed under conditions of nutrient starvation. However, it remains to be clarified whether this autophagy proceeds through process 1 or through process 2, which is typical of autophagy found in mammalian and *Saccharomyces* cells upon nutrient starvation. VVs appear to have the ability to sequester and degrade portions of cytoplasm through microautophagy (process 3). In some cases, the autophagic process 4 seems to occur, when another type of plant vacuole, the protein storage vacuole (PSV), is formed *de novo*. This process proceeds topologically in a manner similar to process 1. In wheat endosperm and some transgenic plants expressing storage proteins, protein bodies in the cytoplasm are also delivered to the vacuole by autophagy, thus also forming PSVs by autophagy (process 5). During seed germination, small PSVs fuse together to make large VVs. Autophagic events occur during this transformation of vacuoles (process 6). In the latter case, these autophagic processes seem to be responsible for the adjustment of the ratio of membrane surface area to vacuolar volume and for the turnover of vacuolar membrane proteins and cytoplasmic constituents, such as organelles and ribosomes. A mechanism similar to process 6 also seems to be used for the reduction of the surface area of the vacuolar membrane at the last stage of process 1, where small VVs fuse together to become large VVs. Abbreviations: ER, endoplasmic reticulum; Golgi, the Golgi apparatus; TGN, *trans*-Golgi network; AP, autophagosome; AL, autolysosome.

Another theory on the genesis of vegetative vacuoles, which also depends on autophagy, has been proposed on the basis of observations made on the testa cells of linseed (*Linum*) and the cells of the vascular cylinder of lettuce (*Lactuca sativa*) roots (Amelunxen and Heinze, 1984; Hilling and Amelunxen, 1985). In this theory, smooth ER (sER) elements rather than the TGN are thought to be the precursor of the vacuole. Vacuolation starts with a segregation of cytoplasm by sER. At first, tubular or lamellar sER aligns itself in a circular configuration at the site where the tonoplast of the vegetative vacuole will eventually form. As a result, the cytoplasm becomes segregated into two parts. Interestingly, the cytoplasm enclosed by sER does not appear to contain any organelles. Although ribosomes can be found inside, their number is always less than outside the sER border. The mechanism of formation of such a zone of exclusion is unknown, and its existence has occasionally been ridiculed as being a fixation artifact. According to Hilling and Amelunxen (1985), the sER elements enclosing the cytoplasm fuse and dilate inwardly until the lumen of the sER builds up a large vegetative vacuole.

Amelunxen and co-workers have presented a further line of evidence showing that the vacuole originates from ER (Hilling and Amelunxen, 1985). Mesophyll cells of a young primary leaf of *Oenothera odorata* accumulate tannins, which can be easily identified as electron dense deposits in the vacuole. In this plant, the same type of vacuolar genesis as seen in linseed can be observed, but the ER enclosing the cytoplasm already contains tannin deposits in its lumen. Root cells of *Althaea officinalis* often contain a very highly contrasted material, which is probably polyphosphate. Here also, a ring of ER vesicles marks the size and position of the future central vacuole. A small vacuole with a convex-concave shape, which is thought to be formed by the dilation of the ER, already appears to contain the same kind of highly contrasted material.

Similar mechanisms have been presented for the genesis of the vegetative vacuole in at least two other cases. In developing wheat endosperm, ER-derived, acid-phosphatase-containing provacuolar structures enclose portions of the cytoplasm from which plastids, mitochondria and other organelles are excluded. A vacuole is formed through fusion of the provacuolar structures, followed by an inwardly directed dilation (Bechtel *et al.*, 1989). In the staminal filaments and lodicules of maize, synchronous cell extension and vacuole genesis occur at anthesis (Keijzer *et al.*, 1996). However, in the filament cells, rough ER (rER) as well as sER fuses to create the zone of exclusion and small membrane-bounded cytoplasmic inclusions remain in the vacuoles formed. In addition, pre-existing vacuoles appear to join with the ER in creating a zone of exclusion in the filaments and the lodicules.

When small vacuoles fuse with each other to form a large vacuole, an adjustment of membrane surface area to volume must occur. This adjustment seems to be brought about by autophagy, that is, through the internalization and degradation of the membrane within vacuoles. In order to trace the ontogeny of vacuoles, Herman *et al.* (1994) have analyzed root tip cells of oat using antibodies against V-ATPase. They observed electron-translucent vesicles of 0.1–0.3 μm in diameter associated with the *trans*-face of the Golgi apparatus, apparently resulting from budding at this face. They named these vesicles 'provacuoles' and suggested that they are precursors of a large vegetative vacuole. Immunolocalization of the peripheral 60 kDa subunit of V-ATPase resulted in labeling of the ER, small vacuoles of 1–2 μm in diameter and the provacuoles. The ER labeling appears to exhibit a spatial variation of gold label on contiguous ER cisternae. Thus, the authors suggested that there might be a domain of the ER that is directly responsible for vacuole formation (Herman *et al.*, 1994). Numerous provacuoles are frequently seen surrounding the small vacuole, and the fusion of these provacuoles appears to contribute to the enlargement of the small vacuole. At this time, the membrane of provacuoles appears to be internalized and partially degraded within the vacuole. This membrane is then stainable with anti-V-ATPase antibody.

4.4.2 Autophagy and genesis of protein storage vacuoles

During seed development in legumes, numerous protein storage vacuoles of 1–10 μm are formed in storage parenchyma cells of cotyledons, which initially have only vegetative vacuoles. It has been widely assumed that the pre-existing vegetative vacuoles are transformed into protein storage vacuoles (PSVs) by deposition of storage proteins into them. Prior to seed dessication, the PSVs are thought to fragment into numerous, small protein bodies. Recently, however, Robinson and co-workers have provided irrefutable evidence that PSVs in maturing pea seeds are formed *de novo* through a dynamic autophagic process (Hoh *et al.*, 1995; Robinson *et al.*, 1995). Before storage protein synthesis begins, there are numerous vegetative vacuoles in the pea cotyledon cell. The deposition of storage proteins, including vicilin and legumin, is first detected at the periphery of groups of numerous vegetative vacuoles. These are contained within a sheet-like, membrane system, which surrounds the pre-existing vegetative vacuoles. Since binding protein (BiP), an ER-resident protein, is present within storage protein deposits, this membrane system may have originated from the ER (Robinson *et al.*, 1995). Subsequently, the tube-like membrane system dilates and the enlargement of its lumen takes place in an inward direction. As a result, the vegetative vacuoles appear to get smaller. While this expansion is occurring, the storage protein deposits remain attached to the inner side of the outer membrane. At this stage, two types of vacuole, the PSV and the vegetative vacuole, coexist in the same cell. Dilation of the tube system continues until it reaches vacuole-like proportions. Finally, only small profiles of internal membranes are visible and the lumen is seen to contain small clumps of storage protein in addition to the larger deposits at the periphery.

Antibodies against α-tonoplast intrinsic protein (α-TIP) label the membrane of the tube-like PSV precursor as well as the membrane of fully developed PSV, but not the vegetative vacuoles. Moreover, antibodies raised against TIP-Ma 27, an integral tonoplast polypeptide of the γ-TIP family, label the membrane of vegetative vacuoles within the dilating PSV, but not the membrane of the PSV. Subcellular fractionation supports the *in situ* labeling data: α- and γ-TIP containing membranes separate clearly on density gradients; there is no evidence for a membrane bearing both. Thus, the formation of the PSV in maturing pea cotyledons is, in a way, similar to the ontogeny of the vegetative vacuole as described by Hilling and Amelunxen (1985).

4.4.3 Transformation of protein storage vacuoles to vegetative vacuoles

When seeds germinate and seedlings grow, storage proteins within the PSV are degraded and, concomitantly, the PSV is transformed into a vegetative vacuole. PSVs in dry seeds contain several kinds of acid hydrolases. In addition, they acquire new hydrolases, which are synthesized *de novo* during germination and, as a result, the degradation of the stored proteins starts to take place. Matile (1975) first suggested that PSVs have a function in degrading cytoplasm in addition to degrading storage proteins, since he noticed that the PSV contains RNase that does not appear to be necessary for the degradation of its reserves. This proposal has been elaborated on by Van der Wilden *et al.* (1980) and Herman *et al.* (1981). These authors proposed two physiological meanings for autophagy during seed germination. One is to degrade components of the cytoplasm, such as ribosomes and ER, which have been extensively produced during seed formation, and the other is the final breakdown of cellular constituents of parenchyma cells that are undergoing senescence.

Numerous protein bodies of 1–10 μm diameter are observed in storage parenchyma cells of mung bean cotyledons 1 day after the start of imbibition. These protein bodies contain legumin and vicilin. Three to five days after the start of imbibition, the storage proteins are degraded by the action of vicilin peptidohydrolase, which is synthesized *de novo* and is transported into the PSV. Concomitantly with the degradation of the storage proteins, membrane-bounded structures of 0.2–1 μm in diameter appear within the PSV (Van der Wilden *et al.*, 1980; Herman *et al.*, 1981). These structures seem to enclose cytoplasm with ribosomes, ER and mitochondria. Serial sectioning has revealed that these vesicles are not cytoplasmic protrusions but are entirely contained within the PSV. Furthermore, the membrane of the PSV was found to have depressions or invaginations, which contain ribosomes, membrane vesicles and ER cisternae. Thus, there is good evidence for microautophagy being performed by the PSV. The cytoplasmic structures within the autophagic vesicles in the PSV seem to disappear together with the digestion of the storage proteins, and eventually the autophagic vesicles become empty.

Two observations are difficult to explain (Herman *et al.*, 1981). Firstly, with regard to acid phosphatase activity, the membrane of the autophagic vesicles, which probably originates from the membrane of the PSV, stains more strongly than the membrane of the PSV. Thus, the property of the membrane must change just before or after it pinches off from the membrane of the PSV. Secondly, the autophagic vesicles often seem to be surrounded by several membranes, whose origin remains unknown.

During the transformation of the PSV into a large vegetative vacuole, dynamic changes in the proteins of the tonoplast must occur. Among these membrane proteins, α-TIP is known to be abundant in the membrane of the PSV, but is absent from the membrane of vegetative vacuoles. Therefore, α-TIP must be removed from the tonoplast during the transformation from PSV to vegetative vacuoles. In fact, in soybean cotyledon cells, the amount of α-TIP does not change for the first 4 days after imbibition but thereafter decreases drastically (Melroy and Herman, 1991). The degradation of α-TIP occurs largely in parallel to the mobilization of the storage proteins and the restitution of the vegetative vacuole, although a functional coupling between these three parameters does not seem to exist. Some of the membranes of the autophagic vesicles are labeled with anti-α-TIP antibodies, but others, apparently empty vesicles, are not. This may be due to the degradation of α-TIP in the vacuole. Thus, microautophagy by the PSV might be an effective means of α-TIP removal and degradation during germination. In this respect, it is interesting to note that Höfte *et al.* (1991) reported the presence of α-TIP on the membranes of multivesicular bodies sequestered in the leaf vacuoles of transgenic tobacco expressing α-TIP. Similar observations were made by Maeshima *et al.* (1994) in 4 day old germinating pumpkin seedlings. They observed PPase-positive vesicles or membrane fragments within the transforming PSV. These vesicles may be autophagic vesicles, which are formed by microautophagy of protein storage vacuoles. Alternatively, they may be formed during the process of fusion of small protein vacuoles into a large vacuole. However, in contrast to α-TIP, PPase increases markedly during germination.

4.4.4 Vacuolar deposition of storage proteins via autophagy

In most cereals, the major storage proteins are alcohol-soluble prolamins. The subcellular localization of these proteins appears to be different between species. The prolamins of maize and rice are shown to aggregate into dense protein bodies formed directly from the rER, and they remain attached to this organelle. In other cereals, such as wheat and oats, the prolamins are known to accumulate in PSVs. A portion of the prolamins in wheat are transported to the PSV via the Golgi apparatus, and Golgi-derived vesicles of about 0.2 μm containing prolamins have been described. However, small protein bodies of 0.1–0.5 μm have also been observed within or in the vicinity of rER in the cytoplasm and, in addition to these, large protein bodies of up to 10 μm are found in the cytoplasm (Levanony *et al.*, 1992). The large protein bodies appear to be formed by the fusion of small protein bodies that are still enclosed by rER membrane.

A pathway for these protein bodies in the cytoplasm to the large central vacuole (then becoming a PSV), which circumvents the Golgi apparatus, has been proposed by Galili and co-workers (Levanony *et al.*, 1992). This route seems to utilize two types of autophagic process. In one type, many electron-lucent vesicles attach to the surface of the protein body in the cytoplasm. These vesicles are apparently provacuoles, which may develop directly from the ER and which already possess acid phosphatase (Galili and Herman, 1997). The provacuoles then appear to fuse with each other around the protein body, eventually enclosing it to form a small vacuole containing the storage protein mass. During this process, the rER membrane surrounding the protein body seems to disappear. However, the surface of some protein bodies in a small vacuole is covered by loosely-attached membranes, which may originate from the membrane of provacuoles. This membrane appears to contain the tonoplast marker enzyme, pyrophosphatase (Galili and Herman, 1997). In some cases, however, a portion of cytoplasm, but not storage proteins, is also detected in the small vacuoles. They may be trapped together with protein bodies during the fusion of the provacuoles. The formation of these vacuoles containing protein bodies is thus a form of macroautophagy.

The other type of autophagic process, which is also observed in developing wheat endosperm cells, relates to how the small protein body-vacuoles in the cytoplasm enter pre-existing vacuoles. In this case, the pre-existing vacuole itself appears to take up the protein bodies by microautophagy (Levanony *et al.*, 1992). Immunoelectron microscopy has shown that the ER lumenal chaperone, BiP, is present in protein bodies within the vacuoles (Levanony *et al.*, 1992). This result is regarded as supporting evidence for the aggregation of prolamins in the rER and the resulting aggregates themselves are transported into vacuoles.

There are possibly two other examples of the involvement of autophagy in the transport of preformed storage protein aggregates/complexes to the vacuoles. The so-called precursor-accumulating (PAC) vesicles in developing pumpkin cotyledons carry storage proteins directly from the ER to the vacuole (Hara-Nishimura *et al.*, 1998). An autophagic uptake of the PAC vesicles into the vacuoles, as seen in the transport of prolamins in wheat endosperm, is conceivable. The second example involves transgenic tobacco expressing α- and γ-zeins (Bagga *et al.*, 1995, 1997; Coleman *et al.*, 1996). Protein bodies (0.1–0.5 μm diameter) containing these proteins are found in the cytoplasm and vacuole of endosperm cells. Some of those in the vacuole still appear to have retained their limiting membranes, whereas others appear to be degraded. In this case, an autophagic uptake also seems likely.

4.4.5 Autophagy and senescence

Senescence is the ultimate developmental process in plants and involves programmed cell death. It is known that the tonoplast ruptures towards the last stage of programmed cell death (Matile, 1975; Fukuda, 1996). As a result, hydrolases are released from the vacuole, which digest the cell contents. The mechanism for the rupture of the tonoplast is totally unknown. An interesting example of this is seen in germinating barley grains. The aleurone cells which have imbibed water undergo a series of dramatic biochemical and cellular changes in response to gibberellic acid (Bethke *et al.*, 1998). Storage proteins are degraded in PSVs, which are formed through the fusion of protein bodies, thereby providing amino acids with which the cells can produce hydrolytic enzymes necessary for mobilization of stored reserves in the endosperm. Some of these enzymes, for example α-amylase, are secreted. Others, for example cysteine protease, appear to be temporarily stored in different vacuoles and are released into the endosperm through autolysis of the aleurone cell. Swanson *et al.* (1998) have characterized these 'secondary vacuoles' and have found them to be smaller than 10 μm, to have an acid pH, and to possess α-TIP in the tonoplast, indicating that these are not true 'lytic' vacuoles.

It is interesting to consider whether or not autophagy contributes to the processes leading to cell death. Many biochemical studies on senescence have shown that various kinds of hydrolytic enzyme accumulate in the vacuole (Matile and Winkenbach, 1971; Baumgartner *et al.*, 1975; Minami and Fukuda, 1995). Such accumulation usually seems to precede the degradation of the substrates of these enzymes, including protein and RNA. Thus, the lytic potential of these vacuoles is intensified in senescent cells in order to achieve a rapid degradation of the remaining cell constituents when vacuoles collapse (Matile, 1997). However, the induction of autophagy may also accompany the increase in the activities of lytic enzymes. Indeed, morphological evidence showing that vacuolar microautophagy takes place during senescence has also been presented (Matile and Winkenbach, 1971; Smith *et al.*, 1992; Inada *et al.*, 1998). There is also a report showing that macroautophagy takes place in senescence, which proceeds during germination in endosperm cells of castor bean (Vigil, 1970).

In the past, it was assumed that vacuolar autophagy plays an important role in the degradation of chloroplast proteins during senescence. Although we still cannot deny the possibility that some chloroplasts are finally broken down within vacuoles by autophagy at late stages of senescence, we now know that vacuolar proteases are unlikely to play a primary role in degradation of chloroplast proteins, such as Rubisco, which is one of the most prominent phenomena in senescence (Matile, 1997).

4.4.6 Selective autophagy in plant cells?

In yeast (*Pichia*) cells, peroxisomes are known to be selectively engulfed and degraded by microautophagy in response to the change of carbon source from methanol to glucose (Tuttle *et al.*, 1993; Sakai *et al.*, 1998). Apparently, a similar kind of selective microautophagy has been reported in plant cells. When an embryo in the seed of *Fraxinus excelsior* that has been subjected to prolonged dormancy is set to germinate by breaking dormancy, vacuoles in the cells of the embryo appear to engulf plastids and mitochondria in an almost selective manner (Villiers, 1971). This is necessary in order to remove organelles that have been damaged through a prolonged dormancy. Indeed, germination in such an embryo is delayed compared to that in a normal embryo.

Spherical, membrane-bounded inclusion bodies, which are composed mainly of lipids, exist in the proplastids of cells around shoot apices of *Bryophyllum* and *Kalanchoe* (Gifford and Stewart, 1968). Similar inclusion bodies are also found in the vacuole of the same cells. Electron-microscopic analysis shows that the inclusion body in the proplastids seems to be specifically transferred into the vacuole. This transfer appears to be brought about by a process in which the evagination of the inclusion body from the plastids is incorporated into the vacuole by microautophagy. A similar incorporation process of part of a chloroplast into the vacuole is observed in leaf cells of *Hypoestes* and *Vigna* (mung bean) (Vaughn and Duke, 1981).

When the cold-sensitive *Arabidopsis* mutant, *fab1*, is grown at 2°C for more than 10 days, chloroplasts in the leaf cells become disrupted and gradually disappear (Wu *et al.*, 1997). Interestingly, the disrupted chloroplasts are always associated with small vacuoles. It is not clear whether this loss is due to autophagy, but it is selective, since mitochondria are unaffected.

4.5 Starvation-induced autophagy

In mammalian and yeast cells, autophagy has been known to be stimulated under conditions of nutrient starvation (Schworer and Mortimore, 1979; Takeshige *et al.*, 1992; Baba *et al.*, 1994; Dunn, 1994). The same situation occurs in plant cells, for which three examples will be given. Firstly, rice cells, when cultured in the presence of sucrose contain many small vacuoles (Chen *et al.*, 1994; Yu, 1999). Vacuolar volume begins to increase 6–12 h after starvation and, concomitantly, starch granules, which initially exist within amyloplasts, appear in the vacuoles. The membranes of the amyloplasts are difficult to identify in the vacuoles, but the amyloplasts were probably nonselectively engulfed into the vacuole by microautophagy. However, the possible formation of an autophagosome and subsequent delivery of the amyloplasts to the vacuole cannot be excluded.

Aubert *et al.* (1996) have performed an ultrastructural analysis on suspension cultured sycamore cells starved of sucrose. Before starvation, plastids contain large starch granules, but after 4 days of starvation, the plastids are starch-free. This response is in contrast to that of rice cells, where some of the starch granules appear to be degraded in vacuoles. The total number of mitochondria in a cell seems to decrease, while the number of peroxisomes seems to increase transiently. Peroxisomes are thought to contribute to the oxidation of an increased level of fatty acids, which originate from the degradation of lipids. After 6 days, the numbers of plastids, mitochondria and peroxisomes decrease, although the cells are still alive. Cell death, heralded by the rupture of the tonoplast, occurs 7–8 days after the start of starvation. Autophagic vacuoles appear as early as 14 h after starvation. These vacuoles are polymorphic, and the authors classify them into two groups. One is a two membrane-bounded vacuole of 3–5 μm, which appears to contain intact or degenerated cytoplasm (Aubert *et al.*, 1996; Marty, 1997). These double membrane-bounded autophagic vacuoles seem to fuse with the central vacuole, because single membrane-bounded vesicles with similar contents are observed inside the vacuole. The second type of autophagic vacuole is detected less frequently. These are single membrane vacuoles and contain a variety of cellular structures at different stages of degradation. The inclusion looks more severely degraded compared with that within the double membrane vacuoles. Whether and how these vacuoles are related to the double membrane-bounded vacuoles is unknown.

Autophagy is also inducible upon sucrose starvation in suspension-cultured tobacco cells (Moriyasu and Ohsumi, 1996). In this study, a coupling between cellular protein degradation and cellular morphology engaging in autophagy was clearly shown. When cells are transferred to sucrose-free medium, 30–50% of the total initial protein is degraded over a 2 days period. Activity of cellular protease, mainly cysteine and aspartic protease, increases three- to seven-fold following 1 day of sucrose starvation. Therefore, protein degradation induced by sucrose starvation is coordinated with an increase in protease activity. This degradation can be inhibited by adding cysteine protease inhibitors, such as E-64, leupeptin and antipain, to the culture medium. Treatment with E-64 completely abolishes the increase in protease activity induced by sucrose starvation. Analysis of cellular proteins by sodium dodecyl sulphate-polyacrylamide gel electrophoresis (SDS-PAGE) shows that the decrease in total protein does not come from the degradation of specific proteins but from the nonselective degradation of all intracellular proteins. Treatment with cysteine protease inhibitors rescues almost all of the polypeptides from degradation.

The cysteine protease inhibitors also cause the accumulation of numerous structures in the perinuclear cytoplasm after 8 h of starvation. These structures are vacuoles of 1–6 μm in diameter and have several osmiophilic particles, including partially degraded mitochondria, within them. They eventually fuse with the large central vacuole, releasing the partially digested contents into the lumen of the vacuole. Like typical lysosomes, these vacuoles are acidic compartments, as evidenced by the fact that they accumulate the acidotrophic dyes, quinacrine and neutral red. Enzyme histochemistry at the electron-microscopic level, using β-glycerophosphate as a substrate, as well as immunogold studies show that these vesicles contain acid phosphatase, a marker enzyme of lysosomes. Isolated vesicles also have cysteine protease activity (Moriyasu, Takatsuka, Hillmer, Robinson, unpublished results). Without a cysteine protease inhibitor, autolysosomes in tobacco cells are difficult to detect. This may imply that the autolysosomes have a short lifetime. In mammalian cells, the half-life of autolysosomes is estimated to be 8–9 min (Mortimore and Schworer, 1977). Thus, plant cells respond to sucrose deprivation by forming a new lytic compartment in addition to the central vacuole. In contrast, the central vacuole is the only lysosomal compartment in yeast cells (Takeshige *et al.*, 1992; Baba *et al.*, 1994; see also Chapter 1 in this volume). Under starvation conditions, autophagosomes are not detected in yeast (Takeshige *et al.*, 1992). Only when proteolysis is inhibited with the serine protease inhibitor, PMSF, do they appear, and undegraded cytoplasm accumulates in the vacuole.

In contrast to mamalian cells, where autophagosomes have been clearly documented as being derived from the ER (Dunn, 1994), the origin of the autolysosomes in suspension-cultured tobacco remains unclear. It is not certain whether the autophagosome, a precursor structure of the autolysosome, exists or not. Macroautophagy induced by nutrient starvation in mammalian and yeast cells is known to prsoceed stepwise, where autophagosomes and autolysosomes can be discriminated on the basis of the presence of acid phosphatase. Vesicles or vacuoles which sequester portion of cytoplasm in plant cells already appear to contain acid phosphatase, and typical autophagosomes without acid hydrolases do not seem to be present. Marty (1999) has suggested that upon sucrose starvation, cultured cells, which already have large vacuoles, revert to the same type of vacuole formation as that occurring in meristematic cells, that is, involving autophagy.

4.6 Does the cytoplasm to vacuole targeting (Cvt) pathway exist in plants?

The Cvt pathway has been well-established in yeast (see Chapter 1 in this volume). In this pathway, aggregates of aminopeptidase I are sequestered into miniature autophagosomes after being synthesized on free ribosomes in the cytoplasm. These structures then fuse with the vacuole to expel the Cvt bodies into the lumen. With the possible exception of the wheat prolamin (see Section 4.4.4), such mini-autophagosomes have not yet been detected in plants. Nevertheless, there are a few intriguing examples for proteins in the plant, which lack a signal sequence but seem to be located in the vacuole, for example, β-amylase in *Arabidopsis*, and the lipoxygenase, vps94, in soybean (Monroe *et al.*, 1991; Tranbarger *et al.*, 1991). Do these proteins gain entry to the vacuole through mini-autophagosomes?

4.7 Endocytosis and autophagy

From animal cells, it is known that the endocytic and autophagic pathways meet at the level of multivesicular bodies (Liou *et al.*, 1997). In plant cells, information about endocytosis is very limited, and receptor-mediated endocytosis is still questioned by some (Hawes *et al.*, 1995). However, Record and Griffing (1988) have shown that multivesicular bodies, the plant equivalent of endosomes (see Chapter 12 in this volume), contain an endocytic marker and also possess acid phosphatase, a vacuolar enzyme. Moreover, Herman and Lamb (1992) have localized arabinogalactan-rich glycoproteins not only to the cell wall but also to multivesicular bodies. They suggested that multivesicular bodies originate at the plasma membrane and are subsequently enclosed within the vacuole by microautophagy. This makes it likely that endocytic and autophagic pathways also converge in plant cells, however, further experiments are certainly needed in this area.

4.8 Future perspective

Up to now, nearly all of the work carried out on autophagy in plants has been descriptive. Considering the importance of autophagy in plant development, it is high time that a more biochemical/molecular biological approach be adopted. With starvation-induced autophagy in suspension-cultured cells we have an exploitable system. Clearly, the autolysosomes formed in these cells need to be characterized. Several questions automatically arise:

- What kind of TIP is contained in their membrane?
- Where do the lytic enzymes come from which are present in them?
- From which organelle do the autolysosomes originate?
- How do the autolysosomes achieve and maintain a low lumenal pH?
- Do the autolysosomes fuse with the central vacuole in order to complete degradation?

A genetic analysis of autophagy in yeast (see Chapter 1 in this volume) has revealed the key function of a number of proteins. It will be interesting to look for plant homologs, to manipulate them, and to prepare antibodies against them. The field is wide open for experimentation along these lines.

References

Abel, S. and Glund, K. (1986) Localization of RNA-degrading enzyme activity within vacuoles of cultured tomato cells. *Physiol. Plant.*, **66** 79-86.

Amelunxen, F. and Heinze, U. (1984) Zur Entwicklung der Vacuole in Testa-Zellen des Leinsamens (On the development of the vacuole in the testa cells of Linum seeds). *Eur. J. Cell Biol.*, **35** 343-54.

Aubert, S., Gout, E., Bligny, R. and Douce, R. (1994) Multiple effects of glycerol on plant cell metabolism: phosphorus-31 nuclear magnetic resonance studies. *J. Biol. Chem.*, **269** 21420-27.

Aubert, S., Gout, E., Bligny, R., Marty-Mazars, D., Barrieu, F., Alabouvette, J., Marty, F. and Douce, R. (1996) Ultrastructural and biochemical characterization of autophagy in higher plant cells subjected to carbon deprivation: control by the supply of mitochondria with respiratory substrates. *J. Cell Biol.*, **133** 1251-63.

Baba, M., Takeshige, K., Baba, N. and Ohsumi, Y. (1994) Ultrastructural analysis of the autophagic process in yeast: detection of autophagosomes and their characterization. *J. Cell Biol.*, **124** 903-13.

Bagga, S., Adams, H., Kemp, J.D. and Sengupta-Gopalan, C. (1995) Accumulation of 15-kilodalton zein in novel protein bodies in transgenic tobacco. *Plant Physiol.*, **107** 13-23.

Bagga, S., Adams, H.P., Rodriguez, F.D., Kemp, J.D. and Sengupta-Gopalan, C. (1997) Coexpression of the maize δ-zein and β-zein genes results in stable accumulation of δ-zein in endoplasmic reticulum-derived protein bodies formed by β-zein. *Plant Cell*, **9** 1683-96.

Baumgartner, B., Kende, H. and Matile, P. (1975) Ribonuclease in senescing morning glory: purification and demonstration of *de novo* synthesis. *Plant Physiol.*, **55** 734-37.

Bechtel, D.B., Frend, A., Kaleikau, L.A., Wilson, J.D. and Shewry, P.R. (1989) Vacuole formation in wheat starchy endosperm. *Food Microstr.*, **8** 183-90.

Bethke, P.C., Swanson, S.J., Hillmer, S. and Jones, R.L. (1998) From storage compartment to lytic organelle: the metamorphosis of the aleurone protein storage vacuole. *Ann. Bot. (London)*, **82** 399-412.

Blommaart, E.F.C., Luiken, J.J.F.P. and Meijer, A.J. (1997) Autophagic proteolysis: control and specificity. *Histochem. J.*, **29** 365-85.

Brouquisse, R., Gaudillère, J.-P. and Raymond, P. (1998) Induction of a carbon-starvation-related proteolysis in whole maize plants submitted to light/dark cycles and to extended darkness. *Plant Physiol.*, **117** 1281-91.

Canut, H., Alibert, G., Carrasco, A. and Boudet, A.M. (1986) Rapid degradation of abnormal proteins in vacuoles from *Acer pseudoplatanus* L. cells. *Plant Physiol.*, **81** 460-63.

Chen, M.-H., Liu, L.-F., Chen, Y.-R., Wu, H.-K. and Yu, S.-M. (1994) Expression of α-amylases, carbohydrate metabolism, and autophagy in cultured rice cells is coordinately regulated by sugar nutrient. *Plant J.*, **6** 625-36.

Coleman, C.E., Herman, E.M., Takasaki, K. and Larkins, B.A. (1996) The maize γ-zein sequesters α-zein and stabilizes its accumulation in protein bodies of transgenic tobacco endosperm. *Plant Cell*, **8** 2335-45.

de Duve, C. (1984) *A Guided Tour of the Living Cell*. Scientific American Books, Inc., New York.

Dorne, A.-J., Bligny, R., Rébeillé, F., Roby, C. and Douce, R. (1987) Fatty acid disappearance and phosphorylcholine accumulation in higher plant cells after a long period of sucrose deprivation. *Plant Physiol. Biochem.*, **25** 589-95.

Douce, R., Bligny, R., Brown, D., Dorne, A.-J., Genix, P. and Roby, C. (1991) Autophagy triggered by sucrose deprivation in sycamore (*Acer pseudoplatanus*) cells, in *Compartmentation of Plant Metabolism in Non-photosynthetic Tissues* (ed. M.J. Emes), Cambridge University Press, Cambridge, New York, Port Chester, Melbourne, Sydney, pp. 127-45.

Douce, R., Pascal, N. and Bligny, R. (1992) Macroautophagy triggered by sucrose starvation in higher plant cells: analysis of a model for prolonged carbon deprivation under water stress, in *Water and Life* (eds. G.N. Somero and C.B. Osmond), Springer-Verlag, Berlin, Heidelberg, pp. 128-40.

Dunn, Jr. W.A. (1994) Autophagy and related mechanisms of lysosome-mediated protein degradation. *Trends Cell Biol.*, **4** 139-43.

Fineran, B.A. (1970) Organization of the tonoplast in frozen-etched root tips. *J. Ultrastruct. Res.*, **33** 574-86.

Fukuda, H. (1996) Xylogenesis: initiation, progression, and cell death. *Annu. Rev. Plant Physiol. Plant Mol. Biol.*, **47** 299-325.

Galili, G. and Herman, E.M. (1997) Protein bodies: storage vacuoles in seeds. *Adv. Bot. Res.*, **25** 113-40.

Genix, P., Bligny, R., Martin, J.-B. and Douce, R. (1990) Transient accumulation of asparagine in sycamore cells after a long period of sucrose starvation. *Plant Physiol.*, **94** 717-22.

Gifford, Jr. E.M. and Stewart, K.D. (1968) Inclusions of the proplastids and vacuoles in the shoot apices of Bryophyllum and Kalanchoe. *Am. J. Bot.*, **55** 269-79.

Hara-Nishimura, I., Shimada, T., Hatano, K., Takeuchi, Y. and Nishimura, M. (1998) Transport of storage proteins to protein storage vacuoles is mediated by large precursor-accumulating vesicles. *Plant Cell*, **10** 825-36.

Hawes, C., Crooks, K., Coleman, J. and Satiat-Jeunemaitre, B. (1995) Endocytosis in plants: fact or artefact? *Plant Cell Environ.*, **18** 1245-52.

Herman, E.M. and Lamb, C.J. (1992) Arabinogalactan-rich glycoproteins are localized on the cell surface and in intravacuolar multivesicular bodies. *Plant Physiol.*, **98** 264-72.

Herman, E.M., Li, X., Su, R.T., Larsen, P., Hsu, H.-T. and Sze, H. (1994) Vacuolar-type H^+-ATPases are associated with the endoplasmic reticulum and provacuoles of root tip cells. *Plant Physiol.*, **106** 1313-24.

Herman, E.M., Baumgartner, B. and Chrispeels, M.J. (1981) Uptake and apparent digestion of cytoplasmic organelles by protein bodies (protein storage vacuoles) in mung bean cotyledons. *Eur. J. Cell Biol.*, **24** 226-35.

Hilling, B. and Amelunxen, F. (1985) On the development of the vacuole. II. Further evidence for endoplasmic reticulum origin. *Eur. J. Cell Biol.*, **38** 195-200.

Höfte, H., Faye, L., Dickinson, C., Herman, E.M. and Chrispeels, M.J. (1991) The protein-body proteins phytohemagglutinin and tonoplast intrinsic protein are targeted to vacuoles in leaves of transgenic tobacco. *Planta*, **184** 431-37.

Hoh, B., Hinz, G., Jeong, B.-K. and Robinson, D.G. (1995) Protein storage vacuoles form *de novo* during pea cotyledon development. *J. Cell Sci.*, **108** 299-310.

Inada, N., Sakai, A., Kuroiwa, H. and Kuroiwa, T. (1998) Three-dimensional analysis of the senescence program in rice (*Oryza sativa* L.) coleoptiles: investigations by fluorescence microscopy and electron microscopy. *Planta*, **206** 585-97.

James, F., Brouquisse, R., Pradet, A. and Raymond, P. (1993) Changes in proteolytic activities in glucose-starved maize root tips: regulation by sugars. *Plant Physiol. Biochem.*, **31** 845-56.

James, F., Brouquisse, R., Suire, C., Pradet, A. and Raymond, P. (1996) Purification and biochemical characterization of a vacuolar serine endopeptidase induced by glucose starvation in maize roots. *Biochem. J.*, **320** 283-92.

Journet, E.-P., Bligny, R. and Douce, R. (1986) Biochemical changes during sucrose deprivation in higher plant cells. *J. Biol. Chem.*, **261** 3193-99.

Keijzer, C.J., Reinders, M.C. and Klooster, H.B.L.-T. (1996) The mechanics of the grass flower: the extension of the staminal filaments and the lodicules of maize. *Ann. Bot.*, **77** 675-83.

Levanony, H., Rubin, R., Altschuler, Y. and Galili, G. (1992) Evidence for a novel route of wheat storage proteins to vacuoles. *J. Cell Biol.*, **119** 1117-28.

Liou, W., Geuze, H.J., Geelen, M.J.H. and Slot, J.W. (1997) The autophagic and endocytic pathways converge at the nascent autophagic vacuoles. *J. Cell Biol.*, **136** 61-70.

Löffler, A., Abel, S., Jost, W., Beintema, J.J. and Glund, K. (1992) Phosphate-regulated induction of intracellular ribonucleases in cultured tomato (*Lycopersicon esculentum*) cells. *Plant Physiol.*, **98** 1472-78.

Maeshima, M., Hara-Nishimura, I., Takeuchi, Y. and Nishimura, M. (1994) Accumulation of vacuolar H^+-pyrophosphatase and H^+-ATPase during reformation of the central vacuole in germinating pumpkin seeds. *Plant Physiol.*, **106** 61-69.

Marty, F. (1978) Cytochemical studies on GERL, provacuoles and vacuoles in root meristematic cells of *Euphorbia*. *Proc. Natl. Acad. Sci. USA*, **75** 852-56.

Marty, F. (1997) The biogenesis of vacuoles: insights from microscopy. *Adv. Bot. Res.*, **25** 1-42.

Marty, F. (1999) Plant vacuoles. *Plant Cell*, **11** 587-99.

Marty, F., Branton, D. and Leigh, R.A. (1980) Plant vacuoles, in *The Biochemistry of Plants*, Vol. 1, pp. 625-58.

Matile, P. (1975) *The Lytic Compartment of Plant Cells, Cell Biology Monographs*, Vol. 1, Springer-Verlag, Wien.

Matile, P. (1997) The vacuole and cell senescence. *Adv. Bot. Res.*, **25** 87-112.

Matile, P. and Winkenbach, F. (1971) Function of lysosomes and lysosomal enzymes in the senescing corolla of the morning glory (*Ipomoea purpurea*). *J. Exp. Bot.*, **22** 759-71.

Melroy, D.L. and Herman, E.M. (1991) TIP, an integral membrane protein of the protein-storage vacuoles of the soybean cotyledon undergoes developmentally regulated membrane accumulation and removal. *Planta*, **184** 113-22.

Minami, A. and Fukuda, H. (1995) Transient and specific expression of a cysteine endopeptidase associated with autolysis during differentiation of *Zinnia* mesophyll cells into tracheary elements. *Plant Cell Physiol.*, **36** 1599-606.

Monroe, J.D., Salminen, M.D. and Preiss, J. (1991) Nucleotide sequence of a cDNA clone encoding β-amylase from *Arabidopsis thaliana*. *Plant Physiol.*, **97** 1599-601.

Moriyasu, Y. (1995) Examination of the contribution of vacuolar proteases to intracellular protein degradation in *Chara corallina*. *Plant Physiol.*, **109** 1309-15.

Moriyasu, Y. and Ohsumi, Y. (1996) Autophagy in tobacco suspension-cultured cells in response to sucrose starvation. *Plant Physiol.*, **111** 1233-41.

Mortimore, G.E. and Schworer, C.M. (1977) Induction of autophagy by amino-acid deprivation in perfused rat liver. *Nature*, **270** 174-76.

Poux, N. (1963) Sur la présence dénclaves cytoplasmiques in voie de dégénerescence dans les vacuoles des cellulles végétales. *C. R. Acad. Sci. (Paris)*, **257** 736-38.

Rébeillé, F., Bligny, R., Martin, J.-B. and Douce, R. (1985) Effect of sucrose starvation on sycamore (*Acer pseudoplatanus*) cell carbohydrate and Pi status. *Biochem. J.*, **226** 679-84.

Record, R.D. and Griffing, L.R. (1988) Convergence of the endocytic and lysosomal pathways in soybean protoplasts. *Planta*, **176** 425-32.

Robinson, D.G., Hoh, B., Hinz, G. and Jeong, B.-K. (1995) One vacuole or two vacuoles: do protein storage vacuoles arise *de novo* during pea cotyledon development? *J. Plant Physiol.*, **145** 654-64.

Robinson, D.G., Galili, G., Herman, E.M. and Hillmer, S. (1998) Topical aspects of vacuolar protein transport: autophagy and prevacuolar compartments. *J. Exp. Bot.*, **49** 1263-70.

Roby, C., Martin, J.-B., Bligny, R. and Douce, R. (1987) Biochemical changes during sucrose deprivation in higher plant cells: phosphorus-31 nuclear magnetic resonance studies. *J. Biol. Chem.*, **262** 5000-5007.

Sakai, Y., Koller, A., Rangell, L.K., Keller, G.A. and Subramani, S. (1998) Peroxisome degradation by microautophagy in *Pichia pastoris*: identification of specific steps and morphological intermediates. *J. Cell. Biol.*, **141** 625-36.

Schworer, C.M. and Mortimore, G.E. (1979) Glucagon-induced autophagy and proteolysis in rat liver: mediation by selective deprivation of intracellular amino acids. *Proc. Natl. Acad. Sci. USA*, **76** 3169-73.

Seglen, P.O. and Bohley, P. (1992) Autophagy and other vacuolar protein degradation mechanisms. *Experientia*, **48** 158-72.

Sikora, A., Hillmer, S. and Robinson, D.G. (1998) Sucrose starvation causes a loss of immunologically detectable pyrophosphatase and V-ATPase in the tonoplast of suspension-cultured tobacco cells. *J. Plant Physiol.*, **152** 207-12.

Smith, M.T., Saks, Y. and Van Staden, J. (1992) Ultrastructural changes in the petals of senescing flowers of *Dianthus caryophyllus* L. *Ann. Bot.*, **69** 277-85.

Swanson, S.J., Bethke, P.C. and Jones, R.L. (1998) Barley aleurone cells contain two types of vacuoles: characterization of lytic organelles by use of fluorescent probes. *Plant Cell*, **10** 685-98.

Takeshige, K., Baba, M., Tsuboi, S., Noda, T. and Ohsumi, Y. (1992) Autophagy in yeast demonstrated with proteinase-deficient mutants and conditions for its induction. *J. Cell Biol.*, **119** 301-11.

Tranbarger, T.J., Franceschi, V.R., Hildebrand, D.F. and Grimes, H.D. (1991) The soybean 94 kilodalton vegetative storage protein is a lipoxygenase that is localized in paraveinal mesophyll cell vacuoles. *Plant Cell*, **3** 973-87.

Tuttle, D.L., Lewin, A.S. and Dunn, Jr. W.A. (1993) Selective autophagy of peroxisomes in methylotrophic yeasts. *Eur. J. Cell Biol.*, **60** 283-90.

Van der Wilden, W., Herman, E.M. and Chrispeels, M.J. (1980) Protein bodies of mung bean cotyledons as autophagic organelles. *Proc. Natl. Acad. Sci. USA*, **77** 428-32.

Vaughn, K.C. and Duke, S.O. (1981) Evaginations from the plastid envelope: a method for transfer of substances from plastid to vacuole. *Cytobios.*, **32** 89-95.

Vigil, E.L. (1970) Cytochemical and developmental changes in microbodies (glyoxysomes) and related organelles of castor bean endosperm. *J. Cell Biol.*, **46** 435-54.

Villiers, T.A. (1967) Cytolysomes in long-dormant plant embryo cells. *Nature*, **214** 1356-57.

Villiers, T.A. (1971) Lysosomal activities of the vacuole in damaged and recovering plant cells. *Nature New Biol.*, **233** 57-58.

Wu, J., Lightner, J., Warwick, N. and Browse, J. (1997) Low-temperature damage and subsequent recovery of *fab1* mutant *Arabidopsis* exposed to 2°C. *Plant Physiol.*, **113** 347-56.

Yu, S.-M. (1999) Cellular and genetic responses of plants to sugar starvation. *Plant Physiol.*, **121** 687-93.

5 Ricinosomes and aleurain-containing vacuoles (ACVs): protease-storing organelles

Christine Gietl, Markus Schmid and David Simpson

5.1 Introduction

Ricinosomes were first described in the senescing endosperm of *Ricinus communis* (Mollenhauer and Totten, 1970; Vigil, 1970) and they have been demonstrated in other tissues, such as senescing *Hemerocallis* petals (Schmid *et al.*, 1999). They store a papain-type cysteine endopeptidase, released in the final stages of programmed cell death. Aleurain-containing vacuoles (ACVs) contain the cysteine protease, aleurain, which has been described in barley aleurone cells, and demonstrated in barley root tip cells and tobacco suspension culture protoplasts. Until ACVs are isolated and their composition determined, their function and their relationship to the protein-storage vacuole (PSV) remains speculative.

Plant cells contain separate protein storage and lytic vacuolar compartments (Rogers, 1998). Lytic vacuoles contain hydrolytic enzymes, such as glycosidases and proteases, that function in an acidic pH environment. Lytic vacuoles have features similar to those of animal lysosomes, and are supposed to digest their substrate within the organelle. PSVs are found in vegetative tissues, but are especially abundant in the storage organs of fruits and seeds (Bethke *et al.*, 1998). During seed maturation, storage proteins can be deposited within the embryo in cotyledons, as in the Fabaceae and Cucurbitaceae families, or they can be localized in the perisperm, which develops from the nucellus, as in the Piperaceae family. Storage proteins can also be deposited in the endosperm, such as the living endosperm of castor bean (*Ricinus communis*) or the endosperm of the Poaceae (includes the cereals), which is a dead tissue in the mature grain.

Storage protein is mobilized during germination. The cotyledons of some species within the Fabaceae, such as soybean (*Glycine max.*), and Cucurbitaceae, become green in the light and perform photosynthesis when storage protein mobilization is complete, with senescence occurring much later in development. In other species of the Fabaceae, such as the hypogeal germinating *Vicia faba*, *Pisum sativum*, and the epigeal germinating *Vigna mungo*, the cotyledons become exhausted and senesce. Castor bean endosperm also senesces during the final stages of storage protein mobilization. In Poaceae grains, the aleurone layer is a secretory tissue that releases hydrolases to digest the dead endosperm into sugars and amino acids, which nourish the embryo. Aleurone cells are not necessary for postembryonic development and senesce when germination is complete. Senescence requires nuclear function, and is an active process also described as 'programmed cell death' (PCD) (Pennell and Lamb, 1997).

5.2 Ricinosomes

5.2.1 Endosperm morphology in germinating castor bean

The endosperm of germinating castor bean (*Ricinus communis*) (Breidenbach and Beevers, 1967) and the cotyledons of Cucurbitaceae (Trelease *et al.*, 1971) represent two major methods for higher plants to store and subsequently mobilize lipids and proteins for developing seedlings. Castor beans store lipid and proteins in a living endosperm, which is laterally attached to the cotyledons. From day 1–5 of germination, glyoxysomes develop rapidly in the endosperm cells and mobilize the triglycerides of the oleosomes (lipid bodies) into fatty acids, acetyl-CoA and succinate (Gietl, 1996). When the storage reserves are depleted by day 6–7, the desiccated endosperm abscises. In these late stages, castor bean endosperm is a senescing tissue.

Endosperm tissue from castor bean contains an additional microbody-like organelle, only slightly larger than glyoxysomes. This organelle was independently discovered in ultrastructural and cytochemical studies 30 years ago, by two groups. It was called 'dilated cisternae', since it seemed to develop from the endoplasmic reticulum (ER) (Vigil, 1970), or 'ricinosome', since it was thought to be unique to castor bean (Mollenhauer and Totten, 1970). From 4 to 6 days after germination, all developmental changes in the fine structure of the endosperm during lipid mobilization can be seen along a cross-section extending from the seed coat inward to the cotyledons in the center of the seed (Vigil, 1970; Schmid *et al.*, 1999). Such a cross-section can be divided into four zones, in which endosperm cells nearest the cotyledons have progressed furthest while those nearest the seed coat resemble cells from imbibed tissue. In zone I, the endosperm is hard and the endosperm cells are still filled with lipid and protein bodies. In Zone II, the endosperm is soft and mitochondria, plastids and microbodies are more numerous. Cells with large vacuoles that restrict most of the cytoplasm to regions along the cell wall are characteristic of Zone III. These vacuoles develop from the limiting membranes of protein bodies. Oleosomes, prominent in Zone I, are not very conspicuous, while ricinosomes and glyoxysomes are most abundant in Zone III. The cells in Zone IV have lost their cytoplasmic contents. The final stages of cellular breakdown are very sudden, resulting in complete disruption of both cytoplasm and cell walls, which produces a line of demarcation between living and non-living cells.

The two organelles bounded by single membranes, glyoxysomes and ricinosomes, can be distinguished by morphology and contents. While glyoxysomes vary in shape from spherical to ellipsoidal, ranging in diameter 0.2–1.7 μm, ricinosomes are spherical, with a diameter averaging 0.9 μm. Glyoxysomes, but not ricinosomes, can be stained for catalase with diaminobenzidine (Mollenhauer and Totten, 1970). Ricinosomes have a rough-surfaced membrane, irregularly studded with densely staining particles resembling ribosomes, whereas glyoxysomes have a smooth membrane. Until recently, attempts to isolate and characterize ricinosomes have not been successful, primarily because of difficulties in recognizing them after isolation and the lack of a marker enzyme. It is most likely that ricinosomes were present as a contaminant in glyoxysome fractions after density gradient purification (Vigil, 1970).

5.2.2 A papain-like cysteine endopeptidase with a C-terminal KDEL motif is a marker enzyme for the ricinosome

A papain-type cysteine endopeptidase (Cys-EP), with a molecular mass of 35 kDa for the mature protein, was originally purified from a crude glyoxysomal preparation from germinating castor bean endosperm (Gietl *et al.*, 1997). Sequencing of cDNA clones from the endosperm of germinating seedlings and from developing seeds established the presence of a presequence for co-translational targeting into the ER, an N-terminal propeptide and a C-terminal KDEL motif (Schmid *et al.*, 1998). The 45 kDa proenzyme was present in isolated, purified ricinosomes. Immunogold labelling, using antibodies raised against the purified Cys-EP, occurred exclusively in the ricinosomes (Figure 5.1),

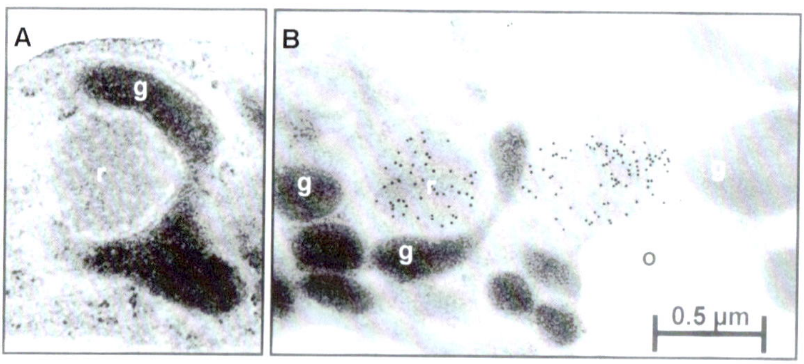

Figure 5.1 Immunocytochemistry of *Ricinus* cysteine endopeptidase (CysEP). Endosperm tissue from 5 day old germinating *Ricinus* seedlings was fixed in aldehyde and post-fixed in OsO_4. Sections were incubated with antibodies against CysEP (B). (A) Ricinosomes (r) are roughly spherical organelles, less electron dense than glyoxysomes (g), with which they are often in close contact. The presence of ribosome-like bodies on the membrane of ricinosomes indicates their origin from the endoplasmic reticulum (ER). (B) Ricinosomes (r) are heavily labelled with the antibody to Cys-EP, with little or no labelling of glyoxysomes (g) or oleosomes (o). (Magnification ×43,000 for both A and B) (Reproduced from Schmid *et al.*, 1998).

while antibodies against gMDH exclusively labelled the glyoxysome. The Cys-EP is thus the first marker enzyme for ricinosomes (Schmid *et al.*, 1998).

The cysteine endopeptidase from senescing castor bean endosperm is homologous to papain-type KDEL cysteine endopeptidases from other senescing tissues (Schmid *et al.*, 1998), such as the cotyledons of germinating mung bean (*Vigna mungo*) and vetch (*Vicia sativa*), the maturing French bean (*Phaseolus vulgaris*) pods and the flowers of daylily (*Hemerocallis*) (Figure 5.2). Ricinosomes containing the 45 kDa Cys-EP are found in the latest stages of seed development in a layer of collapsing cells at the outer edge of the seed; they might represent either the senescing nucellus, which surrounds the embryo sack and collapses as soon as seed maturation is completed, or an integument developing into the seed coat. Furthermore, ricinosome-like organelles containing the 45 kDa Cys-EP are found in senescing *Hemerocallis* petals (Schmid *et al.*, 1999). The existence of other plant KDEL cysteine endoproteases implies that ricinosomes are not unique to *Ricinus* and play an important role in senescing tissues.

The papain-type cysteine endopeptidases with KDEL at their C-terminus fall into a distinct group among the plant cysteine proteases (Figure 5.3). As far as the tissue of expression is known, most of them are localized in senescing tissues. The authors hypothesize, therefore, that these KDEL proteases are directly related to the presence of ricinosomes, so it would be interesting to look for ricinosomes in tissues where papain-type KDEL peptidases are expressed.

5.2.3 *Processing of cysteine endoproteases*

The most extensively studied CysEP with a C-terminal KDEL motif is SH-EP from *Vigna mungo* (Okomoto *et al.*, 1994, 1999). This is synthesized as a 45 kDa inactive precursor, which is co-translationally processed to a 43 kDa intermediate after cleavage of the signal peptide. Inactive intermediates of 39 and 36 kDa have also been observed *in vitro* prior to the formation of the mature 33 kDa protease in *V. mungo* cotyledon extracts. Expression of SH-EP in insect Sf-9 cells revealed the transient presence of a 42 kDa form, which was not recognized by an antibody against the KDEL sequence (Okomoto *et al.*, 1999). Okomoto *et al.*, also showed that the C-terminal KDEL was not required for correct folding and that the mature enzyme was not glycosylated at any of the three potential N-glycosylation sites. In addition, processing of an enzymatically inactive form of SH-EP (C26G, numbering of the mature subunit) to the 33 kDa mature polypeptide occurred much more slowly, without the presence of the 39 and 36 kDa intermediate forms. These results imply that SH-EP maturation involves autocatalytic self-processing. It should be noted, however, that *in vitro* processing experiments were performed with crude extracts, and the possibility of proteolytic attack by proteases and peptidases in other cellular compartments cannot be excluded. The susceptibility of the C-terminal KDEL motif to attack

```
1                                                                                                             100
Hordeum vulgare      ~MWRCILLSA VVVAL.ALAP APALGVPFTE KDLASEESLR GLYERWRSHY TVSR....RG LGADAEE..R RFNVFKQNAR YVHEGNKR.D MPFRLALNKF
Oryza sativa         ~MLRCFLVAA AAVALAAAAA APARAIPFTE SDLSSEESLR ALYERWRSRY TVSRPAASGG VGNDDGEARR RFNVFVENAR YIHEANRRGG RPFRLALNKF
Hemerocallis sp.     ~~MAKPKFIA LALV.ALSFL SIAQSIPFTE KDLASEDSLW NLYEKWRTHH TVAR...... ...DLDEKNR RFNVFKENVK FIHEFNQKKD APYKLALNKF
Hemerocallis sp.     ~~MAKLSYAL LSVVLVLGSV ALAQSIPFDE KDLASEESLW SLYEKWRAHH AVSR...... ...DLDDTDK RFNVFKENVK FIHEFNQKKD ATYKLALNKF
Sandersonia          ~~MASLFPVL LVLALAFGS. ..TLSIPIKE KDLESEDSLW SLYERWRSHH AVSR...... ...DLDQKQK RFNVFKENVK FIHEFNKNKD VTFKLALNKF
Phalaenopsis         ~~~~~MKLFS LILVASFLAS VAATAIDIAD KDLETEDSLW NLYERWRSHH TVSR...... ...DLDEKQK RFNVFKENPR YIHDFNKRKD IPYKLRLNKF
Pisum sativum        MEMKKLLLFS LSLALIFR.. .ATNTFDFNE HDLDSEKSLW DLYERWRSHH TVTR...... ...SLDEKHN RFNVFKANVM HVHNTN.KLD KPYKLKLNKF
Vicia sativa         MEMKKLLFIS LSLALIFT.. .VANTFDFNE HDLESEKSLW NLYERWRSHH TVTR...... ...NLDEKHN RFNVFKANVM HVHNTN.KLD KPYKLKLNKF
Vigna mungo          MAMKKLLWVV LSLSLVLG.. .VANSFDFHE KDLESEESLW DLYERWRSHH TVSR...... ...SLGEKHK RFNVFKANVM HVHNTN.KMD KPYKLKLNKF
Vigna radiata        MAMKKLLWVV LSLSLVLG.. .VANSFDFHE KDLASEESLW DLYERWRSHH TVSR...... ...SLTEKHK RFNVFKENVM HVHNTN.KMD KPYKLKLNKF
Phaseolus vulgaris   MATKKLLWVV LSFSLVLG.. .VANSFDFHD KDLASEESLW DLYERWRSHH TVSR...... ...SLGEKHK RFNVFKANLM HVHNTN.KMD KPYKLKLNKF
Ricinus communis     ~~MQKFILLA LSLALVLA.. .ITESFDFHE KELESEESLW GLYERWRSHH TVSR...... ...SLHEKQK RFNVFKHNAM HVHNAN.KMD KPYKLKLNKF
Arabidopsis          ~~MKRFIVLA LCMLMVLE.. .TTKGLDFHN KDVESENSLW ELYERWRSHH TVAR...... ...SLEEKAK RFNVFKHNVK HIHETN.KKD KSYKLKLNKF
Arabidopsis          ~~MKLFPIVL ISFLSLLLQ.. .ASKGFDFDE KELETEENVW KLYERWRGHH SVSR...... ...ASHEAIK RFNVFRHNVL HVHRTN.KKN KPYKLKINRF
Arabidopsis          ~~MKKLLLIF LFSLVILQ.. .TACGFDYDD KEIESEEGLS TLYDRWRSHH SVPR...... ...SLNEREK RFNVFRHNVM HVHNTN.KKN RSYKLKLNKF
Lycopersicon         ~~~~~~~~~~ ~~~~~~~~~~ ~~~~~~~~~~ ~~~~~~~~~~ ~~~~~~~~~~ ~~~~~~~~~~ ~~~~~~~~~~ ~~~~~~~~~~ ~~~~~~~~~~ ~~~~~~~~~~

                     101                                                                                                      200
Hordeum vulgare      ADMTTDEFRR TYAGSRVRHH LSLSGGRRGD G..G.FRYG. .DADNLPP.A VVWRQKGAVT AIKDQGQC.G SCWVFSTIVA VEGINKIRTG KLVSLSEQEL
Oryza sativa         ADMTTDEFRR TYAGSRARHH RSLSGGRGGE G..GSFRYGG DDEDNLPP.A VDWRERGAVT GIKDQGQC.G SCWAFSTVAA VEGVNKIKTG RLVTLSEQEL
Hemerocallis sp.     GDMTNQEFRS KYAGSKIQHH RSQRGIQKNT GS...FMY.. ENVGSLPAAS IDWRAKGAVT GVKDQGQC.G SCWAFSTIAS VEGINQIKTG ELVSLSEQEL
Hemerocallis sp.     GDMTNQEFRS TYAGSKIDHH MTLRGV.KDA GE...FSY.. EKFHDLPT.S VDWREKGAVT GVKDQGQC.G SCWAFSTVVA VEGINQIKTN ELVSLSEQQL
Sandersonia          GDMTNQEFRA KYAGSKVHHH RTMKGSRHGS GSGAKFMY.. ENAV.APP.S IDWRERGAVA AVKNQGQC.G SCWAFSAIAA VEGINQIVTK ELVPLSEQEL
Phalaenopsis         ADLTNHEFRS TYAGSRINHH RSLRGSRRG. GATNSFMYQS LDSRSLPA.S IDWRQKGAVT AVKDQGQC.G SCWAFSTVAA VEGINQIKTK KLLSLSEQEL
Pisum sativum        ADMTNYEFRR IYADSKVSHH RMFRGMSNEN GT...FMY.. ENVKNVPS.S IDWRKKGAVT DVKDQGQC.G SCWAFSTIVA VEGINQIKTQ KLVSLSEQEL
Vicia sativa         GDMTNYEFRR IYADSKISHH RMFRGMSHEN GT...FMY.. ENAVDVPS.S IDWRNKGAVT GVKDQGQC.G SCWAFSTIAA VEGINQIKTQ KLVSLSEQQL
Vigna mungo          ADMTNHEFRS TYAGSKVNHH KMFRGSQHGS GT...FMY.. EKVGSVPA.S VDWRKKGAVT DVKDQGQC.G SCWAFSTIVA VEGINQIKTN KLVSLSEQEL
Vigna radiata        ADMTNHEFRS TYAGSKVNHH KMFRGTQHGN GT...FMY.. EKVGSVPA.S VDWRKKGAVT DVKDQGQC.G SCWAFSTVVA VEGINQIKTD KLVSLSEQEL
Phaseolus vulgaris   ADMTNHEFRS TYAGSKVNHP RMFRGTPHEN GA...FMY.. EKVVSVPP.S VDWRKKGAVT DVKDQGQC.G SCWAFSTVVA VEGINQIKTN KLVALSEQEL
Ricinus communis     ADMTNHEFRN TYSGSKVKHH RMFRGGPRGN GT...FMY.. EKVDTVPA.S VDWRKKGAVT SVKDQGQC.G SCWAFSTIVA VEGINQIKTN KLVSLSEQEL
Arabidopsis          GDMTSEEFRR TYAGSNIKHH RMFQGEKKAT KS...FMY.. ANVNTLPT.S VDWRKNGAVT PVKNQGQC.G SCWAFSTVVA VEGINQIRTK KLTSLSEQEL
Arabidopsis          ADITHHEFRS SYAGSNVKHH RMLRGPKRGS GG...FMY.. ENVTRVPS.S VDWREKGAVT EVKNQQDC.G SCWAFSTVAA VEGINKIRTN KLVSLSEQEL
Arabidopsis          ADLTINEFKN AYTGSNIKHH RMLQGPKRGS KQ...FMYDH ENLSKLPS.S VDWRKKGAVT EIKNQGKCGG SCWAFSTVAA VEGINKIKTN KLVSLSEQEL
Lycopersicon         ~~~~~~~~~~ ~~~~~~~~~~ ~~~~~~~~~~ ~~~~~~~~~~ ~~~~~~~~~~ ~~~~~~~~~~ ~~~~~~~~~~ ~~~~~~~~~~ ~~~~~~~~~~ ~~~~~~~~~~
```

```
                     201                                                                                                             300
Hordeum vulgare      MDCDNVNNQG CDGGLMDYAF QFIQKN.GIT TESNYPYQGE Q.GSCDQAKE NAQAVTIDGY EDVPANDESA LQKAVAGQPV SVAIDASGQD FQFYSEGVFT
Oryza sativa         VDCDTGDNQG CDGGLMDYAF QFIKRNGGIT TESNYPYRAE Q.GRCNKAKA SSHDVTIDGY EDVPANDESA LQKAVANQPV AVAVEASGQD FQFYSEGVFT
Hemerocallis sp.     VDCDTSYNEG CNGGLMDYAF EFIQKN.GIT TEDSYPYAEQ D.GTCASNLL NSPVVSIDGH QDVPANNENA LMQAVANQPI SVSIEASGYG FQFYSEGVFT
Hemerocallis sp.     VDCDTK.NSG CNGGLMDYAF DFIKNNGGLS SEDSYPYLAE Q.KSCGSE.A NSAVVTIDGY QDVPRNNEAA LMKAVANQPV SVAIEASGYA FQFYSQGVFS
Sandersonia          IDCDTDQNQG CSGGLMDYAF EFIKNNGGIT TEDVYPYQAE D.ATCKK... NSPAVVIDGY EDVPTNDEDA LMKAVANQPV AVAIEASGYV FQFYSEGVFT
Phalaenopsis         IDCDTDENNG CNGGLMDYAF DFIKKNGGIS SEAEYPYAAE D.SYCATEK. KSHVVSIDGH EDVPANDEDS LLKAVANQPV SIAIEASGYD FQFYSEGVFT
Pisum sativum        VDCDTGGNEG CNGGLMEYAF EFIKQN.GIT TESNYPYAAK D.GTCDLKKE DKAEVSIDGY ENVPINNEAA LLKAAAKQPV SVAIDAGGYN FQFYSEGVFS
Vicia sativa         VDCDTEENEG CNGGLMEYAF EFIKQN.GIT TESNYPYAAK D.GTCDVEKE DKA.VSIDGH ENVPINNEAA LLKAAAKQPV SVAIDAGGYN FQFYSEGVFT
Vigna mungo          VDCDKEENQG CNGGLMESAF EFIKQKGGIT TESNYPYTAQ E.GTCDESKV NDLAVSIDGH ENVPVNDENA LLKAVANQPV SVAIDAGGSD FQFYSEGVFT
Vigna radiata        VDCDKEENQG CNGGLMESAF EFIKQKGGIT TESNYPYTAQ E.GTCDASKV NDLAVSIDGH ENVPVNDENA LLKAVANQPV SVAIDAGGSD FQFYSEGVLT
Phaseolus vulgaris   VDCDKEENQG CNGGLMESAF EFIKQKGGIT TESNYPYKAQ E.GTCDASKV NDLAVSIDGH ENVPANDEDA LLKAVANQPV SVAIDAGGSD FQFYSEGVFT
Ricinus communis     VDCDTDQNQG CNGGLMDYAF EFIKQRGGIT TEANYPYEAY D.GTCDVSKE NAPAVSIDGH ENVPENDENA LLKAVANQPV SVAIDAGGSD FQFYSEGVFT
Arabidopsis          VDCDTNQNQG CNGGLMDLAF EFIKEKGGLT SELVYPYKAS D.ETCDTNKE NAPVVSIDGH EDVPKNSEDD LMKAVANQPV SVAIDAGGSD FQFYSEGVFT
Arabidopsis          VDCDTEENQG CAGGLMEPAF EFIKNNGGIK TEETYPYDSS DVQFCRANSI GGETVTIDGH EHVPENDEEE LLKAVAHQPV SVAIDAGSSD FQLYSEGVFI
Arabidopsis          VDCDTKQNEG CNGGLMEIAF EFIKKNGGIT TEDSYPYEGI D.GKCDASKD NGVLVTIDGH EDVPENDENA LLKAVANQPV SVAIDAGSSD FQFYSEGVFT
Lycopersicon         ~~~~~~~~~~ ~~~~~~~~~~ ~~~~~~~~~~ ~~~~~~~~~~ ~~~~~~~~~~ ~~~~~~~~~~ ~~~~~~~~~~ ~~~~~~~~~~ ~~~~~~~~~~ ~~~~~~~~~~

                     301                                                                                               387
Hordeum vulgare      GECSTDLDHG VAAVGYGATR DGTKYWIVKN SWGEDWGEKG YIRMQRGV.S QTEGLCGIAM QASYPTK.SA PHASTVR...E ESHTDEL* EP-A     Z97023
Oryza sativa         GECGTDLDHG VAAVGYGITR DGTKYWIVKN SWGEDWGERG YIRMQRGVSS DSNGLCGIAM EASYPVK.SG ARNAAAS...N RVVKDEM* REP-2    X80876
Hemerocallis sp.     GRCGTELDHG VAIVGYGATR DGTKYWIVKN SWGEEWGESG YIRMQRGI.S DKRGKCGIAM EASYPIK.TS ANPKNS..... .STRDEL* SEN11    X74406
Hemerocallis sp.     GHCGTELDHG VAAVGYGVDD DGKKYWIVKN SWGEGWGESG YIRMERGI.K DKKGKCGIAM EASYPIK.SS PNPKKAE.... .SLKDEL* SEN102   U12637
Sandersonia          GRCGTELDHG VAVVGYGTTQ DGTKYWTVRN SWGADWGESG YVRMQRGI.K ATHGLCGIAM QASYPIK.TS LNP.GMD.... .SLKDEL* PRT5     AF133839
Phalaenopsis         GRSGTELDHG VAIVGYGKTQ QGTKYWIVRN SWGAEWGEKG YIRIS.AA.S DSKRLCGLAM EASYPIK.TS PNPSH...... .KSRDEL* O141     U34747
Pisum sativum        GHCGTDLNHG VAVVGYGVTQ DRTKYWIVKN SWGSEWGEQG YIRMQRGI.S HKEGLCGIAM EASYPIKKSS TNPTES..... STLKDEL* tpE4A    AJ004958
Vicia sativa         GHCDTDLNHG VAIVGYGVTQ DRTKYWIMKN SWGLEF..MG PRMGRTGI.S SREGLCGIAM EASYPIKKSS TKPTES..... SILKDEL* Prot A   Z34895
Vigna mungo          GDCNTDLNHG VAIVGYGTTV DGTNYWIVRN SWGPEWGEQG YIRMQRNI.S KKEGLCGIAM MASYPIKNSS DNPTGS....L SSPKDEL* SH-EP    X15732
Vigna radiata        GDCNTDLNHG VAIVGYGTTV DGTNYWIVRN SWGPEWGEQG YIRMQRNI.S KKEGLCGIAM MASYPIKNSS DNPTGS....F SSPKDEL*          U49445
Phaseolus vulgaris   GDCSTDLNHG VAIVGYGTTV DGTNYWIVRN SWGPEWGEHG YIRMQRNI.S KKEGLCGIAM LPSYPIKNSS DNPTGS....F SSPKDEL* EP-C1    X63102
Ricinus communis     GSCGTELDHG VAIVGYGTTI DGTKYWTVKN SWGPEWGEKG YIRMERGI.S DKEGLCGIAM EASYPIKKSS NNPSGI....K SSPKDEL* CysEP    AF050756
Arabidopsis          GRCGTELNHG VAVVGYGTTI DGTKYWIVKN SWGEEWGEKG YIRMQRGI.R HKEGLCGIAM EASYPLKNSN TNPSRLS...L DSLKDEL* CEP1     AB024031
Arabidopsis          GECGTQLNHG VVIVGYGETK NGTKYWIVRN SWGPEWGEGG YVRIERGI.S ENEGRCGIAM EASYPTKLSS .TPSTHESVVR DDVKDEL* CEP2     AL049659
Arabidopsis          GSCGTELNHG VAAVGYG.SE RGKKYWIVRN SWGAEWGEGG YIKIEREI.D EPEGRCGIAM EASYPIKLSS SNPTPKD.... GDVKDEL* CEP3     AL049659
Lycopersicon         ~~~~~~~~~~ VAIVGYGTTV DGTKYWIVKN SWGAEWGEKG YIRMQRKV.D AEEGLCGIAM QPSYPIKTSS .NPTGSP...A ATPKDEL* E4       S44898
```

Figure 5.2 Amino acid sequence alignment of plant cysteine endoproteases with putative C-terminal endoplasmic reticulum (ER) retention motifs, from *Hordeum vulgare* (Z97023), *Oryza sativa* (X80876), *Hemerocallis* sp. (U12637, X74406), *Sandersonia aurantiaca* (AF133839), *Phalaenopsis* sp. (U34747), *Pisum sativum* (AJ004958), *Vicia sativa* (Z34895), *Vigna mungo* (X15732), *Vigna radiata* (U49445), *Phaseolus vulgaris* (X63102), *Ricinus communis* (AF050756), *Lycopersicon esculentum* (S44898), *Arabidopsis thaliana* (AB024031, AL049659, AL049659), using the GCG program, PILEUP (Devereux *et al.*, 1984).

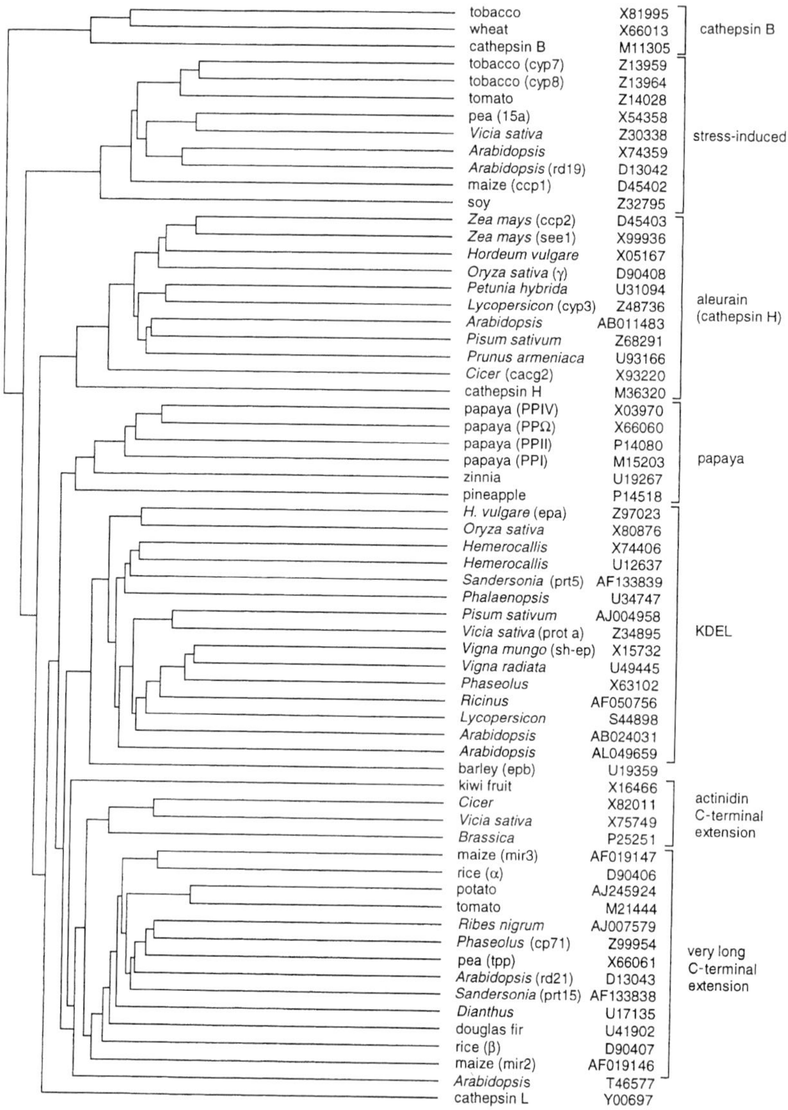

Figure 5.2 Dendrogram of plant cysteine endoproteases produced by the GCG program, PILEUP, which enables them to be placed in various groups. Those similar to aleurain or with KDEL at their C-terminus each fall into a distinct group based on amino acid sequence homology.

by carboxypeptidase A has been demonstrated for the auxin-binding protein (Napier *et al.*, 1992). If the C-terminal processing of SH-EP occurs only *in vitro*, there is no reason to conclude that SH-EP follows the secretory pathway out of the ER.

Apart from the KDEL cysteine endoproteases, there are two groups which have C-terminal extensions (telo sequences). The first is represented by actinidin (X16466), which has a 34 residue C-terminal extension. Tobacco plants expressing an actinidin construct without the C-terminal extension secrete the proteolytically active precursor (Paul *et al.*, 1995). This sequence may, therefore, be a vacuolar sorting determinant, but this has not been tested with foreign proteins. There is only limited homology between the actinidin telo sequence and those of other putative members of the group (X82011, X75749, JQ1121, D41751, T69041, Z99707). The second group has a much longer telo sequence of about 110 residues, with a much greater degree of homology. Two rice proteases, oryzain α (D90406) and β (D90407), are typical for this group, but their C termini have been located only by sequence similarity to actinidin (Watanabe *et al.*, 1991). The 13 members of this group have 13 perfectly conserved cysteine residues in the telo sequence, and the motif NPPXPXP[ST]PP[ST]P after the putative mature C terminus. It is clear, therefore, that for cysteine proteases co-translationally transported into the ER, there are several different targeting sequences that determine their subsequent localization.

5.2.4 *Biogenesis of ricinosomes: indications for the direct development from the endoplasmic reticulum*

Do ricinosomes develop directly from the ER or via the Golgi apparatus? There are several lines of evidence for the direct development of ricinosomes from the endoplasmic reticulum. Vigil (1970) recognized them as ribosome-bearing organelles and suggested, on the basis of electron microscopy, that they arise from the ER, referring to them as ‘dilated cisternae’. They subsequently lose their membrane-bound ribosomes, enabling them to form close associations with glyoxysomes.

The authors have prepared highly enriched ricinosomes by sucrose density centrifugation. Their purity was demonstrated by transmission electron microscopy and immunocytochemistry. Sodium dodecyl sulphate-polyacrylanide gel electrophoresis (SDS-PAGE) and Coomassie-staining of the ricinosome fractions revealed that the matrix consisted almost exclusively of the 45 kDa Cys-EP (manuscript in preparation). The ricinosome Cys-EP is synthesized with a KDEL-motif at the C-terminus, which functions as a Golgi retrieval signal for soluble residents of the ER lumen. Detection of the KDEL-motif of the pro-Cys-EP was possible by western blot analysis of ricinosome fractions, using the anti-KDEL monoclonal antibody, MAC256 (Napier *et al.*, 1992, unpublished observations). Work is in progress to confirm this result by additional, independent methods. Furthermore, the ER lumen protein, binding protein (BiP; Hsp70), was detected within ricinosomes by western blot analysis and by immunofluorescence microscopy of ricinosomes using double labelling of Cys-EP and BiP (manuscript in preparation).

The direct biogenesis of ricinosomes from the ER would appear to be similar to the formation of protein bodies in maize (Lending and Larkins, 1989) and rice (Li *et al.*, 1993), and is consistent with the presence of an ER retention sequence at the C-terminus of the cysteine endoprotease, which is the major soluble polypeptide within ricinosomes. It is suggested that vesicles, containing cargo destined for PSVs in developing seeds and grains, can bud-off from the ER and move directly to the vacuole, thereby bypassing the Golgi. This idea is based mainly on observations of storage protein aggregates in the ER lumen, but there is no direct evidence that such aggregates are subsequently transported out of the ER via precursor accumulating (PAC) vesicles (Levanony *et al.*, 1992; Hara-Nishimura *et al.*, 1998). Indeed, in the barley mutant, Nevsky, hordeins accumulate in the ER lumen in the endosperm but are transported to the protein storage vacuole via the Golgi apparatus in the subaleurone, where the rate of protein synthesis is slower. It has been postulated that if storage proteins are made more rapidly than they can be transported out of the ER, they aggregate, lose their transport competence and are stored in the ER lumen (Rechinger *et al.*, 1993). Biochemical evidence for Golgi-independent transport is lacking in the case of the cereal prolamins, since they are not glycosylated, nor proteolytically cleaved. In the case of maturing *Ricinus* endosperm, PAC vesicles do contain Golgi-derived glycoproteins (Hara-Nishimura *et al.*, 1998). However, ricinosomes differ from PAC vesicles in several respects. They are larger (1 versus 0.2–0.4 μm), are enclosed by a membrane bearing ribosomes, and appear during senescence and not during storage protein accumulation.

5.2.5 *Protein retention within the ER and export from the ER*

The export of soluble proteins from the ER lumen has been postulated to occur either by default or in a regulated manner. The fact that different secretory proteins transverse the ER at different rates indicates regulation rather than bulk flow. Proteins that are to be exported are concentrated in the ER lumen, and are sorted away from ER residents. The latter are a group of proteins which have the intrinsic ability to stay in the ER lumen, and contain a C-terminal motif, using KDEL or HDEL, which functions as part of a salvage system to retrieve the small amounts of ER resident protein that leak out (Vitale and Denecke, 1999). It is possible that secreted proteins have an export signal which binds to (so far unidentified) receptors in COPII vesicles. It is interesting that, in plants, all published sequences of GRP94, protein disulphide isomerase and auxin-binding protein end with KDEL, whereas all HSP70 and calreticulins end with HDEL (Table 5.1).

Fluorescence labelling of maize and onion root cells with monospecific monoclonal antibodies against KDEL and HDEL have shown a different intracellular distribution (Napier *et al.*, 1992). HDEL proteins showed a distribution that was characteristic of plant ER, whereas KDEL proteins were located in discrete parts of the cell cortex. This suggests that the C-terminal KDEL motif

Table 5.1 Plant C-terminal endoplasmic reticulum (ER) retention sequences

Protein	Sequence	Accession no.
Protein disulphide isomerase	KDEL	U11496, L33250/1, M82973, L39014, U41385, Z11499, AC007727, AL04655, AC003033
GRP94	KDEL	X67960, L14594, AL078637, AC002343
Auxin-binding protein	KDEL	X56737, X70902/3, X69901, AB000706/7, U77952, X91839, Z48451, AJ011943, L08425/6
Cysteine endoprotease	KDEL	See Figure 5.2
Myrosinase (β-glucosidase)	KDEL	X89413, U72153, AAC39504
Similar to thyroid hormone binding	KDEL	AB026295, BAA81871
Auxin-conjugate hydrolase	KDEL	AAD48152, AAC32192, CAA73905
Auxin-conjugate hydrolase	KSEL	AAF26972
Immunophilin	KNEL	AAC49390
Immunophilin	NDEL	AAC49391
Auxin-conjugate hydrolase	HDEL	AAC04865
Calreticulin	HDEL	X89813, U74630, L27348/9, Z71395, X85382, U27698, X80756, X78057, AJ002057, AF019376, CAA05161, AAD17490, AAD32207
GRP78 (HSP70)	HDEL	M59449, X60058-62, L08830, L32165, D84414, U58208/9, U96641, L23551/2, AF006825, U08383/4
Putative Ca^{++}-binding protein	HDEL	AL078579, BAA88900, CAB43967
Polygalacturonase	HDEL	AAF04869, AAC98041, AAC98004
Auxin-conjugate hydrolase	HEEL	AAC04866
HSP22	REEL	AAB30525
β-glucosidase	REEL	AF183827
β-glucosidase	RKEL	AAC31962
β-glucosidase	RDEL	U72154, U72155
Cysteine endoprotease	RDEL	X74406, U34747
Ribonuclease I	RDEL	U19923
LEA protein	DAEL	U02966, AJ131342
Putative LEA protein	YAEL	AC006439

NB All have putative signal peptides. Abbreviation: LEA, late embryogenesis abundant.

in plants may target proteins to a subcompartment of the ER. Staehelin (1997) identified 16 different specialized regions of the ER, including one enriched in a vacuolar ATPase subunit, based on subcellular fractionation and immunocytochemistry (Herman *et al.*, 1994).

In addition to KDEL and HDEL, several other C-terminal motifs have been found in putative ER lumen resident proteins (Table 5.1). Some of these, for example HDEF (Kaletta *et al.*, 1998) and RDEL, are sufficient to direct proteins to the ER, while single amino acid substitutions produce sequences that are either partially effective (KEEL) or have a complete loss of function (SDEL, KDDL, KDEI, KDEV) (Denecke *et al.*, 1992). KDEM and TDEL, which are unique to the barley cysteine endoprotease EP-A (Z97023) and the rice homologue (X80876), are not included. Mature EP-A is secreted from isolated aleurone layers (Koehler and Ho, 1990), and this might also be true for the rice homologue, so these sequences are probably not ER retention motifs. The addition of KDEL to the C-terminus of vicilin results in its accumulation within the ER lumen in the form of large aggregates (Wandelt *et al.*, 1992). These closely resemble the protein bodies containing prolamins in tropical cereals, such as rice, sorghum and maize (Lending and Larkins, 1989). In the case of rice, the ribosomes on the cisternal ER membrane surrounding these protein bodies are associated primarily with prolamin transcripts (Li *et al.*, 1993). HDEL will also lead to the retention of foreign proteins in the ER (Gomord *et al.*, 1997), so both of these motifs function as efficient Golgi retrieval signals in plants. Investigations with the maize auxin-binding protein, KDEL (see Table 5.1), showed that less than 15% ever escapes the Golgi, and less than 2% passes further through the secretory pathway (Henderson *et al.*, 1997). There is no reason to suppose that this is not true for other natural proteins with an ER retention motif, including the cysteine endoproteases.

5.2.6 *Ricinosomes are a novel organelle involved in programmed cell death in plants*

Senescence is the final phase of plant vegetative and reproductive development, preceding the widespread death of cells and organs and is a form of programmed cell death (Pennell and Lamb, 1997). The development of ricinosomes is concomitant with nuclear DNA fragmentation, as detected by terminal deoxynucleotidyltransferase-mediated dUTP nick end labelling (TUNEL assay) (Schmid *et al.*, 1999). Nuclear DNA fragmentation in castor bean endosperm was first observed at day 3 in the cells immediately adjacent to the cotyledons, and progressed across to the outermost cell layers by day 5 (Figure 5.4A). Intact ricinosomes accumulate the 45 kDa proform of Cys-EP, as can be demonstrated by western blot analysis and by *in situ* immunofluorescence labelling with antibodies directed specifically against the propeptide of the Cys-EP (Figure 5.4B). Disintegration of ricinosomes during the final cell collapse releases the 35 kDa mature form, giving a diffuse labelling of the collapsed cells near the cotyledons (Figure 5.4C). As long as ricinosomes are intact, the Cys-EP is in the relatively inactive 45 kDa proform, which is approximately 50–100 times less active than the mature form (Schmid *et al.*, 1998). Loss of ricinosome integrity may result from a change in the permeability of the tonoplast, leading to acidification of the cytoplasm. The low pH might possibly trigger autocatalytic cleavage of the Cys-EP propeptide, as for papain and other cysteine endoproteases.

Ricinosomes cannot be defined as lytic vacuoles, since there is no indication that they digest their substrate within the organelle itself. Ricinosomes, therefore, behave as protease-storing vacuoles, which function as suicide bombs in the final stages of programmed cell death.

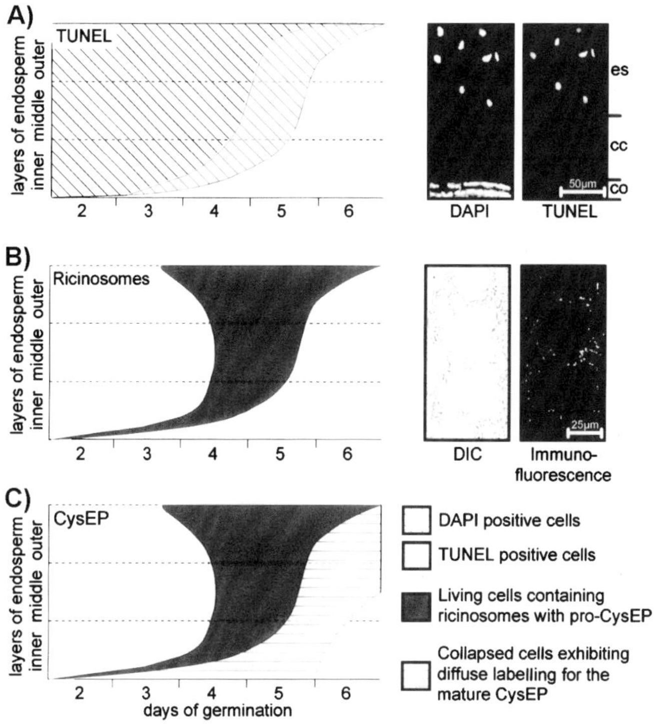

Figure 5.3 Schematic presentation of nuclear DNA fragmentation and development of ricinosomes in germinating *Ricinus* endosperm from inner, middle and outer layers of the endosperm in 2–6 day old seedlings. (A) Nuclei were stained with 4,6-diamidino-2-phenylindole (DAPI) and DNA fragmentation was detected by terminal deoxynucleotidyl transferase-mediated dUTP nick end labelling (TUNEL); cotyledon nuclear DNA does not fragment. The collapsed cells (cc) developing in the layer between the cotyledons (co) and the still living endosperm cells (es) do not label with DAPI or TUNEL. (B) Development of ricinosomes: cells were visualized using differential interference contrast (DIC) optics; ricinosomes can be detected by immunofluorescence microscopy using antibodies directed either against the propeptide or against the mature subunit of cysteine endopeptidase (CysEP) and fluorescein isothiocyanate (FITC)-labelled secondary antibodies giving rise to a punctate labelling. (C) Cellular distribution of CysEP: collapsed cells show a diffuse labelling, with antibodies recognizing the mature enzyme, and no labelling is observed over the collapsed cells with the anti-propeptide antibody. Living cells contain intact ricinosomes showing a punctate labelling with antibodies against either the propeptide or the mature subunit of CysEP.

5.3 Aleurain containing vacuoles

5.3.1 Aleurain structure and processing

Aleurain is a cysteine protease that was first isolated from barley aleurone cells (Rogers *et al.*, 1985). It is the plant homologue of the mammalian lysosomal protease, cathepsin H. Although aleurain shows proteolytic activity toward azocasein, it is regarded as an aminopeptidase, because it has a higher level of activity with non-blocked substrates (Holwerda and Rogers, 1992). This is due to the presence of a 'mini-chain' of eight amino acid residues, which is covalently bound to the major chain through a disulphide bond. The mini-chain occupies the S_2 subsite and the resulting steric hindrance is responsible for the low endoprotease activity and the slow reaction rate with the inhibitor, E64 (Rothe *et al.*, 1994). Aleurain is a glycoprotein with one complex-type and one high-mannose type oligosaccharide (Holwerda *et al.*, 1990), with peak aminopeptidase activity at pH 6.5–7.0 and 25–35°C but unstable in neutral or higher pH and at $\geq 40^{\circ}$C.

Aleurain undergoes a series of processing steps after the initial cleavage of the signal peptide at the N-terminus, resulting in a 42 kDa glycosylated proaleurain. Pulse-chase experiments show that this is followed by the cleavage of a 9 kDa peptide, to produce a 33 kDa protein, which is slowly 'trimmed' to a stable 32 kDa form (Holwerda *et al.*, 1990). The initial cleavage can occur *in vitro* at pH 5 or 6 and is inhibited by the thiol protease inhibitor, E-64, but is not autocatalytic, whereas trimming, which probably involves the removal of the 'mini-chain' may be autocatalytic (Holwerda and Rogers, 1993). The mini-chain is covalently bound to the main chain via a disulphide bond; removal of the mini-chain should result in an increased endoprotease activity (Rothe *et al.*, 1994). One can speculate that aleurain, by analogy with cathepsin H, may act as aminopeptidase on specific protein substrates and/or may perform highly specific endoproteolytic processing reactions on specific substrates (Holwerda and Rogers, 1993). When aleurain is expressed in *Xenopus* oocytes, it apparently does not come into contact with a processing thiol protease, and is secreted as proaleurain. It has been concluded that aleurain processing occurs in an acidic post-Golgi compartment *in planta*, based on the inhibition of aleurain processing by monensin and brefeldin A and the acidic pH under which processing takes place *in vitro* (Holwerda *et al.*, 1990). There is no evidence for C-terminal processing.

5.3.2 Aleurain homologues

The first aleurain homologue (see Table 5.2) to be found was obtained by screening a cDNA library from germinating rice grains (Watanabe *et al.*, 1991). The rice homologue (D90408, oryzain-γ) is expressed only in the aleurone layer of germinating grains, and not in roots or shoots from grains germinated for up to 8 days. The same technique has been used to isolate a homologue (Z68291) from germinating pea seeds (Jones *et al.*, 1996), and germinating maize grains (D45403), which showed maximal expression after 3 days (Domoto *et al.*, 1995).

Aleurain homologues have been detected as upregulated genes in senescing leaves from several plant species, including tomato (Drake *et al.*, 1996), where

Table 5.2 Plant aleurain homologues

Species	Name	Accession no.	Where expressed
Hordeum vulgare	Aleurain	X05167	Germinating aleurone, leaves
Zea mays	CCP2	D45403	Germinating grains
Zea mays	See1	X99936	Senescing leaves
Oryza sativa	Oryzain-γ	D90408	Germinating aleurone
Petunia hybrida	PeTh3 (P21)	U31094	Low cytokinin callus
Lycopersicon esculentum	SENU3 (cyp-3)	Z48736	Senescing leaves, germinating seeds
Arabidopsis thaliana	Sag2	PQ0650 AB011483	Senescing leaves
Pisum sativum	PCTP	Z68291	Germinating seeds
Prunus armeniaca		U93166	Ripe fruit
Cicer arietinum	cacG2	X93220	Cotyledons
Gossypium hirsutum		AI727443	Developing fibres
Glycine max.		AI437657	Seedling roots

the SENU3 gene (Z48736) is constitutively expressed throughout the plant, but is induced during senescence and seed germination. The maize homologue, See1 (X99936), is also expressed in senescing leaves (but not in detached yellowing leaves). It is different from the grain homologue (CCP2) and is encoded by two genes on chromosomes 2 and 7 (Griffiths *et al.*, 1997). The *Arabidopsis* homologue (Sag2) is a single copy gene, whose expression is increased fourfold during senescence, perhaps due to the decline in photosynthesis (Hensel *et al.*, 1993). In *Petunia* callus, an aleurain homologue (PeTh3) was identified as an early marker of senescence triggered by low concentrations of cytokinin (Tournaire *et al.*, 1996). Other aleurain homologues in the database have been reported as ESTs from chickpea cotyledons (X93220), ripening apricot fruit (U93166), developing cotton fibres (AI727443) and roots of soybean seedlings (AI437657). Aleurain and its homologues form a distinct class within the plant cysteine proteases (Figure 5.3) and a common function might be assumed. It is noteworthy that the aleurain homologues are found mainly in senescing tissue, which raises the possibility that aleurains might be involved in PCD, similar to the ricinosome KDEL-cysteine peptidases.

5.3.3 Aleurain targeting

Aleurain contains two glycosylation sites, both of which are used, but glycosylation is not essential for targeting or processing, since aleurain is not secreted in the presence of tunicamycin (Holwerda *et al.*, 1990). Targeting is mediated by sequences (vacuolar sorting determinants) in the N-terminal propeptide. Aleurain sequences were substituted into the propeptide of EP-B, a secreted barley thiol protease, with a similar C-terminal sequence to aleurain. The sequence, SNPIRPVTDRAASTL, targeted EP-B with 90% efficiency, whereas SSSSFADSNPIRP was 62% efficient and SNPIRP was only 3.3% efficient (Holwerda *et al.*, 1992). Nevertheless, it is the NPIR motif, occurring at position 32–35, which is conserved in aleurain homologues, except for cotton, where it is NPIQ (Figure 5.5). The same sequence occurs at position 26–29 in sporamin precursors, the major soluble protein of sweet potato tuberous roots, and the barley aspartic protease, HvAP, contains the sequence NPLR, which is an efficient targeting determinant (K. Matsuoka, cited in Paris *et al.*, 1996).

In contrast to the aleurains, where there is a high degree of sequence conservation around the NPIR motif in the prosequence, the aspartic proteases show neither conservation of the NPLR motif, nor any consensus in the residues surrounding it (Figure 5.5). A possible explanation is that there are many different aspartic proteases, and those from the other species are targeted to other parts of the cell. A database search reveals this motif at the N-terminus of two protein kinase-like sequences (AL049660) and a hypothetical protein sequence (AL050300) in *Arabidopsis*, but both lack putative signal peptides. On the other hand, *Arabidopsis* cysteine endoprotease (Z97343), which is not an aleurain homologue, contains the NPIR motif and does have a signal peptide. The cathepsin D inhibitors, which are homologous to the protease inhibitor, sporamin (Figure 5.5), have a conserved NPIN motif, which is contained within the prosequence, but this does not appear to be used by other proteins in the database. The Kunitz trypsin/α-amylase inhibitors contain members with the NPIR and NPLR motifs, with some degree of conservation, but this is contained within the mature protein sequence, after cleavage of the signal peptide. It is not known if these motifs function as vacuolar targeting sequences, but the cathepsin D inhibitor sequence, NPIE (S70766), is found in the mature coding sequence of the 19 kDa zein protein, which is located in protein bodies in the ER lumen of maize endosperm (Lending and Larkins, 1989).

The aleurain-based vacuolar targeting sequence, SSSSFADSNPIRPVTDRA-ASTYC, was used to make a peptide affinity column, which bound an 80 kDa protein from lysates of pea clathrin-coated vesicles (Kirsch *et al.*, 1994). This protein, BP-80 (U79958), has the properties of a potential vacuolar targeting receptor, and homologues have been found in maize and *A. thaliana* (see also chapter 7 in this volume). It remains to be discovered whether this putative receptor also binds to other potential vacuolar sorting determinants, such as those found in aleurain homologues, aspartic proteases, cathepsin D inhibitors and trypsin/α-amylase inhibitors (Figure 5.5). The high conservation of the targeting signal and the sequence conservation around the NPIR motif of aleurains compared to other proteins, may indicate that aleurains are targeted to a special type of vacuole, which may contain only a very limited number of other proteins.

5.3.4 Aleurain localization

Aleurain has been localized to cell compartments described as aleurain-containing vacuoles (ACVs), based on immunolabelling at both the light-microscopic and ultrastructural levels. The antibody, which was specific in immunoprecipitation and western blots, recognized numerous small, round organelles in aleurone cells, which under the electron microscope were revealed as elliptical to round bodies, about 1–2 μm in diameter, and enclosed by a single membrane (Holwerda *et al.*, 1990). Similar organelles were demonstrated by confocal laser-scanning microscopy in barley root tip cells (Paris *et al.*, 1996; Rogers *et al.*, 1997) and tobacco suspension culture protoplasts (Jiang and Rogers, 1998). These studies have been pivotal in demonstrating the presence of more than one type of vacuole in plant cells (Rogers, 1998). Localization of aleurain in barley aleurone cells also supports the idea that this enzyme is not present in protein storage vacuoles but is located in a separate organelle. This observation is supported by biochemical evidence showing that, although the barley aspartic protease, HvAP, and cysteine proteases are present in barley aleurone PSVs (Swanson *et al.*, 1998), aleurain is absent from this organelle (Bethke *et al.*, 1996). Proteins, such as barley lectin, chitinase, a glucanase, PR-4, thaumatin and an osmotin, are targeted by a C-terminal sequence to PSVs in a wortmannin-sensitive manner (Neuhaus and Rogers, 1998). Since PSVs contain their own cysteine proteases and aspartic proteases, there is no need to involve the aleurain in protein mobilization (Bethke *et al.*, 1996, 1998).

The aleurain-containing vacuoles also label for the barley aspartic protease (HvAP) (Paris *et al.*, 1996), and have been isolated from tobacco protoplasts, where they show a modest enrichment (2.8 times) in aleurain and α-mannosidase compared with the unfractionated whole cell lysate (Holwerda *et al.*, 1992). The presence of other proteins was not determined.

The availability of vacuolar membrane markers, the tonoplast integral proteins (TIPs), allows vacuoles to be identified and different types of vacuole to be distinguished. Although TIPs may function as aquaporins, they are so abundant that they are thought to be structural proteins, and specific isoforms are required for the biogenesis of a specific type of vacuole (Neuhaus and Rogers, 1998; see also Chapter 2 in this volume). At least three TIPs are known: α-TIP defines a protein storage vacuole, γ-TIP defines a lytic vacuole and δ-TIP defines a vegetative storage vacuole (Jauh *et al.*, 1998). Aleurain is located in organelles which label for γ-TIP alone, for δ-TIP alone and for both δ- and γ-TIP (Paris *et al.*, 1996; Jauh *et al.*, 1999).

5.3.5 *Vacuole fusion*

Some plant vacuoles contain two different TIP isoforms, and it has been suggested that they may arise by fusion of two different vacuoles. For example, gibberellic acid-induced fusion of lytic vacuoles with PSVs in aleurone cells may be a means to mobilize the contents of the PSVs, giving rise to vacuoles containing both α- and γ-TIP (Paris *et al.*, 1996). The cells of soybean leaves contain large central vacuoles, which label for γ-TIP, but when they begin to accumulate vegetative storage proteins after de-podding, these vacuoles label

A) Aleurain homologues

```
                                1                                                                                70
Hordeum     (X05167)    MAHARVLLLA LAVLATAAVA V..ASSSSFA D.SNPIRPV. .TDRAASTLE SAVLGALGRT RHALRFARFA
Zea mays    (D45403)    MVPRR..LFV LAVVVLADTA A..VVNSGFA D.SNPIRPV. .TDRAASALE STVFAALGRT RDALRFARFA
Zea mays    (X99936)    MAPRR..LLV LAVVALAATA A..AANSGFA D.SNPIRPV. .TDRAASALE STVFAALGRT RDALRFARFA
Oryza       (D90408)    MAHRRIILL. LAVAAVAATS AVAAASSGFD D.SNPIRSV. .TDHAASALE STVIAALGRT RGALRFARFA
Petunia     (U31094)    ~~~MSRLSLL LVLVAGLFAV AFARTA.NFA D.ENPIRQVV .SD.SFHELE SGILHVVGQT RHALSFARFA
Lycopersicon (Z48736)~~MSRLSLV LILVAGLFAT ALAGPA.TFA D.KNPIRQVV FPD....ELE NGILQVVGQT RSALSFARFA
Arabidopsis (AB011483)~MSAKTIL SSVVLVVLVA ASAAANIGFD E.SNPIRMV. .SD.GLREVE ESVSQILGQS RHVLSFARFT
Pisum       (Z68291)    ~~~~~~~MAQ WSLLIVLFCV ASAAAGFSFH D.SNPIRMV. .SD.....VE EQLLQVIGES RHAVSFARFA
Prunus      (U93166)    ~~~MARVTLV LSAALVLVAI SCGAAASSFD E.SNPIRLV. .SD.GLRELE QQVVQVLGNS RRALHFARFA
Cicer       (X93220)    ~~~~~~~MAQ WSLLIVLFCV ATAAAGLSFE DNTNPIRMV. .SD.....ME EQLLQVIGES HHALSFVRFA
Gossypium (AI727443)~~~~~~MAGL TLMSSVIFLL CCLASASTFE D.TNPIQMV. .SD....GLD SSVLRVIGHT RRAISFARFA
```

B) Sporamin and cathepsin D inhibitors

```
                                               ↓                       ↓
            1                                                                              60
A30835      ~~~MKA..LALF FLLSLYLLPN PAHSKFNPIR LRPA.HETASS ETPVLDINGD EVRAGENYYI
X13510      ~~~MKA..LALL FVLSLYLLPN PAHSKFNPIR LRPA.HETASS ETPVLDINGD EVRAGENYYI
X15094      ~~~MKA..FALF FVLSLYLLPN PTHSKFNPIR LRPA.HETASS ETPVLDINGD EVRAGENYYM
X15091      ~~~MKALTLALF LALSLYLLPN PAHSRFNPIR L.PTTHEPASS ETPVLDINGD EVRAGGNYYM
U12436      ~~~MKAFTLALF LALFLYLLPN PAHSRFNPIR L.PTTHEPASS ETPVLDINGD EVRAGGNYYM
X13509      ~~~MKALTLALF LALSLYLLPN PAHSRFNPIR L.PTTHEPASS ETPVLDINGD EVRAGGNYYM
X64370      ~~~MKCLFL.LC LCLVPIVVFS STFTSKNPIN L.PSD...A.. .TPVLDVAGK ELDSRLSYRI
U30814      ~~~MKCLFL.LC LCLFPIVVFS STFTSQNPIN L.PSD...A.. .TPVLDVTGK ELDPRLSYHI
D17332      ~~~MKCLFL.LC LCLVPIVVFS STFTSQNPIN L.PSD...A.. .TPVLDVTGK ELDPRLSYRI
D17328      ~~MMKCLFF.LC LCLFPILVFS STFTSQNPIN L.PSE...SPV PKPVLDTDGK KLNPNSSYRI
D17331      ~~MMKCLFF.LC LCLFPILVFS STFTSQNPIN L.PSE...SPV PKPVLDTNGK KLNPNSSYRI
M96257      ~~MMKCLFL.LC LCLFPILVFS STFTSQNPIN L.PSE...SPV PKRVLDTNGK KLNPNSSYRI
X73986      ~~MMKCLFL.LC LCLFPIVVFS SSFTSQNPIE L.PSA...SPK PNPVLDTNGN ELNPNSSYRI
X56874      MKSINILSFLLL SSTLSLVAFA RSFSSENPIV L.PST..CHDD DNLVLPEVYD QDGHPLR GQ
                                      ↑                     ↑
```

C) Kunitz trypsin, α-amylase inhibitors

```
                1                                                              100
S45092      ~~MKSTIFFL FLFCAFTTSY LPSAIADFVL DNEGNPLEN GGTYYILSDI TA.FGG
X64448      ~MKSTIFFAL FLFCAFTTSY LPSAIADFVL DNEGNPLDS GGTYYILSDI TA.FGG
TISYB  ~                               DFVL DNEGNPLSN GGTYYILSDI TA.FGG
S45035      ~MKSTIFFAL FLVCAFTISY LPSATAQFVL DTDDDPLQN GGTYYMLPVM RGKGGG
X80039      ~MKSTTSLAL FLLCALTSSY QPSATADIVF DTEGNPIRN GGTYYVLPVI RGKGGG
X65637      ~MKT...FFL VTL..LLAAA VCTHGREQVK DSNGNPVKR GAKYFIQPAK ..SNAG
Z25770      ~MMSSFPLVS FLITLMLASR VCTHGENRVD DTDGNPLRT TAQYLILPL. SPRSNG
Z29648      ~~~~~~~~~~ ~~~~~~~~~~ ~~~~~LEEVK DSNGNPVRV GAQYFIQPVK TESNNG
X78586      ~MKATISITT IFLVVALAAP SLARPDNHVE DSVGRLLRP GQTYHIVPAN PE..TG
U20592      PLALSSTFSS DLLLPSDEVV PNGKTYASVV DSDGNPVKA GAKYFVLPSL RGSGGG
Z12961      MGSRRAGLLL LSLILASTAL SRSADPPPVH DTDGHELRA DANYYVLSAN RAHGGG
S38955      ~~~~~~~~~~ ~~~~~~~~~~ ~~~~DPPPVH DTDGNELRA DANYYVLPAN RAHGGG
```

D) Aspartic proteases

```
                            36                                                  ↓                                 100
Cucurbita    (AB002695) LKKIKLDPEN ..RLAARVES KDAEILKAAF RKYNPKGNL GESSDTD..IV ALKNYLDAQY YGEIA
Vigna          (U61396) LKKIKLDPNN ..RLAARIGS ND.DSFRASI RKFHLQNNF AGTGETD..IV ALKNYLDAQY YGEIS
Brassica       (U55032) LKKLKFDPRS RFTIAAPVGS KQL....... ...KPLRGY GLGDSGDADIV TLKNYLDAQY YGEIA
Cynara         (X81984) LKKRKVDQIN ..QLSGHGVS MEAKARKDFG FG.GALR.. ..DSGSD..II ALKNYMDAQY YGEIG
Hordeum        (X56136) LKKRPIDRNS ..RVATGLSG GEEQPLLSGA ...NPLR.. .SEEEGD..IV ALKNYMNAQY FGEIG
Lycopersicon   (L46681) LKKHRLDVDS ..IKAARVA. KLQDRYGKHV ...NGIEKK ..SSDSDIYKV PLKNYLDAQY YGEIG
Oryza          (D12777) LNKKRLDKED ..LTAAKLA. QQG....... ...NRLL.K TGSSDSD..PV PLVDYLNTQY YGVIG
Oryza          (D32165) LKKRPIDENS ..RVAARLSG EEGARRLG L RGANSLG.. GGGGEGD..IV ALKNYMNAQY FGEIG
```

Figure 5.5 Alignment of plant polypeptides containing a signal peptide and a putative vacuolar targeting sequence at the N-terminus. A) In aleurain (X05167) and homologues, the NPIR motif is conserved in all sequences, except cotton, which has NPIQ. B) Alignment of sporamin from sweet potato (*Ipomoea batatas*) (A30835, X13510, X15094, X15091, U12436, X13509) and cathepsin D inhibitors from potato (*Solanum tuberosum*) (X64370, U30814, D17332, D17328, D17331, M96257, X56874) and tomato (*Lycopersicon esculentum*) (X73986). Predicted signal peptide cleavage site and known propeptide cleavage site are indicated by arrows. The NPIR motif is conserved in the sporamin sequences but not in two other sporamins (X15516 and X60930). The NPIN sequence in the homologous cathepsin D inhibitors might be involved in vacuolar targeting, but this motif is missing from most of the members of this family. C) Alignment of Kunitz trypsin inhibitors from soybean (S45092, X64448, TISYB, S45035, X80039), *Brassica napus* (X65637, Z25770), *Raphanus sativus* (Z29648), *Arabidopsis thaliana* (X78586) and *Lycopersicon esculentum* (U20592) and α-amylase/subtilisin inhibitors from *Hordeum vulgare* (Z12961) and *Triticum aestivum* (S38955). The N-terminus of the mature protein is underlined, where known, so that the putative vacuolar targeting motifs (NPIR, NPLR) are part of the mature protein. D) Alignment of plant aspartic proteases reveals that the NPIR motif in barley is poorly conserved in other species, although the mature protein sequences are highly conserved.

for δ-TIP (Jauh *et al.*, 1998). In this case, a lytic vacuole has been converted to a vegetative protein storage vacuole. Neuhaus and Rogers (1998) have report unpublished data concerning vacuoles which label for both α- and δ-TIP.

5.3.6 Possible function for aleurain-containing vesicles

The protein and lipid reserves of the aleurone are used in the synthesis of the hydrolases that are secreted into the starchy endosperm. Some cysteine endoproteases are not secreted, and have been localized to the PSVs of the aleurone using a fluorogenic protease substrate that is transported into the vacuole and then cleaved (Swanson *et al.*, 1998). Aleurone cells undergo extensive vacuolation and die a few days after hydrolase secretion is complete. Aleurone cells die by PCD, which is promoted by gibberellin and inhibited by abscisic acid (Bethke *et al.*, 1999).

The synthesis of aleurain is induced by gibberellin and inhibited by abscisic acid in aleurone cells of germinating barley grains, although it is constitutively expressed in roots and leaves (Rogers *et al.*, 1985). Aleurain-containing vacuoles have been isolated from tobacco protoplasts (Holwerda *et al.*, 1992) but not purified. Aleurain-containing vacuoles may be lytic vacuoles, the plant equivalent of lysosomes, but the hydrolysis of substrates within the aleurain-containing vacuole needs to be demonstrated. Perhaps aleurain-containing vesicles are protease-storing vacuoles similar to ricinosomes, but act by fusion with protein storage vacuoles during senescence and programmed cell death. Does aleurain lose its mini-chain and gain endoproteolytic activity during this process? The role of aleurain-containing vacuoles in other plant tissues which are not undergoing PCD has not been elucidated. Isolation of the purified organelle will enable its composition to be determined and its possible function to be deduced.

References

Bethke, P.C., Hillmer, S. and Jones, R.L. (1996) Isolation of intact storage vacuoles from barley aleurone: identification of aspartic and cysteine proteases. *Plant Physiol.*, **110** 521-29.

Bethke, P.C., Swanson, S.S., Hillmer, S. and Jones, R.L. (1998) From storage compartment to lytic organelle: the metamorphosis of the aleurone protein storage vacuole. *Ann. Bot.*, **82** 399-412.

Bethke, P.C., Lonsdale, J.E., Fath, A. and Jones, R.L. (1999) Hormonally-regulated programmed cell death in barley aleurone cells. *Plant Cell*, **11** 1033-45.

Breidenbach, R.W. and Beevers, H. (1967) Association of the glyoxylate cycle enzymes in a novel subcellular particle from castor bean endosperm. *Biochem. Biophys. Res. Commun.*, **27** 462-69.

Denecke, J., De Rycke, R. and Botterman, J. (1992) Plant and mammalian sorting signals for protein retention in the endoplasmic reticulum contain a conserved epitope. *EMBO J.*, **11** 2345-55.

Devereux, J., Haeberli, P. and Smithies, O. (1984) A comprehensive set of sequence-analysis programs for the VAX. *Nucleic Acids Res.*, **12** 387-95.

Domoto, C., Watanabe, H., Abe, M., Abe, K. and Arai, S. (1995) Isolation and characterization of two distinct cDNA clones encoding corn seed cysteine proteinases. *Biochim. Biophys. Acta.*, **1263** 241-44.

Drake, R., John, I., Farrell, A., Cooper, W., Schuch, W. and Grierson, D. (1996) Isolation and analysis of cDNAs encoding tomato cysteine proteases expressed during leaf senescence. *Plant Mol. Biol.*, **30** 755-67.

Gietl, C. (1996) Protein targeting and import into plant peroxisomes. *Physiol. Plant*, **97** 599-608.

Gietl, C., Wimmer, B., Adamec, J. and Kalousek, F. (1997) A cysteine endopeptidase from castor bean endosperm microbodies processes the glyoxysomal malate dehydrogenase precursor protein. *Plant Physiol.*, **113** 863-71.

Gomord, V., Denmat, L.A., Fitchette-Laine, A.C., Satiat-Jeunemaitre, B., Hawes, C. and Faye, L. (1997) The C-terminal HDEL sequence is sufficient for retention of secretory proteins in the endoplasmic reticulum (ER) but promotes vacuolar targeting of proteins that escape the ER. *Plant J.*, **11** 313-25.

Griffiths, C.M., Hosken, S.E., Oliver, D., Chojecki, J. and Thomas, H. (1997) Sequencing, expression pattern and RFLP mapping of a senescence-enhanced cDNA from *Zea mays* with high homology to oryzain-gamma and aleurain. *Plant Mol. Biol.*, **34** 815-21.

Hara-Nishimura, I., Shimada, T., Hatano, K., Takeuchi, Y. and Nishimura, M. (1998) Transport of storage proteins to protein-storage vacuoles is mediated by large precursor-accumulating vesicles. *Plant Cell*, **10** 825-36.

Henderson, J., Bauly, J.M., Ashford, D.A., Oliver, S.C., Hawes, C.R., Lazarus, C.M., Venis, M.A. and Napier, R.M. (1997) Retention of maize auxin-binding protein in the endoplasmic reticulum: quantifying escape and the role of auxin. *Planta*, **202** 313-23.

Hensel, L.L., Grbic, V., Baumgarten, D.A. and Bleecker, A.B. (1993) Developmental and age-related processes that influence the longevity and senescence of photosynthetic tissues in *Arabidopsis. Plant Cell*, **5** 553-64.

Herman, E.M., Li, X., Su, R.T., Larsen, P., Hsu, H.-T. and Sze, H. (1994) Vacuolar-type H^+-ATPases are associated with the endoplasmic reticulum and provacuoles of root tip cells. *Plant Physiol.*, **106** 1313-24.

Holwerda, B.C. and Rogers, J.C. (1992) Purification and characterization of aleurain: a plant thiol protease functionally homologous to mammalian cathepsin H. *Plant Physiol.*, **99** 848-55.

Holwerda, B.C. and Rogers, J.C. (1993) Structure, functional properties and vacuolar targeting of the barley thiol protease, aleurain. *J. Exp. Bot.*, **44** 321-29.

Holwerda, B.C., Galvin, N.J., Baranski, T.J. and Rogers, J.C. (1990) *In vitro* processing of aleurain, a barley vacuolar thiol protease. *Plant Cell*, **2** 1091-106.

Holwerda, B.C., Padgett, H.S. and Rogers, J.C. (1992) Proaleurain vacuolar targeting is mediated by short contiguous peptide interactions. *Plant Cell*, **4** 307-18.

Jauh, G.Y., Fischer, A.M., Grimes, H.D., Ryan, C.A. and Rogers, J.C. (1998) δ-Tonoplast intrinsic protein defines unique plant vacuole functions. *Proc. Natl. Acad. Sci. USA*, **95** 12995-99.

Jauh, G.Y., Phillips, T.E. and Rogers, J.C. (1999) Tonoplast intrinsic protein isoforms as markers for vacuolar functions. *Plant Cell*, **11** 1867-82.

Jiang, L. and Rogers, J.C. (1998) Integral membrane protein sorting to vacuoles in plant cells: evidence for two pathways. *J. Cell Biol.*, **143** 1183-99.

Jones, C.G., Tucker, G.A. and Lycett, G.W. (1996) Pattern of expression and characteristics of a cysteine proteinase cDNA from germinating seeds of pea (*Pisum sativum* L.). *Biochim. Biophys. Acta.*, **1296** 13-15.

Kaletta, K., Kunze, I., Kunze, G. and Kock, M. (1998) The peptide, HDEF, as a new retention signal is necessary and sufficient to direct proteins to the endoplasmic reticulum. *FEBS Lett.*, **434** 377-81.

Kirsch, T., Paris, N., Butler, J.M., Beevers, L. and Rogers, J.C. (1994) Purification and initial characterization of a potential plant vacuolar targeting receptor. *Proc. Natl. Acad. Sci. USA*, **91** 3403-407.

Koehler, S.M. and Ho, T.H. (1990) Hormonal regulation, processing, and secretion of cysteine proteinases in barley aleurone layers. *Plant Cell*, **2** 769-83.

Lending, C.R. and Larkins, B.A. (1989) Changes in the zein composition of protein bodies during maize endosperm development. *Plant Cell*, **1** 1011-23.

Levanony, H., Rubin, R., Altschuler, Y. and Galili, G. (1992) Evidence for a novel route of wheat storage proteins to vacuoles. *J. Cell Biol.*, **119** 1117-28.

Li, X., Franceschi, V.R. and Okita, T.W. (1993) Segregation of storage protein mRNAs on the rough endoplasmic reticulum membranes of rice endosperm cells. *Cell*, **72** 869-79.

Mollenhauer, H.H. and Totten, C. (1970) Studies on seeds. V. Microbodies, glyoxysomes and ricinosomes of castor bean endosperm. *Plant Physiol.*, **46** 794-99.

Napier, R.M., Fowke, L.C., Hawes, C., Lewis, M. and Pelham, H.R. (1992) Immunological evidence that plants use both HDEL and KDEL for targeting proteins to the endoplasmic reticulum. *J. Cell Sci.*, **102** 261-71.

Neuhaus, J.M. and Rogers, J.C. (1998) Sorting of proteins to vacuoles in plant cells. *Plant Mol. Biol.*, **38** 127-44.

Okomoto, T., Nakayama, H., Seta, K., Isobe, T. and Minamikawa, T. (1994) Post-translational processing of a carboxyterminal propeptide containing a KDEL sequence of plant vacuolar cysteine endopeptidase (SH-EP). *FEBS Lett.*, **35** 31-34.

Okomoto, T., Minamikawa, T., Edward, G., Vakharia, V. and Herman, E. (1999) Post-translational removal of the carboxyterminal KDEL of the cysteine protease SH-EP occurs prior to maturation of the enzyme. *J. Biol. Chem.*, **274** 11390-98.

Paris, N., Stanley, C.M., Jones, R.L. and Rogers, J.C. (1996) Plant cells contain two functionally distinct vacuolar compartments. *Cell*, **85** 563-72.

Paul, W., Amiss, J., Try, R., Praekelt, U., Scott, R. and Smith, H. (1995) Correct processing of the kiwi fruit protease, actinidin, in transgenic tobacco requires the presence of the C-terminal propeptide. *Plant Physiol.*, **108** 261-68.

Pennell, R.I. and Lamb, C. (1997) Programmed cell death in plants. *Plant Cell*, **9** 1157-68.

Rechinger, K.B., Simpson, D.J., Svendsen, I. and Cameron-Mills, V. (1993) A role for γ3-hordein in the transport and targeting of prolamin polypeptides to the vacuole of developing barley endosperm. *Plant J.*, **4** 841-53.

Rogers, J.C. (1998) Compartmentation of plant cell proteins in separate lytic and protein storage vacuoles. *J. Plant Physiol.*, **152** 653-58.

Rogers, J.C., Dean, D. and Heck, G.R. (1985) Aleurain a barley thiol protease closely related to mammalian cathepsin H. *Proc. Natl. Acad. Sci. USA*, **82** 6512-16.

Rogers, S.W., Burks, M. and Rogers, J.C. (1997) Monoclonal antibodies to barley aleurain and homologs from other plants. *Plant J.*, **11** 1359-68.

Rothe, M., Zichner, A., Auerswald, E.A. and Dodt, J. (1994) Structure/function implications for the aminopeptidase specificity of aleurain. *Eur. J. Biochem.*, **224** 559-65.

Schmid, M., Simpson, D., Kalousek, F. and Gietl, C. (1998) A cysteine endopeptidase with a C-terminal KDEL motif isolated from castor bean endosperm is a marker enzyme for the ricinosome, a putative lytic compartment. *Planta*, **206** 466-75.

Schmid, M., Simpson, D. and Gietl, C. (1999) Programmed cell death in castor bean endosperm is associated with the accumulation and release of a cysteine endopeptidase from ricinosomes. *Proc. Nat. Acad. Sci. USA*, **96** 14159-64.

Staehelin, L.A. (1997) The plant ER: a dynamic organelle composed of a large number of discrete functional domains. *Plant J.*, **11** 1151-65.

Swanson, S.J., Bethke, P.C. and Jones, R.L. (1998) Barley aleurone cells contain two types of vacuoles: characterization of lytic organelles by use of fluorescent probes. *Plant Cell*, **10** 685-98.

Tournaire, C., Kushnir, S., Bauw, G., Inze, D., Teyssendier de la Serve, B. and Renaudin, J.P. (1996) A thiol protease and an anionic peroxidase are induced by lowering cytokinins during callus growth in Petunia. *Plant Physiol.*, **111** 159-68.

Trelease, R.N., Becker, W.M., Gruber, P.J. and Newcomb, E.H. (1971) Microbodies (glyoxysomes and peroxisomes) in cucumber cotyledons: correlative biochemical and ultrastructural study in light and dark grown seedlings. *Plant Physiol.*, **48** 461-75.

Vigil, E.L. (1970) Cytochemical and developmental changes in microbodies (glyoxysomes) and related organelles of castor bean endosperm. *J. Cell Biol.*, **46** 435-54.

Vitale, A. and Denecke, J. (1999) The endoplasmic reticulum—gateway of the secretory pathway. *Plant Cell*, **11** 615-28.

Wandelt, C.I., Khan, M.R., Craig, S., Schroeder, H.E., Spencer, D. and Higgins, T.J. (1992) Vicilin with carboxyterminal KDEL is retained in the endoplasmic reticulum and accumulates to high levels in the leaves of transgenic plants. *Plant J.*, **2** 181-92.

Watanabe, H., Abe, K., Emori, Y., Hosoyama, H. and Arai, S. (1991) Molecular cloning and gibberellin-induced expression of multiple cysteine proteinases of rice seeds (oryzains). *J. Biol. Chem.*, **266** 16897-902.

6 Vacuoles and plant defense

Alejandro Iglesias and Fred Meins Jr

6.1 Introduction

Plants have evolved effective defenses against pests and pathogens. Some of these defenses are constitutive, including physical barriers and the production of toxic or antimicrobial compounds, such as the phytoanticipins (Osbourn, 1999). Others are induced locally in response to infestation or infection. These include the production of phytoalexins (Dixon *et al.*, 1994; Clerivet *et al.*, 1996), the induction of pathogenesis-related (PR) proteins (Linthorst, 1991), the strengthening of cell walls (Brisson *et al.*, 1994), and the hypersensitive response (HR) (Dangl *et al.*, 1996). In some cases, the HR following infection of a resistant host can result in systemic acquired resistance (SAR) against subsequent pathogen attacks, when salicylate plays a key role (Ryals *et al.*, 1996). The defense system of plants also includes a variety of relatively small, basic, cysteine-rich polypeptides with antimicrobial activities (Garcia-Olmedo *et al.*, 1998). Among these are lipid transfer proteins (Garcia-Olmedo *et al.*, 1995), plant defensins (Terras *et al.*, 1995) and thionins (Bohlmann *et al.*, 1994).

Toxic secondary products and proteins implicated in defense responses often accumulate or are stored in vacuoles, raising the possibility that vacuoles play a role in plant defense. This review deals with the structure, mode of action, and vacuolar transport of these substances. Phytoanticipins and phytoalexins are touched on briefly and have been reviewed from a different perspective by others (Bailey and Mansfield, 1982; Kuc, 1982, 1995; Dixon, 1986; Barz *et al.*, 1990; Dixon and Harrison, 1990; Ye *et al.*, 1992; Sharma and Salunkhe, 1999). This chapter focuses on well-characterized PR proteins.

6.2 Phytoanticipins and phytoalexins

Many plants produce antimicrobial secondary metabolites. Phytoanticipins are found in healthy plants, and are believed to help defend against potential pathogens, while the phytoalexins are synthesized as part of the plant's defense response to pathogen attack.

Phytoalexins are low molecular weight, lipophilic compounds that accumulate rapidly around sites of incompatible pathogen infections and in response to an extensive array of biotic and abiotic elicitors (Smith, 1996). More than

350 phytoalexins have been chemically characterized from approximately 30 plant families, including dicotyledons and monocotyledons, with the greatest number (130) found in Leguminosae (Kuc, 1995). They include a variety of structural types, e.g. phenylpropanoids, isoflavonoids and sesquiterpenoids. Little is known about the relationship between their structure and biological activity (Maestro *et al.*, 1993).

Phytoalexins are derived from a number of major biosynthetic pathways. Diversion of these pathways toward phytoalexin production is regulated by key, branch-point enzymes, such as phenylalanine ammonia lyase (PAL), as well as by regulation of downstream metabolic transformations, which appear to be targets of highly-coordinated signaling pathways (Graham and Graham, 1991; Dixon and Paiva, 1995; Ranansky *et al.*, 1996). Common sequence elements have been identified in the promoters of several genes encoding enzymes in the biosynthesis of flavonoid phytoalexins (Hahlbrock and Schell, 1989; Hahlbrock *et al.*, 1995). Interestingly, in beneficial mycorrhizal interactions, induction of genes early in this pathway is uncoupled from that of later, phytoalexin-specific genes (Dixon *et al.*, 1995). A number of molecular species can function as elicitors of phytoalexin synthesis, including poly- and oligosaccharides, proteins and polypeptides, and fatty acids. Several receptors have been identified localized on the plasma membrane (Dixon and Paiva, 1995). Transgenic and mutant plants, with specific alterations in one or more of the elements necessary for the plant to respond to the signals for phytoalexin synthesis, are beginning to aid resolution of the complex pattern of signaling events and the respective roles of the inducible defense mechanisms in resistance (Smith, 1996).

Compartmentation and conjugation also appear to be important in regulating phytoalexin accumulation (Kuc, 1995). Chickpea cell suspension cultures treated with a yeast glucan elicitor accumulate the phytoalexins, medicarpin and maackiain, and their malonate conjugates (Mackenbrock *et al.*, 1993). Both conjugates are localized in the vacuoles (Weidemann *et al.*, 1991); there appears to be specific release of these conjugates. Whereas isoflavone and isoflavone conjugates of medicarpin and maackiain remain in the vacuoles, the glucoside malonate conjugates are released (Barz and Welle, 1992).

Although phytoalexins have antimicrobial activity *in vitro*, their contribution to disease resistance in the incompatible response is less clear (Mauch-Mani and Slusarenko, 1996). It is presumed that they halt, or at least reduce, the growth of pathogens and reduce the severity of secondary infections. This view is supported by the observation that successful pathogens can avoid eliciting phytoalexin synthesis, suppress phytoalexin synthesis, or detoxify phytoalexins (Mackintosh *et al.*, 1989). Increased resistance to pests, for example, is correlated with increased phytoalexin accumulation. Alfalfa varieties selected for increased antibiosis-based resistance to the root-lesion nematode, *Pratylenchus penetrans*, exhibit higher levels of the phytoalexin, medicarpin, in their roots as well as higher expression of a key enzyme in its biosynthesis (Baldridge *et al.*, 1998). More direct evidence is provided by experiments with tobacco plants transformed with a constitutive expressed chimeric gene encoding stilbene synthetase, a key enzyme in the production of the phytoalexin, resveratol (Hain *et al.*, 1993). These plants overproduce resveratol and exhibit enhanced resistance to the necrotrophic fungus, *Botrytis cinerea*.

The biosynthesis of many phytoanticipins is well understood and several types, for example, cinnamic acids, flavonoids, isoflavonoids, deoxianthocyanins, stilbenes, coumarins and chromones, have been shown to have antifungal and antibacterial activity *in vitro* (Maestro *et al.*, 1993). Nevertheless, as is the case for phytoalexins, their role in disease resistance is still unclear (Clerivet *et al.*, 1996). For example, in the mutualistic association of legumes with nitrogen-fixing bacteria, isoflavonoids, believed to be phytoalexins and phytoanticipins, can serve as chemoattractants, promote microbial growth, and induce expression of nodulation genes in *Rhizobium* and *Bradyrhizobium* bacteria (Dakora and Phillips, 1996).

6.3 Pathogenesis-related (PR) proteins

PR proteins were first described as a group of acidic proteins induced in tissues as part of the hypersensitive response of tobacco to tobacco mosaic virus (TMV) (Gianinazzi *et al.*, 1970; van Loon and van Kammen, 1970). Later, it was found that similar proteins were induced more generally by a variety of viral, bacterial and fungal pathogens and potential pathogens, following attack by nematodes, phytophagous insects, and herbivores and in response to certain stresses (van Loon, 1999). So far, 14 families of PR proteins (PR-1 to PR-14), defined as 'proteins coded by the host plant but induced only in pathological or related situations' (Antoniw *et al.*, 1980), are recognized.

Some PR proteins are also systemically induced by infection as part of SAR (Ryals *et al.*, 1996). Similarly, the PR-6 proteins, which are inhibitors of serine proteases, are systemically induced by wounding and accompany infections resulting in tissue damage in tomato and potato (Green and Ryan, 1972). A subclass of PR proteins is also induced by chemicals that mimic the effect of pathogen infection or induce some aspects of the host response, as has been well documented for salicylic acid and its derivatives, which are involved in the signal-transduction pathway leading to SAR in tobacco (Gaffney *et al.*, 1993).

It is intriguing that most, and possibly all, PR proteins exist as vacuolar and extracellular isoforms. The vacuolar isoforms are often basic proteins, whereas the extracellular isoforms are often acidic proteins. The present discussion focuses on the vacuolar isoforms of well-characterized PR-proteins—in most cases of tobacco origin—which can have pronounced antifungal or antibacterial activity. A comprehensive overview of PR proteins was recently published (Datta and Muthukrishnan, 1999).

6.3.1 PR-1

Genes encoding PR-1 proteins are found in many plant species (White *et al.*, 1987; Jung *et al.*, 1993). Homologous sequences have also been found in fungi, insects, mammals and yeast (Fang *et al.*, 1988; Schuren *et al.*, 1993; Miosga *et al.*, 1995; Murphy *et al.*, 1995), suggesting that the PR-1 gene family arose rather early in eukaryote evolution.

PR-1 of tobacco is one of the most abundant PR proteins, and can accumulate to concentrations of 1–2% of the total protein in infected leaves (Alexander *et al.*, 1993). Studies of the isolated proteins and cDNA clones indicate there are several expressed acidic and basic isoforms (Cornelissen *et al.*, 1987; Payne *et al.*, 1989; Eyal *et al.*, 1992; Niderman *et al.*, 1995). Acidic and basic isoforms have putative N-terminal signal sequences and are approximately 61% similar in amino acid sequence. The basic isoforms carry, in addition, a C-terminal extension believed to be important for targeting.

The basic isoform, PRB-1b, accumulates in response to TMV infection, treatment with ethylene, salicylic acid and α-aminobutyric acid, and when plants are kept in darkness (Eyal *et al.*, 1992). Subcellular localization studies show that PRB-1b accumulates predominantly in vacuoles isolated from tobacco protoplasts and is not secreted into the medium of suspension cultures (Sessa *et al.*, 1995). PRB-1b also accumulates in the vacuoles of ethylene-treated tobacco leaves (Sessa and Fluhr, 1995); whereas, another basic isoform, PR-1g, accumulates in vacuoles and in extracellular space (Niderman *et al.*, 1995). Although the tobacco acidic isoforms, PR-1a, -1b and -1c, are usually secreted into the extracellular spaces of leaves in response to pathogen infection or chemical treatment, they are also localized in the vacuoles of specialized leaf cells known as 'crystal idioblasts' (Dixon *et al.*, 1991).

Because PR-1 proteins were first discovered in TMV infected plants, it was widely assumed that the protein functions in virus resistance. Evidence supporting this view is weak. Although certain interspecific *Nicotiana* hybrids showing constitutive expression of PR-1 as well as PR-2, PR-3 and PR-9 are resistant to TMV infection (Ahl and Gianinazzi, 1982; Ahl-Goy *et al.*, 1992), transgenic tobacco plants constitutively expressing PR-1 proteins do not show higher levels of local or systemic resistance to TMV or alfalfa mosaic virus (Cutt *et al.*, 1989; Linthorst *et al.*, 1989; Alexander *et al.*, 1993).

The evidence implicating PR-1 in fungal resistance is more convincing. Transgenic tobacco plants expressing the acidic PR-1a protein show increased tolerance to infection by *Peronospora tabacina* and *Phytophthora parasitica* var. *Nicotianae* (Alexander *et al.*, 1993). Studies with purified tobacco and tomato PR-1 proteins have shown that the basic vacuolar PR-1 proteins of tobacco,

PR-1g, and tomato, P14c, effectively inhibit growth of *Phytophthora infestans in vitro*, while acidic PR-1 proteins were only slightly inhibitory (Niderman *et al.*, 1995).

6.3.2 *β-1,3-Glucanases (PR-2)*

β-1,3-Glucanases (PR-2) are abundant, highly-regulated enzymes, widely distributed in plant species that catalyze the endo-type hydrolysis of 1,3-β-D-glucosidic linkages in β-1,3-glucans (Meins *et al.*, 1992; Stone and Clarke, 1992; Simmons, 1994; Høj and Fincher, 1995; Leubner-Metzger and Meins, 1999). β-1,3-Glucanases exist in multiple structural isoforms that differ in size, isoelectric point, primary structure and cellular localization (Meins *et al.*, 1992). They have been classified into three structural classes based on comparison of the amino acid sequences of the tobacco enzymes (Payne *et al.*, 1990; Ward *et al.*, 1991; Meins *et al.*, 1992). The approximately 33 kDa class I enzymes (GLU I) are basic proteins localized in the cell vacuole (Van den Bulcke *et al.*, 1989; Keefe *et al.*, 1990; Sticher *et al.*, 1992a). In contrast, the class II and III isoforms, which are often acidic proteins, are secreted into the extracellular space and lack the N-glycosylated C-terminal extension characteristic of the class I enzyme (Meins *et al.*, 1992; Stintzi *et al.*, 1993; Simmons, 1994). The three-dimensional structure of a barley β-1,3-glucanase has been determined (Varghese *et al.*, 1994).

There is compelling evidence that β-1,3-glucanases are important in pathogenesis. β-1,3-Glucanases and chitinases, acting alone or, in particular, in combination, can help defend plants against fungal infection. Tomato plants expressing vacuolar tobacco class I β-1,3-glucanase and chitinase transgenes show reduced susceptibility to infection by *Fusarium oxysporum* f.sp. *lycopersici*, whereas expression of either gene alone is not effective (Jongedijk *et al.*, 1995). Similarly, transgenic tobacco expressing alfalfa acidic β-1,3-glucanase, Aglul, combined with basic rice chitinase, Rch10, showed substantially increased protection when infected with *Cercospora nicotianae*, the causal agent of frogeye disease, relative to plants expressing either transgene alone (Zhu *et al.*, 1994). In contrast, susceptibility of *N. sylvestris* to *C. nicotianae* was not affected by high level expression of tobacco class I chitinase (Neuhaus *et al.*, 1991a) or deficiencies in host β-1,3-glucanase generated by antisense transformation (Neuhaus *et al.*, 1992). This suggests that expression of these enzymes is not a limiting step in the defense reaction of *N. sylvestris* to this pathogen. However, the finding that novel 'ersatz' forms of β-1,3-glucanase are specifically induced by pathogen infection in antisense plants raises the possibility that plants can also compensate for deficiencies in this enzyme (Beffa *et al.*, 1993).

β-1,3-Glucanases are believed to exert their antifungal effects either directly, by degrading the β-1,3-glucan-containing cell walls of the pathogens, or indirectly, by promoting the release of cell-wall derived materials that can act as elicitors of defense reactions (Boller, 1988, 1995; Bowles, 1990). Purified β-1, 3-glucanases and chitinases have pronounced antifungal activity *in vitro* (Mauch *et al.*, 1988; Ludwig and Boller, 1990; Leah *et al.*, 1991; Sela-Buurlage *et al.*, 1993; Lawrence *et al.*, 1996; Kim and Hwang, 1997). Secreted and vacuolar isoforms differ dramatically in activity. For example, the tobacco class I vacuolar isoforms effectively inhibited growth *in vitro* of *Fusarium solani*, *Alternaria solani*, *Trichoderma viride* and *Phytophthora infestans* (Sela-Buurlage *et al.*, 1993; Lawrence *et al.*, 1996). In contrast, secreted class II β-1,3-glucanases did not exhibit antifungal activity either alone or in any of the combinations tested.

Following fungal infection, soybean β-1,3-glucanases release β-glucans from fungal cell wall, which then induce accumulation of the phytoalexin, glyceollin. Released β-glucans showed structural requirements for eliciting activity (Sharp *et al.*, 1984; Takeuchi *et al.*, 1990; Okinaka *et al.*, 1995; Ham *et al.*, 1997).

In addition to their role in defense against fungal pathogens, β-1,3-glucanases appear to promote virus infection. Antisense transformation of tobacco and *Nicotiana sylvestris* with sequences of the tobacco GLU I *Gla* gene, specifically blocks expression of the class I, but not the class II or class III isoforms (Neuhaus *et al.*, 1992; Beffa *et al.*, 1993, 1996; Beffa and Meins, 1996). These GLU I-deficient mutants showed markedly reduced lesion size, lesion number and virus yield in local response of tobacco to TMV and of *N. sylvestris* to tobacco necrosis virus. Moreover, GLU I-deficient *N. sylvestris* mutants showed decreased severity and spread of mosaic symptoms in the susceptible response to TMV (Beffa *et al.*, 1996). Recent studies have established a plausible hypothesis for these effects on virus susceptibility (Iglesias and Meins, 2000). Relative to wild-type plants, a GLU I-deficient mutant of tobacco showed delayed intercellular trafficking via plasmodesmata of a tobamovirus (TMV), of a potexvirus (recombinant potato virus X expressing GFP), and of the movement protein (MP) *3a* of a cucumovirus (cucumber mosaic virus), and a reduced plasmodesmatal size exclusion limit defined by cell-to-cell movement of dextrans or peptides in non-infected plants. Therefore, GLU I-deficiency has a broad effect on plasmodesmatal movement, which is not limited to a particular virus type. Callose deposition was increased in the GLU I-deficient mutant, suggesting that GLU I could have an important function in regulating callose turnover at plas-modesmata and, hence, plasmodesmatal size exclusion limit. The role of vacuolar GLU I in callose degradation, which occurs in the extracellular compartment, suggests that GLU I, like the PR-1 proteins (Niderman *et al.*, 1995), may exhibit dual localization. These observations raised the possibility that viruses can use a defense response of the host against fungal infection—production of β-1,3-glucanases—to promote their own replication and spread (Beffa *et al.*, 1996).

In addition to their role in pathogenesis, β-1,3-glucanases have been implicated in diverse developmental and physiological processes, including microsporogenesis, pollen germination, fruit ripening and seed germination

(Leubner-Metzger and Meins, 1999). The function of GLU I in the germination of tobacco seeds has been studied in detail (Vögeli-Lange *et al.*, 1994; Leubner-Metzger *et al.*, 1995, 1996). Measurements of β-1,3-glucanase activity, GLU I protein and mRNA and reporter gene experiments have established that practically all of β-1,3-glucanase activity accumulated during germination is due to induction of the GLU I isoforms in the micropylar region of the endosperm, where the radicle will penetrate. These results led to the hypothesis that GLU I helps in the digestion of cell-wall material, leading to weakening of the endosperm and radicle protrusion.

6.3.3 Plant chitinases (PR-3, PR-4, PR-8 and PR-11)

Chitinases are abundant hydrolases found in bacteria, fungi, animals and plants. They hydrolyze the β-1,4-linkage of N-acetylglucosamine residues of chitin and, in some cases, exhibit variable lysozymal activity as well (Beintema and Terwisscha, 1996). Several structural types of plant chitinases have been purified and cloned from at least 33 plant species (Neuhaus, 1999). The PR nomenclature has placed plant chitinases into several groups (PR-3, PR-4, PR-8 and PR-11) (van Loon *et al.*, 1994), while a classification based on sequence comparison, distinguishes six classes (Meins *et al.*, 1994; Neuhaus, 1999). The three-dimensional structure of a class II chitinase from barley has been determined (Hart *et al.*, 1995).

Like the PR-1 and PR-2 proteins, the PR-3 chitinases exist as class I vacuolar isoforms, which are often basic proteins, and class II extracellular isoforms, which are often acidic proteins and lack the N-terminal lectin domain, spacer region and C-terminal extension present in class I isoforms (Neuhaus, 1999).

Chitinases can inhibit the growth *in vitro* of fungi with chitin as a major cell-wall component. Although some fungi, for example, *Trichoderma viride* and *T. longibfachiatum*, are susceptible to chitinases alone (Schlumbaum *et al.*, 1986; Arlorio *et al.*, 1992), in most cases, inhibition of growth depends on combination of chitinases with β-1,3-glucanases (Mauch *et al.*, 1988; Sela-Buurlage *et al.*, 1993) or compounds that alter membrane structure (Lorito *et al.*, 1996). Class I vacuolar chitinases with both an N-terminal chitin-binding domain and a chitinolytic domain are particularly active. Nevertheless, other chitinases are also active, and there is evidence that chitinases may carry more than one antifungal domain with different modes of action (Broekaert *et al.*, 1989; van Parijs *et al.*, 1991; Melchers *et al.*, 1994; Ponstein *et al.*, 1994; Neuhaus, 1999).

Expression of chitinases alone or in combination with β-1,3-glucanases in transgenic plants has been shown to reduce susceptibility to certain fungi. In addition to the examples cited in Section 6.3.2, transgenic tobacco expressing bean vacuolar class I chitinase was less susceptible to *Rhizoctonia solani*

(Broglie *et al.*, 1991), and in field trials carried out at two different geographical locations, transgenic winter-type oilseed rape expressing a chimeric vacuolar class I chitinase was less susceptible to *Cylindrosporium concentricum*, *Phoma lingam* and *Sclerotinia sclerotiorum* (Grison *et al.*, 1996).

6.3.4 Thaumatin-like proteins (PR-5)

The thaumatin-like (TL) proteins were first identified in tobacco leaves infected with TMV (van Loon *et al.*, 1987; Kauffmann *et al.*, 1990). Although they are similar in sequence to the sweet-tasting protein, thaumatin, present in *Thaumatococcus danielli*, none of the TL proteins tested have been found to have a sweet taste. Like the PR-2 and PR-3 families, the PR-5 family includes acidic and basic isoforms, localized in the extracellular compartment and vacuole, respectively (Dudler *et al.*, 1993).

Originally, the TL protein family was divided into three distinct groups: TL proteins induced by pathogens (the bona fide PR-5 proteins); TL proteins induced by osmotic stress (e.g. tobacco osmotin and tomato NP24); and the TL proteins developmentally regulated (e.g. TL proteins expressed in floral organs, seeds and fruits) (Rodrigo *et al.*, 1991a; Stintzi *et al.*, 1991; Dudler *et al.*, 1993). Later, it was found that osmotin and tomato NP24 were also induced by pathogens and were included in the PR-5 family (Rodrigo *et al.*, 1991b, 1993; Stintzi *et al.*, 1991). These proteins are thought to be localized in vacuoles and have been designated vacuolar TLs (VTLs) (King *et al.*, 1986, 1988; Singh *et al.*, 1987, 1989). Although there are exceptions (King *et al.*, 1986; LaRosa *et al.*, 1989), VTL proteins are not usually found in healthy plants. They accumulate rapidly in response to pathogen attack, elicitor treatment, or abiotic stress (Xu *et al.*, 1994; Chang *et al.*, 1995).

Zeamatin, a protein isolated from maize seeds with sequence similarity to tobacco osmotin and PR-R, has antifungal activity against *Candida albicans*, *Neurospora crassa* and *Trichoderma reesei*. This has led to the hypothesis that TL proteins might have a role in defense against fungi (Roberts and Selitrennikoff, 1990; Vigers *et al.*, 1991). Several *in vitro* assays against numerous fungi have confirmed the antifungal activity of osmotin (Hejgaard *et al.*, 1991; Vigers *et al.*, 1992; Stintzi *et al.*, 1993) and the activity of an NP24 against *Phytophthora infestans* (Woloshuk *et al.*, 1991). Abad and co-workers carried out an extensive study with tobacco osmotin and 31 fungal isolates representing 18 fungal genera (Abad *et al.*, 1996). Their evidence suggests that osmotin acts through specific receptors on the plasma membrane, and juxtaposition of the cell wall and the plasma membrane might be required for susceptibility of fungi to the protein. TL proteins have also been found to be effective in transgenic plants. Potato plants constitutively expressing the pA13 osmotin-like protein show an increased tolerance to the late-blight fungus, *Phytophthora infestans*, with a greater resistance at an early phase of fungal infection (Zhu *et al.*, 1996).

One possible explanation for the antifungal activity of these PR-5 proteins is the recent finding that two tomato fruit osmotins have β-1,3-glucanase activity (Grenier *et al.*, 1999). This suggests that they could act by hydrolyzing β-1,3-glucan-components of fungal cell walls.

6.3.5 Thionins (PR-13)

Thionins are basic, cysteine-rich peptides with a molecular weight of approximately 5 kDa. They are found in numerous plant species, including barley, wheat, *Aegilops squarrosa*, *Tulipa gesneriana*, mistletoes and several Brassicaceae (Bohlmann, 1999). They have been classified into five different classes (Bohlmann and Apel, 1994) according to the number of cysteine residues and the pattern of disulfide bridges. Three-dimensional structures of several thionins have been determined. In all cases, a compact L-shaped structure has been revealed, which is stabilized by three or four disulfide bridges (Hase *et al.*, 1978).

Thionins have been found in several cellular compartments. For example, in barley they accumulate in both the vacuole and walls of leaf cells (Bohlmann *et al.*, 1988; Reimann-Philipp *et al.*, 1989), whereas viscotoxins accumulate in storage vacuoles of leaf cells, and in both the storage vacuole and cytoplasm of cells in seeds (Schrader-Fischer and Apel, 1993).

Most thionins show pronounced toxic effects against yeast, bacteria, fungi and mammals (reviewed by Bohlmann *et al.*, 1994). In the case of mammals, thionins had no effect when eaten but only when injected intravenously. The precise mechanism for toxicity is not known, but is likely to involve the destruction of membranes, since treatment with thionins results in the leakage of low molecular weight compounds and ions from cells (Carrasco *et al.*, 1981; Thevissen *et al.*, 1996). Barley leaf thionins lower the viability of tobacco and barley protoplasts (Reimann-Philipp *et al.*, 1989), posing the problem of how plants protect themselves from auto-intoxication. One possibility is that the thionins are stored as inactive complexes. This speculation is supported by the finding that thionins have a phospholipid binding site, and that the antimicrobial activity of thionins is reduced by phospholipids (Woolley and Krampitz, 1942; Rao *et al.*, 1995). It has also been proposed that the long, acidic C-terminal region of the thionin proprotein might have a protective function (Bohlmann *et al.*, 1994).

Evidence for defense-related functions of thionins comes from ectopic expression studies with transgenic plants. High-level expression of a viscotoxin, a class II thionin from *Viscum album*, and of endogenous thionins in transgenic *Arabidopsis* resulted in significantly enhanced resistance against *Plasmodiopora brassicae* and *Fusarium oxysporum*, respectively (Epple *et al.*, 1997). Expression of barley thionin in transgenic tobacco enhanced resistance to the bacterial plant pathogens, *Pseudomonas syringae* pv. tabaci 153 and *P. syringae* pv. *syringae* (Carmona *et al.*, 1993).

6.4 Post-translational modification, intracellular transport and vacuolar targeting of PR proteins

Proteins destined for the vacuole and the extracellular space carry an N-terminal signal peptide that targets the nascent polypeptides to the lumen of the endomembrane system. In plants, the default pathway of polypeptides targeted to the endomembrane system is secretion. Positive sorting information carried in the polypeptide is required to direct proteins to the vacuole (Beevers and Raikhel, 1998). The mechanisms underlying intracellular transport and sorting are discussed in Chapter 11 of this volume. This section focuses on PR proteins and thionins known to exist as vacuolar and extracellular homologs or to show dual localization in these compartments.

The processing, intracellular transport and vacuolar targeting of PR-3 chitinases and PR-2 β-1,3-glucanases of tobacco have been studied in detail. The two vacuolar class I chitinases, CHN A and CHN B, are multidomained polypeptides, consisting of an N-terminal signal sequence followed by a highly folded cysteine-rich lectin domain. This domain is attached by a threonine- and proline-rich spacer (of variable length) to the catalytic domain and a short C-terminal extension. CHN A and CHN B were the first example of a new type of intracellular, hydroxyproline-containing proteins. They contain a small number of hydroxyprolyl residues, restricted to the short spacer region (Sticher *et al.*, 1992b). The results of pulse-chase experiments indicate that these class I chitinases undergo two distinct steps in post-translational processing (Sticher *et al.*, 1993; Freydl *et al.*, 1995). In the first step, the nascent protein undergoes an increase in apparent M_r of approximately 1500 Da due to hydroxylation of prolyl residues. This step occurs in the endomembrane system before sorting for secretion and vacuolar transport, and does not appear to be required for correct targeting of CHN A to the vacuole. The second step is a proteolytic removal of the C-terminal heptapeptide, which appears to occur late in intracellular transport but before the polypeptide reaches the vacuole.

Studies with stably expressed recombinant CHN A and transient expression of reporter-gene constructs have established that the C-terminal extension is necessary and sufficient for vacuolar localization of chitinases and, therefore, comprises a targeting signal for plant vacuoles (Neuhaus *et al.*, 1991b; Di Sansebastiano *et al.*, 1998). Exchanges of amino acids in the C-terminal heptapeptide, as well as substitution with random sequences, had intermediary effects that covered the whole range of a gradual transition from vacuolar retention to secretion (Neuhaus *et al.*, 1994).

The tobacco vacuolar class I β-1,3-glucanases are synthesized as preproproteins with an N-terminal signal peptide and a C-terminal extension carrying a single N-glycan (Shinshi *et al.*, 1988). Pulse-chase labelling and cell fractionation studies have shown that the nascent polypeptide is N-glycosylated in the endoplasmic reticulum and then transported via the Golgi to the vacuole, where the C-terminal N-glycopeptide is removed to produce the mature enzyme (Sticher *et al.*, 1992a). N-glycosylation is not required for either proteolytic processing or correct targeting to the vacuole. The C-terminal extension characteristic of vacuolar isoforms is absent from the otherwise highly homologous class II and class III enzymes that are usually localized in the extracellular compartment (Meins *et al.*, 1992). This observation suggests that the C-terminal extension could contain vacuolar targeting information. This is consistent with the finding that recombinant forms of tobacco GLU I, with the C-terminal extension deleted, are secreted into the extracellular space, indicating that the peptide is necessary for proper transport to the vacuole (Worrall *et al.*, 1992; Melchers *et al.*, 1993). It has not, however, been demonstrated that C-termination is sufficient for targeting, thus raising the possibility that the main domain of the polypeptide also carries sorting information.

Thionins are synthesized as preproproteins with an N-terminal signal peptide followed by the thionin domain and a long C-terminal acidic domain with six highly conserved cysteine residues (Ponz *et al.*, 1983; Schrader and Apel, 1991). The acidic domain is proteolytically removed during intracellular transport (Ponz *et al.*, 1983). Studies in barley leaves, with antibodies against the thionin and the acidic domains, have shown that the mature thionin accumulates in vacuoles but that the acidic domain is not detected in any of the cell fractions. This and the finding that a proteinase that cleaves the acidic domain is present in vacuoles suggests that proteolytic processing to the mature form occurs in a vacuole (Romero *et al.*, 1997).

As expected for proteins that traverse the secretory pathway, extracellular as well as vacuolar PR-1 isoforms have N-terminal signal sequences (Payne *et al.*, 1989). The special feature of the vacuolar isoforms is the presence of a C-terminal extension (Cornelissen *et al.*, 1987). By analogy to β-1,3-glucanase (PR-2) and chitinase (PR-3), this peptide could function as a signal for targeting to the vacuole (Neuhaus *et al.*, 1991b; Melchers *et al.*, 1993). On the other hand, the dual localization of isoforms with and without C-terminal extensions, in the vacuole and extracellular compartment (Dixon *et al.*, 1991; Sessa *et al.*, 1995), suggests that additional information contributes to targeting.

6.5 Biotechnological applications

There is, at present, a strong interest in the use of phytoalexins and PR proteins for biotechnological purposes. In leguminous plants, such as the forage legume, alfalfa, products of the phenylpropanoid pathway of secondary metabolism are involved in interactions with beneficial microorganisms (flavonoid inducers of the *Rhizobium* symbiosis), and in defense against pathogens (isoflavonoid phytoalexins). The recent isolation of genes encoding key enzymes in the various phenylpropanoid branch pathways opens up the possibility of engineering important crop plants, such as alfalfa, for increased or broader-spectrum disease resistance, by introducing novel phytoalexins or structural variants of the naturally occurring phytoalexins, or by modifying expression of transcriptional regulators of phytoalexin pathways (Dixon *et al.*, 1996). Probably the best illustration of the role of vacuolar phytoalexins in improved disease resistance is presented by certain varieties of *Medicago sativa* (alfalfa) produced by conventional breeding. Varieties were obtained by recurrent selection that showed antibiosis-based resistance to the root-lesion nematode, *Pratylenchus penetrans* (Baldridge *et al.*, 1998). Resistant plants showed higher expression of key genes in the phenylpropanoid pathway and of a key enzyme in biosynthesis of the phytoalexin, medicarpin. Constitutive levels of medicarpin were highest in roots of the most resistant plants.

Expression of β-1,3-glucanase and chitinase transgenes, alone or in combination, has been used to increase the resistance of plants to fungal infection. Tobacco plants expressing vacuolar bean class I beta chitinase gene showed decreased susceptibility to the root pathogen, *Rhyzoctonia solani* (Broglie *et al.*, 1991). Similarly, transgenic rice constitutively expressing chitinase showed enhanced resistance to blast disease (Nishizawa *et al.*, 1999), and transgenic rose plants expressing chitinase were tolerant to blackspot disease (Marchant *et al.*, 1998). A transgenic winter-type oilseed rape (*Brassica napus* var. oleifera) inbred line expressing a hybrid endochitinase gene containing the 3′-end of tobacco vacuolar chitinase was challenged with three different fungal pathogens, *Cylindrosporium concentricum*, *Phoma lingam* and *Sclerotinia sclerotiorum*, in field trials at two different geographical locations. Transgenic plants exhibited an increased tolerance to disease as compared with the non-transgenic parental plants (Grison *et al.*, 1996). A combination of chitinase and β-1,3-glucanase transgenes in wheat resulted in enhanced resistance to powdery mildew disease (Blifeld *et al.*, 1999).

Osmotin and thionins have also been engineered in transgenic plants for pathogen defense. Transgenic potato (*Solanum commersonii* Dun.) plants constitutively expressing the pA13 osmotin-like protein showed an increased tolerance to the late-blight fungus, *Phytophthora infestans*, with a greater resistance at an early phase of fungal infection (Zhu *et al.*, 1996). Transgenic *A. thaliana* lines expressing the thionin, viscotoxin A3, from *Viscum album* were more resistant to infection by clubroot pathogen, *Plasmodiophora brassicae*, than the parental line (Holtorf *et al.*, 1998). Similarly, expression of barley alpha-thionin conferred to transgenic tobacco enhanced resistance to the bacterial plant pathogens, *Pseudomonas syringae* pv. tabaci 153 and *P. syringae* pv. syringae (Carmona *et al.*, 1993).

6.6 Concluding remarks

A number of substances with antimicrobial or cytotoxic activity accumulate in vacuoles, suggesting that this organelle has a role in plant defense. Differentiation into vacuolar and extracellular isoforms is an evolutionarily conserved feature of several PR proteins; in some cases, the same isoform shows dual localization in both compartments, a finding that strongly suggests this compartmentation is functionally significant. Vacuolar isoforms of PR-1 and the glucanohydrolases, β-1,3-glucanases and PR-3 chitinase, have been shown to have appreciably stronger antifungal activity than the extracellular isoforms (Meins *et al.*, 1992; Fritig *et al.*, 1998). The different glucanohydrolase isoforms can also differ in specific activity and specificity (Kauffmann *et al.*, 1987; Legrand *et al.*, 1987; Brunner *et al.*, 1998; Schultze *et al.*, 1998). In the case of the highly active vacuolar β-1,3-glucanases, the major endogenous substrate, callose, is localized in the extracellular compartment, which is also the site where these enzymes act in microsporogenesis, seed germination, and regulation of plasmodesmatal trafficking (Leubner-Metzger and Meins, 1999; Iglesias and Meins, 2000). Thus, vacuolar compartmentation may be a means of isolating proteins with dual defense and non-defense related functions. This implies that targeting to the two compartments might be a regulated process. Indirect evidence for secretion of nominally vacuolar proteins supports this view (Kunze *et al.*, 1998). The mechanisms for this dual localization and its regulation are not known.

References

Abad, L.R., D'Urzo, M.P., Liu, D., Narasimhan, M., Reuveni, M., Zhou, J., Niu, X., Singh, N.K., Hasegawa, K. and Bressan, R.A. (1996) Antifungal activity of tobacco osmotin has specificity and involves plasma membrane permeabilization. *Plant Sci.*, **118** 11-19.

Ahl, P. and Gianinazzi, S. (1982) b-Protein as a constitutive component in highly (TMV) resistant interspecific hybrids of *Nicotiana glutinosa* × *Nicotiana debneyi*. *Plant Sci. Lett.*, **26** 173-79.

Ahl-Goy, P., Felix, G., Métraux, J.-P. and Meins, F. Jr. (1992) Resistance to disease in the hybrid, *Nicotiana glutinosa* × *Nicotiana debneyi*, is associated with high constitutive levels of β-1,3-glucanase, chitinase, peroxidase and polyphenoloxidase. *Physiol. Mol. Plant Pathol.*, **41** 11-21.

Alexander, D., Goodman, R.M., Gut-Rella, M., Glascock, C., Weymann, K., Friedrich, L., Maddox, D., Ahl-Goy, P., Luntz, T. and Ward, E. (1993) Increased tolerance to two oomycete pathogens in transgenic tobacco expressing pathogenesis-related protein 1a. *Proc. Natl. Acad. Sci. USA*, **90** 7327-31.

Antoniw, J.F., Ritter, C.E., Pierpoint, W.S. and van Loon, L.C. (1980) Comparison of three pathogenesis-related proteins from plants of two cultivars of tobacco infected with TMV. *J. Gen. Virol.*, **47** 79-89.

Arlorio, M., Ludwig, A., Boller, T. and Bonfante, P. (1992) Inhibition of fungal growth by plant chitinases and β-1,3-glucanases: a morphological study. *Protoplasma*, **171** 34-44.

Bailey, J. and Mansfield, J. (eds.) (1982) *Phytoalexins*, Blackie, Glasgow.

Baldridge, G.G., O'Neill, N.R. and Samac, D.A. (1998) Alfalfa (*Medicago sativa* L.) resistance to the root-lesion nematode, *Pratylenchus penetrans*: defense-response gene mRNA and isoflavonoid phytoalexin levels in roots. *Plant Mol. Biol.*, **38** 999-1010.

Barz, W. and Welle, R. (1992) Biosynthesis and metabolism of isoflavones and pterocarpan phytoalexins in chickpea, soybean and phytopathogenic fungi. *Recent Adv. Phytochem.*, **26** 139-64.

Barz, W., Bless, W., Boerger-Papendorf, G., Gunia, W., Mackenbrock, U., Meier, D., Otto, C. and Sueper, E. (1990) Phytoalexins as part of induced defense reactions in plants: their elicitation, function and metabolism, in *CIBA Foundation Symposium No. 154. Bioactive Compounds from Plants* (eds. D.J. Chadwick and J. Marsh), John Wiley & Sons, Chichester, pp. 140-56.

Beevers, L. and Raikhel, N.V. (1998) Transport to the vacuole: receptors and *trans* elements. *J. Exp. Bot.*, **49** 1279-89.

Beffa, R. and Meins, F. Jr. (1996) Pathogenesis-related functions of plant β-1,3-glucanases investigated by antisense transformation: a review. *Gene*, **179** 97-103.

Beffa, R.S., Neuhaus, J.-M. and Meins, F. Jr. (1993) Physiological compensation in antisense transformants: specific induction of an ersatz glucan endo-1,3-β-glucosidase in plants infected with necrotizing viruses. *Proc. Natl. Acad. Sci. USA*, **90** 8792-96.

Beffa, R.S., Hofer, R.-M., Thomas, M. and Meins, F. Jr. (1996) Decreased susceptibility to virus disease of β-1,3-glucanase-deficient plants generated by antisense transformation. *Plant Cell*, **8** 1001-11.

Beintema, J.J. and Terwisscha van Scheltinga, A.C. (1996) Plant lysozymes, in *Lysozymes: Model Enzymes in Biochemistry and Biology* (ed. P. Jolles), Birkhaeuser Verlag, Basel, pp. 75-86.

Blifeld, M., Mundy, J., Potrykus, I. and Fuetterer, J. (1999) Genetic engineering of wheat for increased resistance to powdery mildew disease. *Theor. Appl. Genet.*, **99** 1079-86.

Bohlmann, H. (1999) The role of thionins in the resistance of plants, in *Pathogenesis-Related Proteins in Plants* (eds. S.K. Datta and S. Muthukrishnan), CRC Press, Boca Raton, pp. 207-34.

Bohlmann, H. and Apel, K. (1994) The role of thionins in plant protection. *Crit. Rev. Plant Sci.*, **13** 1-13.

Bohlmann, H., Clausen, S., Behnke, S., Giese, H., Hiller, C., Schrader, G., Barkholt, V. and Apel, K. (1988) Leaf-thionins of barley—a novel class of cell wall proteins toxic to plant-pathogenic fungi and possibly involved in the defense mechanism of plants. *EMBO J.*, **7** 1559-69.

Bohlmann, H., Apel, K. and Garcia-Olmedo, F. (1994) Thionins. *Plant Mol. Biol. Rep.*, **12** 75-85.

Boller, T. (1988) Ethylene and the regulation of antifungal hydrolases in plants. *Oxf. Surv. Plant Mol. Cell Biol.*, **5** 145-74.

Boller, T. (1995) Chemoperception of microbial signals in plant cells. *Annu. Rev. Plant Physiol. Plant Mol. Biol.*, **46** 189-214.

Bowles, D.J. (1990) Defense-related proteins in higher plants. *Annu. Rev. Biochem.*, **59** 873-907.

Brisson, L.F., Tenhaken, R. and Lamb, C. (1994) Functions of oxidative cross-linking of cell wall structural proteins in plant disease resistance. *Plant Cell*, **6** 1703-13.

Broekaert, W.F., van Parijs, J., Leyns, F., Joos, H. and Peumans, W.J. (1989) A chitin-binding lectin from stinging nettle rhizomes with antifungal properties. *Science*, **245** 1100-10.

Broglie, K., Chet, I., Holliday, M., Cressman, R., Biddle, P., Knowlton, S., Mauvis, C.J. and Broglie, R. (1991) Transgenic plants with enhanced resistance to the fungal pathogen, *Rhizoctonia solani*. *Science*, **254** 1194-97.

Brunner, F., Stintzi, A., Fritig, B. and Legrand, M. (1998) Substrate specificities of tobacco chitinases. *Plant J.*, **14** 225-34.

Carmona, M.J., Molina, A., Fernandez, J.A., Lopez-Fando, J.J. and Garcia-Olmedo, F. (1993) Expression of the alpha-thionin gene from barley in tobacco confers enhanced resistance to bacterial pathogens. *Plant J.*, **3** 457-62.

Carrasco, L., Vazquez, D., Hernandez-Lucas, C., Carbonero, P. and Garcia-Olmedo, F. (1981) Thionins: plant peptides that modify membrane permeability in cultured mammalian cells. *Eur. J. Biochem.*, **116** 185-89.

Chang, P.F.L., Cheah, K.T., Narasimhan, M., Hasgawa, P.M. and Bressan, R.A. (1995) Osmotin gene expression is controlled by elicitor synergism. *Physiol. Plant*, **95** 620-30.

Clerivet, A., Alami, I., Breton, F., Garcia, D. and Sanier, C. (1996) Phenolic compounds and plant resistance to pathogenic microorganisms. *Acta Bot. Gallica*, **143** 531-38.

Cornelissen, B.J., Horowitz, J., van Kan, J.A., Goldberg, R.B. and Bol, J.F. (1987) Structure of tobacco genes encoding pathogenesis-related proteins from the PR-1 group. *Nucleic Acids Res.*, **15** 6799-811.

Cutt, J.R., Harpster, M.H., Dixon, D.C., Dunsmuir, P. and Klessig, D.F. (1989) Constitutive expression of pathogenesis-related proteins, PR-1, GRP and PR-S, in tobacco has no effect on virus infection. *Virology*, **173** 89-97.

Dakora, F.D. and Phillips, D.A. (1996) Diverse functions of isoflavonoids in legumes transcend antimicrobial definitions of phytoalexins. *Physiol. Mol. Plant Pathol.*, **49** 1-20.

Dangl, J.L., Dietrich, R.A. and Richberg, M.H. (1996) Death don't have no mercy: cell death programs in plant-microbe interactions. *Plant Cell*, **8** 1793-807.

Datta, S.K. and Muthukrishnan, S. (eds.) (1999) *Pathogenesis-Related Proteins in Plants*, CRC Press, Boca Raton, FL.

Di Sansebastiano, G.P., Paris, N., Marc-Martin, S. and Neuhaus, J.M. (1998) Specific accumulation of GFP in a non-acidic vacuolar compartment via a C-terminal propeptide-mediated sorting pathway. *Plant J.*, **15** 449-57.

Dixon, D.C., Cutt, J.R. and Klessig, D.F. (1991) Differential targeting of the tobacco PR-1 pathogenesis-related proteins to the extracellular space and vacuoles of crystal idioblasts. *EMBO J.*, **10** 1317-24.

Dixon, R.A. (1986) The phytoalexin response: elicitation, signalling and control of host gene expression. *Biol. Rev.*, **61** 239-91.

Dixon, R.A. and Harrison, M.J. (1990) Activation, structure and organization of genes involved in microbial defense in plants. *Adv. Genet.*, **28** 165-233.

Dixon, R.A. and Paiva, N.L. (1995) Stress-induced phenylpropanoid metabolism. *Plant Cell*, **7** 1085-97.

Dixon, R.A., Harrison, M.J. and Paiva, N.L. (1995) The isoflavonoid phytoalexin pathway: from enzymes to genes to transcription factors. *Physiol. Plant*, **93** 385-92.

Dixon, R.A., Harrison, M.J. and Lamb, C.J. (1994) Early events in the activation of plant defense responses. *Annu. Rev. Phytopathol.*, **32** 479-501.

Dixon, R.A., Lamb, C.J., Masoud, S., Sewalt, V.J. and Paiva, N.L. (1996) Metabolic engineering: prospects for crop improvement through the genetic manipulation of phenylpropanoid biosynthesis and defense responses: a review. *Gene*, **179** 61-71.

Dudler, R., Mauch, F. and Reimmann, C. (1993) Thaumatin-like proteins, in *Thaumatin* (eds. M. Witty and J.D. Higginbotham), CRC Press, Boca Raton, FL, pp. 193-209.

Epple, P., Apel, K. and Bohlmann, H. (1997) Overexpression of an endogenous thionin enhances resistance of *Arabidopsis* against *Fusarium oxysporum*. *Plant Cell*, **9** 509-20.

Eyal, Y., Sagee, O. and Fluhr, R. (1992) Dark-induced accumulation of a basic pathogenesis-related (PR-1) transcript and a light requirement for its induction by ethylene. *Plant Mol. Biol.*, **19** 589-99.

Fang, K.S.F., Vitale, M., Fehlner, P. and King, T.P. (1988) cDNA cloning and primary structure of a white-face hornet venom allergen, antigen 5. *Proc. Natl. Acad. Sci. USA*, **85** 895-99.

Freydl, E., Meins, F. Jr., Boller, T. and Neuhaus, J.-M. (1995) Kinetics of prolyl hydroxylation, intracellular transport and C-terminal processing of the tobacco vacuolar chitinase. *Planta*, **197** 250-56.

Fritig, B., Heitz, T. and Legrand, M. (1998) Antimicrobial proteins in induced plant defense. *Curr. Opin. Immunol.*, **10** 16-22.

Gaffney, T., Friedrich, L., Vernooij, B., Negrotto, D., Nye, G., Uknes, S., Ward, E., Kessmann, H. and Ryals, J. (1993) Requirement of salicylic acid for the induction of systemic acquired resistance. *Science*, **261** 754-56.

Garcia-Olmedo, F., Molina, A., Segura, A. and Moreno, M. (1995) The defensive role of nonspecific lipid-transfer proteins in plants. *Trends Microbiol.*, **3** 72-74.

Garcia-Olmedo, F., Molina, A., Alamillo, J.M. and Rodriguez-Palenzuela, P. (1998) Plant defense peptides. *Biopolymers*, **47** 479-91.

Gianinazzi, S., Martin, C. and Vallee, J.C. (1970) Hypersensibilité aux virus, températures et proteines solubles chez le *Nicotiana* Xanthi-nc: apparition de nouvelles macromolécules lors de la répression de la synthèse virale. *C. R. Acad. Sci. Paris D*, **270** 2383-86.

Graham, T.L. and Graham, M.Y. (1991) Gyceollin elicitors induce major but distinctly different shifts in isoflovonoid metabolism in proximal and distal soybean cell populations. *Mol. Plant Microbe Interact.*, **4** 60-68.

Green, T.E. and Ryan, C.A. (1972) Wound-induced proteinase inhibitor in plant leaves: a possible defense mechanism against insects. *Science*, **175** 776-80.

Grenier, J., Potvin, C., Trudel, J. and Asselin, A. (1999) Some thaumatin-like proteins hydrolyse polymeric β-1,3-glucans. *Plant J.*, **19** 473-80.

Grison, R., Grezes-Besset, B., Schneider, M., Lucante, N., Olsen, L., Leguay, J.J. and Toppan, A. (1996) Field tolerance to fungal pathogens of *Brassica napus* constitutively expressing a chimeric chitinase gene. *Nat. Biotechnol.*, **14** 643-46.

Hahlbrock, K. and Schell, D. (1989) Physiology and molecular biology of phenylpropanoid metabolism. *Annu. Rev. Plant Physiol.*, **40** 347-69.

Hahlbrock, K., Scheel, D., Logemann, E., Nürnberger, T., Parniske, M., Reinold, S., Sacks, W.R. and Schmelzer, E. (1995) Oligopeptides elicitor-mediated defense gene activation in cultured parsley cells. *Proc. Natl. Acad. Sci. USA*, **92** 4150-57.

Hain, R., Reif, H., Krause, E., Langebartels, R., Kindl, H., Vornam, B., Wiese, W., Schmelzer, E., Schreier, P.H., Stocker, R.H. and Stenzel, K. (1993) Disease resistance results from foreign phytoalexin expression in a novel plant. *Nature*, **361** 153-56.

Ham, K.S., Wu, S., Darvill, A.G. and Albersheim, P. (1997) Fungal pathogens secrete an inhibitor protein that distinguishes isoforms of plant pathogenesis-related endo-β-1,3-glucanases. *Plant J.*, **11** 169-78.

Hart, P.J., Pfluger, H.D., Monzingo, A.F., Hollis, T. and Robertus, J.D. (1995) The refined crystal structure of an endochitinase from *Hordeum vulgare* L. seeds at 1.8 A resolution. *J. Mol. Biol.*, **248** 402-13.

Hase, T., Matsubara, H. and Yoshizumi, H. (1978) Disulfide bonds of purothionins, a lethal toxin for yeasts. *J. Biochem.*, **83** 1671-79.

Hejgaard, J., Jacobsen, S. and Svendsen, I. (1991) Two antifungal thaumatin-like proteins from barley grain. *FEBS Lett.*, **291** 127-31.

Holtorf, S., Ludwig-Muller, J., Apel, K. and Bohlmann, H. (1998) High-level expression of a viscotoxin in *Arabidopsis thaliana* gives enhanced resistance against *Plasmodiophora brassicae*. *Plant Mol. Biol.*, **36** 673-80.

Høj, P.B. and Fincher, G.B. (1995) Molecular evolution of plant β-glucan endohydrolases. *Plant J.*, **7** 367-79.

Iglesias, V.A. and Meins, F. Jr. (2000) Movement of plant viruses is delayed in a β-1,3-glucanase-deficient mutant showing a reduced plasmodesmatal size exclusion limit and enhanced callose deposition. *Plant J.*, **21** 157-66.

Jongedijk, E., Tigelaar, H., van Roeckel, J.S.C., Bres-Vloemans, S.A., Dekker, I., van den Elzen, P.J.M., Cornelissen, B.J.C. and Melchers, L.S. (1995) Synergistic activity of chitinases and β-1,3-glucanases enhances fungal resistance in transgenic tomato plants. *Euphytica*, **85** 173-80.

Jung, J.L., Fritig, B. and Hahne, G. (1993) Sunflower (*Helinathus annuus* L.) pathogenesis-related proteins: induction by aspirin (acetylsalicylic acid) and characterization. *Plant Physiol.*, **101** 873-80.

Kauffmann, S., Legrand, M., Geoffroy, P. and Fritig, B. (1987) Biological function of 'pathogenesis-related' proteins: four PR proteins of tobacco have 1,3-β-glucanase activity. *EMBO J.*, **6** 3209-12.

Kauffmann, S., Legrand, M. and Fritig, B. (1990) Isolation and characterization of six pathogenesis-related (PR) proteins of Samsun NN tobacco. *Plant Mol. Biol.*, **14** 381-90.

Keefe, D., Hinz, U. and Meins, F. Jr. (1990) The effect of ethylene on the cell-type-specific and intracellular localization of β-1,3-glucanase and chitinase in tobacco leaves. *Planta*, **182** 43-51.

Kim, Y.J. and Hwang, B.K. (1997) Isolation of a basic 34 kiloDalton β-1,3-glucanase with inhibitory activity against *Phytophthora capsici* from pepper stems. *Physiol. Mol. Plant Pathol.*, **50** 103-15.

King, G.J., Hussey, C.E. and Turner, V.A. (1986) A protein induced by NaCl in suspension cultures of *Nicotiana tabacum* accumulates in whole plant roots. *Plant Mol. Biol.*, **7** 441-49.

King, G.J., Turner, V.A., Hussey, C.E. Jr., Wurtele, E.S. and Lee, S.-M. (1988) Isolation and characterization of a tomato cDNA clone which codes for a salt-induced protein. *Plant Mol. Biol.*, **10** 401-10.

Kuc, J. (1982) Induced immunity to plant disease. *Bioscience*, **32** 854-60.

Kuc, J. (1995) Phytoalexins, stress metabolism and disease resistance in plants. *Annu. Rev. Phytopathol.*, **33** 275-97.

Kunze, I., Kunze, G., Broker, M., Manteuffel, R., Meins, F.J. and Müntz, K. (1998) Evidence for secretion of vacuolar alpha-mannosidase, class I chitinase and class I β-1,3-glucanase in suspension cultures of tobacco cells. *Planta*, **205** 92-99.

LaRosa, P.C., Singh, N.K., Hasegawa, P.M. and Bressan, R.A. (1989) Stable NaCl tolerance of tobacco cells is associated with enhanced accumulation of osmotin. *Plant Physiol.*, **91** 855-59.

Lawrence, C., Joosten, M.H.A.J. and Tuzun, S. (1996) Differential induction of pathogenesis-related proteins in tomato by *Alternaria solani* and the association of a basic chitinase isozyme with resistance. *Physiol. Mol. Plant Pathol.*, **48** 361-70.

Leah, R., Tommerup, H., Svendsen, I. and Mundy, J. (1991) Biochemical and molecular characterization of three barley seed proteins with antifungal properties. *J. Biol. Chem.*, **266** 1564-73.

Legrand, M., Kauffmann, S., Geoffroy, P. and Fritig, B. (1987) Biological function of pathogenesis-related proteins: four tobacco pathogenesis-related proteins are chitinases. *Proc. Natl. Acad. Sci. USA*, **84** 6750-54.

Leubner-Metzger, G. and Meins, F. Jr. (1999) Function and regulation of plant β-1,3-glucanases (PR-2), in *Pathogenesis-Related Proteins in Plants* (eds. K. Datta and S. Muthukrishnan), CRC Press, Boca Raton, FL, pp. 49-76.

Leubner-Metzger, G., Fründt, C., Vögeli-Lange, R. and Meins, F. Jr. (1995) Class I β-1,3-glucanases in the endosperm of tobacco during germination. *Plant Physiol.*, **109** 751-59.

Leubner-Metzger, G., Fründt, C. and Meins, F. Jr. (1996) Effects of gibberellins, darkness and osmotica on endosperm rupture and class I β-1,3-glucanase induction in tobacco seed germination. *Planta*, **199** 282-88.

Linthorst, H.J.M. (1991) Pathogenesis-related proteins of plants. *Crit. Rev. Plant Sci.*, **10** 123-50.

Linthorst, H.J., Meuwissen, R.L., Kauffmann, S. and Bol, J.F. (1989) Constitutive expression of pathogenesis-related proteins, PR-1, GRP and PR-S, in tobacco has no effect on virus infection. *Plant Cell*, **1** 285-91.

Lorito, M., Woo, S.L., Dambrosio, M., Harman, G.E., Hayer, C.K., Kubicek, C.P. and Scala, F. (1996) Synergistic interaction between cell-wall-degrading enzymes and membrane-affecting compounds. *Mol. Plant Microbe Interact.*, **9** 206-13.

Ludwig, A. and Boller, T. (1990) A method for the study of fungal growth inhibition by plant proteins. *FEMS Microbiol. Lett.*, **57** 61-66.

Mackenbrock, U., Gunia, W. and Barz, W. (1993) Accumulation and metabolism of medicarpin and maackiain malonylglucosides in elicited chickpea (*Cicer prietinum* L.) cell supension cultures. *J. Plant Physiol.*, **142** 385-91.

Mackintosh, S., Matthews, D. and VanEtten, H. (1989) Two additional genes for pisatin demethylation and their relationship to the pathogenicity of *Nectaria haematococca* metabolites released by damaged bean cells. *Mol. Plant Microbe Interact.*, **2** 354-62.

Maestro, D.R., Leon, R. and Ruiz, G.V. (1993) Phenolic compounds as vegetables' autoprotection. *Grasas y Aceites*, **44** 365-69.

Marchant, R., Davey, M.R., Lucas, J.A., Lamb, C., Dixon, R.A. and Power, J.B. (1998) Expression of a chitinase transgene in rose (*Rosa hybrida*, L.) reduces development of blackspot disease (*Diplocarpon rosea*, Wolf). *Mol. Breed.*, **4** 187-94.

Mauch-Mani, B. and Slusarenko, A.J. (1996) Production of salicylic acid precursors is a major function of phenylalanine ammonia-lyase in the resistance of *Arabidopsis* to *Peronospora parasitica*. *Plant Cell*, **8** 203-12.

Mauch, F., Mauch-Mani, B. and Boller, T. (1988) Antifungal hydrolases in pea tissue. II. Inhibition of fungal growth by combinations of chitinase and β-1,3-glucanase. *Plant Physiol.*, **88** 936-42.

Meins, F. Jr., Neuhaus, J.-M., Sperisen, C. and Ryals, J. (1992) The primary structure of plant pathogenesis-related glucanohydrolases and their genes, in *Genes Involved in Plant Defense* (eds. T. Boller and F. Meins Jr.), Springer Verlag, Vienna/New York, pp. 245-82.

Meins, F. Jr., Fritig, B., Linthorst, H.J.M., Mikkelsen, J.D., Neuhaus, J.-M. and Ryals, J. (1994) Plant chitinase genes. *Plant Mol. Biol. Rep.*, **12** 22-28.

Melchers, L.S., Sela-Buurlage, M.B., Vloemans, S.A., Woloshuk, C.P., van Roekel, J.S., Pen, J., van den Elzen, P.J. and Cornelissen, B.J. (1993) Extracellular targeting of the vacuolar tobacco proteins, AP24, chitinase and beta-1,3-glucanase, in transgenic plants. *Plant Mol. Biol.*, **21** 583-93.

Melchers, L.S., Apotheker-de Groot, M., van der Knaap, J.A., Ponstein, A.S., Sela-Buurlage, M.B., Bol, J.F., Cornelissen, B.J., van den Elzen, P.J. and Linthorst, H.J. (1994) A new class of tobacco chitinases, homologous to bacterial exo-chitinases, displays antifungal activity. *Plant J.*, **5** 469-80.

Miosga, T., Schaaff-Gerstenschlager, I., Chalwatzis, N., Baur, A. and Boles, E. (1995) Sequence analysis of a 33.1 kb fragment from the left arm of *Saccharomyces cervisiae* chromosome X, including putative proteins with leucine zippers, a fungal Zn(II)2-Cys-6 binuclear cluster domain and a putative alpha-2-SCB-alpha-2 binding site. *Yeast.*, **11** 681-89.

Murphy, E.V., Zhang, Y., Zhu, W.J. and Biggs, J. (1995) The human glioma pathogenesis-related protein is structurally related to plant pathogenesis-related proteins and its gene is expressed specifically in brain tumors. *Gene*, **159** 131-35.

Neuhaus, J.-M. (1999) Plant chitinases, in *Pathogenesis-Related Proteins in Plants* (eds. K. Datta and S. Muthukrishnan), CRC Press, Boca Raton, FL, pp. 77-105.

Neuhaus, J.M., Ahl-Goy, P., Hinz, U., Flores, S. and Meins, F.J. (1991a) High-level expression of a tobacco chitinase gene in *Nicotiana sylvestris*: susceptibility of transgenic plants to *Cercospora nicotianae* infection. *Plant Mol. Biol.*, **16** 141-51.

Neuhaus, J.M., Sticher, L., Meins, F.J. and Boller, T. (1991b) A short C-terminal sequence is necessary and sufficient for the targeting of chitinases to the plant vacuole. *Proc. Natl. Acad. Sci. USA*, **88** 10362-66.

Neuhaus, J.M., Flores, S., Keefe, D., Ahl-Goy, P. and Meins, F.J. (1992) The function of vacuolar beta-1,3-glucanase investigated by antisense transformation: susceptibility of transgenic *Nicotiana sylvestris* plants to *Cercospora Nicotianae* infection. *Plant Mol. Biol.*, **19** 803-13.

Neuhaus, J.M., Pietrzak, M. and Boller, T. (1994) Mutation analysis of the C-terminal vacuolar targeting peptide of tobacco chitinase: low specificity of the sorting system, and gradual transition between intracellular retention and secretion into the extracellular space. *Plant J.*, **5** 45-54.

Niderman, T., Genetet, I., Bruyere, T., Gees, R., Stintzi, A., Legrand, M., Fritig, B. and Mosinger, E. (1995) Pathogenesis-related PR-1 proteins are antifungal: isolation and characterization of three 14 kilodalton proteins of tomato and of a basic PR-1 of tobacco with inhibitory activity against *Phytophthora infestans*. *Plant Physiol.*, **108** 17-27.

Nishizawa, Y., Nishio, Z., Nakazono, K., Soma, M., Nakajima, E., Ugaki, M. and Hibi, T. (1999) Enhanced resistance to blast (*Magnaporthe grisea*) in transgenic Japonica rice by constitutive expression of rice chitinase. *Theor. Appl. Genet.*, 383-90.

Okinaka, Y., Mimori, K., Takeo, K., Kitamura, S., Takeuchi, Y., Yamaoka, N. and Yoshikawa, M.A. (1995) A structural model for the mechanisms of elicitor release from fungal cell walls by plant β-1,3-endoglucanase. *Plant Physiol.*, **109** 839-45.

Osbourn, A.E. (1999) Antimicrobial phytoprotectants and fungal pathogens: a commentary. *Fungal Genet. Biol.*, **26** 163-68.

Payne, G., Middlesteadt, W., Desai, N., Williams, S., Dincher, S., Carnes, J. and Ryals, J. (1989) Isolation and sequence of a genomic clone encoding the basic form of pathogenesis-related protein 1 from *Nicotiana tabacum*. *Plant Mol. Biol.*, **12** 595-99.

Payne, G., Ward, E., Gaffney, T., Goy, P.A., Moyer, M., Harper, A., Meins, F.J. and Ryals, J. (1990) Evidence for a third structural class of β-1,3-glucanase in tobacco. *Plant Mol. Biol.*, **15** 797-808.

Ponstein, A.S., Bres-Vloemans, S.A., Sela-Buurlage, M.B., van den Elzen, P.J., Melchers, L.S. and Cornelissen, B.J. (1994) A novel pathogen- and wound-inducible tobacco (*Nicotiana tabacum*) protein with antifungal activity. *Plant Physiol.*, **104** 109-18.

Ponz, F., Paz-Ares, J., Hernandez-Lucas, C., Carbonero, P. and Garcia-Olmedo, F. (1983) Synthesis and processing of thionin precursors in developing endosperm from barley (*Hordeum vulgare* L.). *EMBO J.*, **2** 1035-39.

Ranansky, J., Erdelsky, K., Ziegler, W. and Pavlovkin, J. (1996) Fungal elicitors: active biological substances of fungi attacking cells of higher plants. *Biologia*, **51** 471-77.

Rao, U., Stec, B. and Teeter, M.M. (1995) Refinement of purothionins reveals solute particles important for lattice formation and toxicity. Part 1. 1-Purothionins revisited. *Acta Cryst.*, **51** 904-13.

Reimann-Philipp, U., Schrader, G., Martinoia, E., Barkholt, V. and Apel, K. (1989) Intracellular thionins of barley: a second group of leaf thionins closely related to but distinct from cell-wall-bound thionins. *J. Biol. Chem.*, **264** 8978-84.

Roberts, W.K. and Selitrennikoff, C.P. (1990) Zeamatin, an antifungal protein from maize with membrane-permeabilizing activity. *J. Gen. Microbiol.*, **136** 1771-79.

Rodrigo, I., Vera, P., Frank, R. and Conejero, V. (1991a) cDNA cloning of viroid-induced tomato pathogenesis-related protein, P23: characterization as a vacuolar antifungal factor. *Plant Physiol.*, **103** 939-45.

Rodrigo, I., Vera, P., Frank, R. and Conejero, V. (1991b) Identification of the viroid-induced tomato pathogenesis-related (PR) protein, P23, as the thaumatin-like tomato protein, NP24, associated with osmotic stress. *Plant Mol. Biol.*, **16** 931-34.

Rodrigo, I., Vera, P., Tornero, P., Hernandez, Y.J. and Conejero, V. (1993) cDNA cloning of viroid-induced tomato pathogenesis-related protein, P23: characterization as a vacuolar antifungal factor. *Plant Physiol.*, **3** 39-45.

Romero, A., Alamillo, J.M. and Garcia-Olmedo, F. (1997) Processing of thionin precursors in barley leaves by a vacuolar proteinase. *Eur. J. Biochem.*, **243** 202-208.

Ryals, J.A., Neuenschwander, U.H., Willits, M.G., Molina, A., Steiner, H.-Y. and Hunt, M.D. (1996) Systemic acquired resistance. *Plant Cell*, **8** 1809-19.

Schlumbaum, A., Mauch, F. and Boller, T. (1986) Plant chitinases are potent inhibitors of fungal growth. *Nature*, **324** 365-69.

Schrader, G. and Apel, K. (1991) Isolation and characterization of cDNAs encoding viscotoxins of mistletoe (*Viscum album*). *Eur. J. Biochem.*, **198** 549-53.

Schrader-Fischer, G. and Apel, K. (1993) cDNA-derived identification of novel thionin precursors in *Viscum album* that contain highly divergent thionin domains but conserved signal and acidic polypeptide domains. *Plant Mol. Biol.*, **23** 1233-42.

Schultze, M., Staehelin, C., Brunner, F., Genetet, I., Legrand, M., Fritig, B., Kondorosi, E. and Kondorosi, A. (1998) Plant chitinase/lysozyme isoforms show distinct substrate specificity and cleavage site preference towards lipochito-oligosaccharide Nod signals. *Plant J.*, **16** 571-80.

Schuren, F.H., Asgeirsdottic, S.A., Kothe, E.M., Scheer, J.M.J. and Wessels, J.G.H. (1993) The Sc7/Sc14 gene family of *Schizophyllum commune* codes for extracellular proteins specifically expressed during fruit-body formation. *J. Gen. Microbiol.*, **139** 2083-90.

Sela-Buurlage, M.B., Ponstein, A.S., Bres-Vloemans, S.A., Melchers, L.S., van den Elzen, P.J.M. and Cornelissen, B.J.C. (1993) Only specific tobacco (*Nicotiana tabacum*) chitinases and β-1,3-glucanases exhibit antifungal activity. *Plant Physiol.*, **101** 857-63.

Sessa, G. and Fluhr, R. (1995) The expression of an abundant transmitting tract-specific endoglucanase (Sp41) is promoter-dependent and not essential for the reproductive physiology of tobacco. *Plant Mol. Biol.*, **29** 969-82.

Sessa, G., Yang, X.Q., Raz, V., Eyal, Y. and Fluhr, R. (1995) Dark induction and subcellular localization of the pathogenesis-related, PRB-1b, protein. *Plant Mol. Biol.*, **28** 537-47.

Sharma, R. and Salunkhe, D. (1999) *Mycotoxins and Phytoalexins*, CRC Press, Boca Raton, FL, USA.

Sharp, J.K., Valent, B. and Albersheim, P. (1984) Purification and partial characterization of a β-glucan fragment that elicits phytoalexin accumulation in soybean. *J. Biol. Chem.*, **259** 11312-20.

Shinshi, H., Wenzler, H., Neuhaus, J.M., Felix, G., Hofsteenge, J. and Meins, F.J. (1988) Evidence for N- and C-terminal processing of a plant defense-related enzyme: primary structure of tobacco prepro-β-1,3-glucanase. *Proc. Natl. Acad. Sci. USA*, **85** 5541-45.

Simmons, C.R. (1994) The physiology and molecular biology of plant 1,3-β-D-glucanases and 1,3;1,4-β-D-glucanases. *Crit. Rev. Plant Sci.*, **13** 325-87.

Singh, N.K., Bracker, C.A., Hasegawa, K., Handa, A.K., Buckel, S., Hermodson, M.A., Pfankock, E., Regnier, F.E. and Bressan, R.A. (1987) Characterization of osmotin. *Plant Physiol.*, **85** 529-35.

Singh, N.K., Nelson, D.E., Kuhn, D., Hasegawa, K. and Bressan, R.A. (1989) Molecular cloning of osmotin and regulation of its expression by ABA and adaptation to low water potential. *Plant Physiol.*, **90** 1096-102.

Smith, C.J. (1996) Accumulation of phytoalexins: defence mechanism and stimulus response systems. *New Phytol.*, **132** 1-45.

Sticher, L., Hinz, U., Meyer, A.D. and Meins, F. Jr. (1992a) Intracellular transport and processing of a tobacco vacuolar β-1,3-glucanase. *Planta*, **188** 559-65.

Sticher, L., Hofsteenge, J., Milani, A., Neuhaus, J.M. and Meins, F.J. (1992b) Vacuolar chitinases of tobacco: a new class of hydroxyproline-containing proteins. *Science*, **257** 655-57.

Sticher, L., Hofsteenge, J., Neuhaus, J.M., Boller, T. and Meins, F.J. (1993) Post-translational processing of a new class of hydroxyproline-containing proteins: prolyl hydroxylation and C-terminal cleavage of tobacco (*Nicotiana tabacum*) vacuolar chitinase. *Plant Physiol.*, **101** 1239-47.

Stintzi, A., Heitz, T., Kauffmann, S., Legrand, M. and Fritig, B. (1991) Identification of a basic pathogenesis-related, thaumatin-like protein in virus-infected tobacco as osmotin. *Physiol. Mol. Plant Pathol.*, **38** 137-43.

Stintzi, A., Heitz, T., Prasad, V., Wiedemann-Merdinoglu, S., Kauffmann, S., Geoffroy, P., Legrand, M. and Fritig, B. (1993) Plant 'pathogenesis-related' proteins and their role in defense against pathogens. *Biochimie*, **75** 687-706.

Stone, B.A. and Clarke, A.E. (1992) *Chemistry and Biology of (1,3)-β-Glucans*, La Trobe University Press, Melbourne, Victoria, Australia.

Takeuchi, Y., Yoshikawa, M.A., Takeba, G., Tanaka, K., Shibata, D. and Horino, O. (1990) Molecular cloning and ethylene induction of mRNA encoding a phytoalexin elicitor-releasing factor, β-1,3-glucanase, in soybean. *Plant Physiol.*, **93** 673-80.

Terras, F.R., Eggermont, K., Kovaleva, V., Raikhel, N., Osborn, R.W., Kester, A., Rees, S.B., Torrekens, S.76, Van Leuven, F., Vanderleyden, J., Cammue, B.P. and Broekaert, W.F. (1995) Small cysteine-rich antifungal proteins from radish: their role in host defense. *Plant Cell*, **7** 573-88.

Thevissen, K., Ghazi, A., De Samblanx, G.W., Brownlee, C., Osborn, R.W. and Broekaert, W.F. (1996) Fungal membrane responses induced by plant defensins and thionins. *J. Biol. Chem.*, **271** 15018-25.

Van den Bulcke, M., Bauw, G., Castresana, C., van Montagu, M. and Vandekerckhove, J. (1989) Characterization of vacuolar and extracellular β(1,3)-glucanases of tobacco: evidence for a strictly compartmentalized plant defense system. *Proc. Natl. Acad. Sci. USA*, **86** 2673-80.

van Loon, L.C. (1999) Occurrence and properties of plant pathogenesis-related proteins, in *Pathogenesis-Related Proteins in Plants* (eds. S.K. Datta and S. Muthukrishnan), CRC Press, Boca Raton, FL, USA, pp. 1-19.

van Loon, L.C. and van Kammen, A. (1970) Polyacrylamide disc electrophoresis of the soluble leaf proteins from *Nicotiana tabacum* var 'Samsun' and 'Samsun NN'. II. Changes in protein constitution after infection with tobacco mosaic virus. *Virology*, **40** 199-211.

van Loon, L.C., Gerritsen, Y.A.M. and Ritter, C.E. (1987) Identification, purification, and characterization of pathogenesis-related proteins from virus-infected Samsun NN tobacco leaves. *Plant Mol. Biol.*, **9** 593-609.

van Loon, L.C., Pierpoint, W.S., Boller, T. and Conejero, V. (1994) Recommendations for naming plant pathogenesis-related proteins. *Plant Mol. Biol. Rep.*, **12** 245-64.

van Parijs, J., Broekaert, W.F., Goldstein, I.J. and Peumans, W.J. (1991) Hevein: an antifungal protein from rubber-tree (*Hevea brasiliensis*) latex. *Planta*, **183** 258-64.

Varghese, J.N., Garrett, T.P.J., Colman, P.M., Chen, L., Hoej, P.B. and Fincher, G.B. (1994) Three-dimensional structures of two plant β-glucan endohydrolases with distinct substrate specificities. *Proc. Natl. Acad. Sci. USA*, **91** 2785-89.

Vigers, A.J., Roberts, W.K. and Selitrennikoff, C.P. (1991) A new family of plant antifungal proteins. *Mol. Plant Microbe Interact.*, **4** 315-23.

Vigers, A.J., Wiedemann, S., Roberts, W.K., Legrand, M., Selitrennikoff, C.P. and Fritig, B. (1992) Thaumatin-like pathogen-related proteins are antifungal. *Plant Sci.*, **83** 155-63.

Vögeli-Lange, R., Fründt, C., Hart, C.M., Beffa, R., Nagy, F. and Meins, F. Jr. (1994) Evidence for a role of β-1,3-glucanase in dicot seed germination. *Plant J.*, **5** 273-78.

Ward, E.R., Payne, G.B., Moyer, M.B., Williams, S.C., Dincher, S.S., Sharkey, K.C., Beck, J.J., Taylor, H.T., Ahl-Goy, P., Meins, F. Jr. and Ryals, J. (1991) Differential regulation of β-1,3-glucanase messenger RNAs in response to pathogen infection. *Plant Physiol.*, **96** 390-97.

Weidemann, C., Tenhaken, R., Hohe, U. and Barz, W. (1991) Medicarpin and maackiain-3-*o*-glucoside-6′-*o* malonate conjugates are constitutive compounds of chickpea (*Cicer arietinum* L.) cell cultures. *Plant Cell Rep.*, **10** 371-74.

White, R.F., Rybicki, E.P., von Wechmar, M.B., Dekker, J.L. and Antoniw, J.F. (1987) Detection of PR-1 type proteins in Amaranthaceae, Chenopodiaceae, Gramineae and Solanaceae by immunoelectroblotting. *J. Gen. Virol.*, **68** 2043-50.

Woloshuk, C.P., Meulenhoff, J.S., Sela-Buurlage, M., van den Elzen, P.J. and Cornelissen, B.J. (1991) Pathogen-induced proteins with inhibitory activity toward *Phytophthora infestans*. *Plant Cell*, **3** 619-28.

Woolley, D. and Krampitz, L.O. (1942) Reversal by phosphatides of the antimicrobial action of a crystalline protein from wheat. *J. Biol. Chem.*, **146** 273-80.

Worrall, D., Hird, D.L., Hodge, R., Paul, W., Draper, J. and Scott, R. (1992) Premature dissolution of the microsporocyte callose wall causes male sterility in transgenic tobacco. *Plant Cell*, **4** 759-71.

Xu, Y., Chang, P.F.L., Liu, D., Narasimhan, M., Raghothama, K.G., Hasegawa, P.M. and Bressan, R.A. (1994) Plant defense genes are synergistically induced by ethylene and methyl jasmonate. *Plant Cell*, **6** 1077-84.

Ye, X.S., Pan, S.Q. and Kuc, J. (1992) Specificity of induced systemic resistance as elicited by ethephon and tobacco mosaic virus in tobacco. *Plant Sci.*, **84** 1-9.

Zhu, B., Chen, T.H. and Li, P.H. (1996) Analysis of late-blight disease resistance and freezing tolerance in transgenic potato plants expressing sense and antisense genes for an osmotin-like protein. *Planta*, **198** 70-77.

Zhu, Q., Maher, E.A., Masoud, S., Dixon, R.A. and Lamb, C.J. (1994) Enhanced protection against fungal attack by constitutive co-expression of chitinase and glucanase genes in transgenic tobacco. *Biotechnol.*, **12** 807-12.

7 Energizing the tonoplast

Rafael Ratajczak and Thea A. Wilkins

7.1 Introduction

More than 95% of the total volume of a mature plant cell can be occupied by the large central vacuole. This fact already indicates that the vacuole represents a cellular compartment that is of major importance. In the past, the vacuole was wrongly assumed to be a sump for waste and toxic molecules. However, intensive research performed during the last decade has shown that the vacuole is an essential compartment for maintaining cytosolic pH and ion homeostasis, as well as for the short- and long-term storage of metabolites. Molecules stored in the vacuole include: inorganic ions, such as sodium, potassium, calcium, magnesium, heavy metals, chloride and nitrate (for review, see Blumwald and Gelli, 1997); organic molecules, such as carbohydrates, organic and amino acids, polyamines and peptides (for review, see Martinoia and Ratajczak, 1997); and products of secondary plant metabolism (for review, see Wink, 1997). These substances have to be transported into and out of the vacuole across the membrane surrounding the vacuolar lumen, the tonoplast. Transport of these ions and metabolites is mediated by numerous secondary transport systems, comprised of integral membrane proteins in the tonoplast that include ion channels and carriers (for reviews, see Allen and Sanders, 1997; Martinoia and Ratajczak, 1997; and Chapter 9 in this volume). However, since transport in most cases has to be driven against a concentration gradient, energization of the tonoplast is required. This is achieved by the action of two electrogenic primary-active proton pumps (H^+-phosphohydrolases) located at the tonoplast, that is, the vacuolar-type H^+-transporting adenosine triphosphatase (V-ATPase; EC 3.6.1.3) and the vacuolar-type H^+-transporting inorganic pyrophosphatase (V-PPase; EC 3.6.1.1). Together, the V-ATPase and the V-PPase establish an electrochemical proton gradient across the tonoplast by pumping protons into the vacuolar lumen using the energy released by hydrolyzing phosphate bonds of ATP and inorganic pyrophosphate, respectively.

This chapter summarizes the status of investigation of the two pumps, including their molecular structure and gene organization, protein targeting, assembly and turnover, the mechanism and the regulation of gene expression and enzymatic activity. It must be stressed that we still do not know why there are two primary-active proton pumps present at one membrane, and if and how both pumps might co-operate in vacuolar functions. The present chapter concentrates on recent findings in the field including results obtained by investigation of the yeast and animal V-ATPase, and considers future strategies to study the proton pumps.

7.2 V-ATPase

7.2.1 Structure, biochemistry and gene organization

Electron microscopy studies performed with negatively-stained tonoplast vesicles revealed the presence of lollipop-shaped structures at the rim of the membrane, consisting of a globular head (9.4–11.9 nm in diameter) and a stalk (2.3–3.1 nm in diameter and 4.5–6.6 nm in length) connecting the head with the membrane (Klink and Lüttge, 1991; Lee Taiz and Taiz, 1991; Dschida and Bowman, 1992; Emig, 1995; Getz and Klein, 1995). In addition to the central stalk, inspection of electron micrographs, using more sophisticated methods of image analysis, has recently led to the visualization of a second peripheral stalk. This peripheral stalk was first observed for the *Clostridium fervidus* Na^+-translocating V-ATPase (Boekema *et al.*, 1997), and later in the H^+-translocating V-ATPase from bovine brain clathrin-coated vesicles (Wilkens *et al.*, 1999). It was also found in the chloroplastic F_oF_1-ATP synthase (F-ATPase) (Böttcher *et al.*, 1998), underscoring the long known structural similarity of V-ATPases and F-ATPases. These head-and-stalk structures represent the membrane-peripheral V_1-domain of the V-ATPase. The stalks connect the head with the membrane-integral V_o-domain, which can be visualized by freeze-fracture electron microscopy as spherical structures exhibiting diameters of 6.5–9.1 nm (Klink *et al.*, 1990; Klink and Lüttge, 1992; Mariaux *et al.*, 1994).

In yeast, V-ATPase (*VMA*) genes encode 13 known subunits, disruption of which leads to the loss of V-ATPase activity and to a lethal phenotype, that is, genes encoding for subunits A, B, C, D, E, F, G, H, a, d, c, c', and c" (see Table 7.1) (Nelson and Nelson, 1990; Nelson and Klionsky, 1996; Stevens and Forgac, 1997; Forgac, 1999). The V-ATPase holoenzyme contains three copies of subunits A and B, six copies of subunit c, and a single copy of all other subunits (Arai *et al.*, 1988; Kaestner *et al.*, 1988) (see Figure 7.1 for a structural model of the V-ATPase). In 1997, plant genes for only four V-ATPase subunits (A, B, E and c) had been cloned (see Lüttge and Ratajczak, 1997, and references therein), although up to 10 different subunits had been proposed to constitute the V-ATPase holoenzyme in some plant species (DuPont and Morrissey, 1992; Ward and Sze, 1992a; Hosaka *et al.*, 1994). The assessment of other proteins to be V-ATPase subunits was based on comparison of molecular mass and immunological evidence. During the last few years, four additional genes coding for plant V-ATPase subunits have been characterized: subunits C (Tavakoli *et al.*, 1999), D (Kluge *et al.*, 1999) and G (Rouquié *et al.*, 1998). The completed sequence of an *Arabidopsis* EST clone (Accession No. Z34714) for

Table 7.1 Nomenclature and properties of vacuolar-type H^+-transporting adenosine triphosphatase (V-ATPase) subunits

Subunit	Yeast V-ATPase gene	Molecular mass of yeast V-ATPase subunit (kDa)	Function	Plant gene cloned	Isoforms found	Molecular mass of plant V-ATPase subunit (kDa)
V_1-domain						
A	*VMA1*	69	ATP-binding; catalytic	Yes	Yes	69
B	*VMA2*	57	ATP-binding; non-catalytic	Yes	Yes	54–55
H	*VMA13*	54	V_1-stabilization	No		50–54[a]
C	*VMA5*	42	V_1-stabilization, activity	Yes	No	42
D	*VMA8*	32	Central stalk, coupling	Yes		29
E	*VMA4*	27	V_1-assembly, activity	Yes	Yes	26–27
F	*VMA7*	14	Bridge V_1-V_o contacts	No		14[a]
G	*VMA10*	13	Peripheral stalk, coupling	Yes	Yes	13
V_o-domain						
a	*VPH1/STV1*	100	Peripheral stalk, assembly; Targeting, activity	No		95–115[a]
d	*VMA6*	36	V_o-assembly, activity	No		32–36[a]
c''	*VMA16*	23	Proton translocation	No		
c	*VMA3*	17	Proton translocation	Yes	Yes	16
c'	*VMA11*	17	proton translocation	No		

[a]Molecular masses of proteins which might correspond to the yeast V-ATPase subunits based on apparent molecular mass comparison and/or immunological evidence. For information about yeast V-ATPase subunits and putative functions, see Stevens and Forgac (1997). See text for relevant references pertaining to plant V-ATPase subunits.

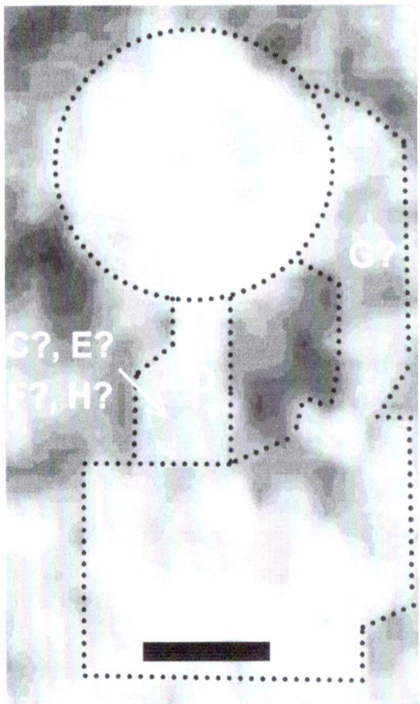

Figure 7.1 (Topological model of the vacuolar-type H^+-transporting adenosine triphosphatase (V-ATPase) holoenzyme. The electron micrograph (provided by Dr I. Emig) shows a negatively-stained V-ATPase of *Mesembryanthemum crystallinum*. It should be mentioned that the resolution of the electron micrograph is not sufficient to unequivocally identify the second, peripheral stalk indicated on the righthand side. Subunits, which have not been cloned in plants and the location of which is unclear are marked by '?'. (Internal scale bar 5 nm).

subunit F (T.A. Wilkins, unpublished) is identical to the sequence deposited in GenBank from the *Arabidopsis* genome sequencing project, indicating that the gene is located on chromosome IV. Subunit F is also a remarkably well conserved polypeptide of 14 kDa that shares 51 and 54.5% amino acid identity with the corresponding subunits of bovine and yeast, respectively (Figure 7.2). However, there are still some subunits with homologs in yeast that have yet to be isolated from plants.

One overriding goal of V-ATPase research has been to determine the localization of V-ATPase subunits in the holoenzyme in order to relate structure to function. Various strategies have been used to determine the subunit composition of V_0 and V_1 and to investigate nearest-neighbor relationships. These strategies include: dissociation of the V_1-domain from the membrane by incubation with chaotropic agents (Lai *et al.*, 1988; Ward and Sze, 1992a; Lüttge *et al.*, 1995a); covalent cross-linking of V-ATPase subunits (Adachi *et al.*, 1990); co-immunoprecipitation (Holzenberg *et al.*, 1993); and expression of a limited number of the V-ATPase subunits in yeast (Tomashek *et al.*, 1997). It has been postulated that the V-ATPase head contains three copies each of subunits A

```
Arabidopsis : MAGRATIPARNSALIAMIADEDTVVGFLMAGVGNVD-IRRKTNYLIV-DSKTTVRQIEDAFKEFS-ARDDIAIILLSQYIANMIRFLV
Bovine      : ~~~~~~~~~~~~~~~~VIGDEDTVTGFLLGGIGELN-KNRHPNFLVV-EKDTTINEIEDTFRQFL-NRDDIGIILINQYIAEMVRHAL
Yeast       : ~~~~~~~MAEKRTLIAVIADEDTTTGLLLAGIGQITPETQEKNFFVYQEGKTTKEEITDKFNHFTEERDDIAILLINQHIAENIRARV

Arabidopsis : DSYNKPVPAILEIPSKDHPYDPAHDSVLSRVKYLFSAESVSQR
Bovine      : DAHQRSIPAVLEIPSKEHPYDAAKDSILRRARGMFTAEDLR~~
Yeast       : DSFTNAFPAILEIPSKDHPYDPEKDSVLKRVRKLFGE~~~~~~
```

Figure 7.2 Amino acid sequence alignment of subunit F of the *Arabidopsis thaliana* vacuolar-type H^+-transporting adenosine triphosphatase (V-ATPase) (Accession No. Z34714; Wilkins, unpublished) with subunits F from the bovine and yeast V-ATPase.

and B, as shown for the β- and α-subunits of the F-ATPase by X-ray crytallography (Abrahams *et al.*, 1994). Although such detailed structural data are not yet available for the V-ATPase, inspection of top views of negatively stained *Mesembryanthemum crystallinum* V-ATPase heads after rotational image analysis (Kramer *et al.*, 1995), and of the *Manduca sexta* midgut V-ATPase heads after image averaging (Radermacher *et al.*, 1999), has revealed a hexameric structure which might represent three A- and three B-subunits. However, pentameric structures have also been found (Kramer *et al.*, 1995), and were discussed as states of V-ATPase assembly and/or turnover.

Although both subunits A and B contain nucleotide binding sites, for several reasons subunit A seems to be the catalytic V-ATPase subunit. Subunit A contains all the consensus loops, including a glycine-rich loop region, which has been shown to be crucial for ATP-hydrolytic activity of the catalytic F-ATPase subunit β (Hirata *et al.*, 1990). Covalent binding of N-ethylmaleimide (NEM) (Wang and Sze, 1985; Randall and Sze, 1986), fluorescein 5′-isothiocyanate (FITC) (Tzeng *et al.*, 1992) and 7-chloro-4-nitrobenzo-2-oxa-1,3-diazole (Nbd-Cl) (Mandala and Taiz, 1986; Feng and Forgac, 1992; 1994; Yamanishi and Kasamo, 1992) to critical amino acid residues in the ATP-binding site of subunit A inhibit ATP-hydrolysis activity. Nbd-Cl was postulated to bind to a critical cysteine residue at or near the catalytic site, since inhibition of V-ATPase activity can be prevented by addition of its substrate, MgATP (Mandala and Taiz, 1986). This was supported by site-directed mutagenesis of a single cysteine residue in the glycine-rich loop, leading to a fully active V-ATPase resistant to the cysteine modifying reagents, Nbd-Cl and NEM (Zhang *et al.*, 1995; Liu *et al.*, 1997). Finally, photolabeling of subunit A by 2-azido-[^{32}P]-ATP correlates directly with inhibition of V-ATPase activity (Zhang *et al.*, 1995).

The first hint of the existence of a nucleotide binding site in subunit B came from labeling studies with the ATP analog [α-^{32}P]3′-O-(4-benzoyl benzoyl) adenosine 5′-triphosphate (Manolson *et al.*, 1985). This was later supported by the identification of a putative nucleotide binding sequence in subunit B (Manolson *et al.*, 1988). However, in contrast to subunit A, the binding of the ATP analog did not inhibit ATP hydrolysis but changed the kinetic properties of the V-ATPase.

The main subunit of the central stalk appears to be subunit D. Although it does not show high sequence similarity to the central-stalk γ-subunit of the F-ATPase, structural similarity indicates that both subunits have similar functions in F- and V-ATPases (Nelson *et al.*, 1995). Subunits C, E, F and H seem to be stalk subunits, however, at present, there is no unequivocal evidence as to whether they belong to the central or the peripheral stalk (see Forgac, 1999). Although it is shorter than its F-ATPase b-subunit counterpart and does not contain an apparent transmembrane domain at its N-terminus, subunit G exhibits significant sequence similarity to the F-ATPase b-subunit (Supekova *et al.*, 1996). Thus, subunit G is a good candidate to be part of the peripheral V-ATPase stalk. It appears to be in tight contact with the membrane peripheral V_0-domain: a 12 kDa polypeptide which might represent subunit G of the *Avena sativa* V-ATPase remains with the V_0-domain after dissociation of the V_1 domain (Ward and Sze, 1992a). In addition, a 12 kDa *Me. crystallinum* polypeptide that cross-reacts with an antiserum against subunit G of *Nicotiana tabacum* has been found in the V-ATPase holoenzyme, but not in a fraction of dissociated V_1-subunits (I. Emig and R. Ratajczak, unpublished).

At least five subunits are suggested to form the V_0-domain of the yeast V-ATPase, namely, subunits a, d and the proteolipids c, c′ and c″ (Stevens and Forgac, 1997). In plants, however, only genes encoding subunit c have so far been cloned. One characteristic structural component of V_0 seems to be a ring of, probably, six proteolipid subunits (Arai *et al.*, 1988; Kaestner *et al.*, 1988). While the position of subunit d in V_0 is not known, *in vitro* proteolysis experiments have provided some insight into the position of subunit a of the yeast V-ATPase (Landoldt-Marticorena *et al.*, 1999). Subunit a is the only V-ATPase subunit present in the yeast genome as two isoforms, *Vph1* (Manolson *et al.*, 1992) and *Stv1* (Manolson *et al.*, 1994). The Vph1p subunit a contains a hydrophobic C-terminal domain containing 6–8 putative transmembrane segments and a large N-terminal hydrophilic domain (Manolson *et al.*, 1992). Since some regions of subunit a have been shown to be exposed to the cytosol by proteolytic digestion and labeling by membrane impermeable agents (Arai *et al.*, 1988; Adachi *et al.*, 1990), it is a good candidate for linking V_1 and V_0, possibly representing one component of the peripheral stalk. *In vitro* susceptibility of Vph1p to trypsin was decreased by the addition of ATP and increased by the addition of Mg^{2+} to the reaction mixture, that is, by two molecules suggested to bind to V_1, indicating that structural changes in V_1 have an impact on Vph1p structure (Landoldt-Marticorena *et al.*, 1999). On the other hand, incubation with bafilomycin A_1, which possibly binds to subunit a (see following paragraph for details), decreases the trypsin susceptibility of subunit A (Landoldt-Marticorena *et al.*, 1999).

V-ATPase actually has two activities: ATP hydrolysis and proton transport. Both activities are maximal at pH values ranging 7.0–8.5 (Sze, 1985), are stimulated by millimolar concentrations of chloride, and inhibited by millimolar concentrations of nitrate (Bennett and Spanswick, 1983; Sze, 1985), and nanomolar concentrations of macrolide antibiotics, such as bafilomycin A_1

and concanamycin A, the latter being 10-fold more effective (Bowman *et al.*, 1988; Dröse *et al.*, 1993). Compared to nitrate, bafilomycin seems to be a much more specific and effective inhibitor of V-ATPase activity. However, there are two examples of V-ATPases exhibiting no or little bafilomycin sensitivity: the V-ATPase of the *Citrus limon* fruit tonoplast appears to be bafilomycin A_1-insensitive (Müller *et al.*, 1996), and senescence induction of *Hordeum vulgare* leaves with methyljasmonate decreases bafilomycin A_1-sensitivity of V-ATPase (Ratajczak *et al.*, 1998), possibly due to a special structure of the V_o-domain. Patch-clamp studies of whole vacuoles have indicated that nitrate interacts with the cytoplasmic side of the V-ATPase, since nitrate applied to the lumenal side does not inhibit V-ATPase activity (Hedrich *et al.*, 1989). For the *Neurospora crassa* V-ATPase, it has been suggested that inhibition of activity by nitrate is due to oxidation of essential cysteine residues in V_1 (Dschida and Bowman, 1995). The site of bafilomycin inhibition seems to be at the V_o-domain, as indicated by studies of proteoliposomes containing reconstituted V_o-subunits. The fact that the F_0-domain of F-ATPase is not inhibited by bafilomycin seems to be due to structural and functional differences of the V- and F-ATPase integral membrane domains. In contrast to the F_0-domain, the native V-ATPase V_0-domain does not exhibit passive proton conductivity after dissociation of the V_1-domain (Zhang *et al.*, 1992). This difference might be due, at least in part, to the lipid environment *in vivo*, since proton conduction is measurable after reassembly of partially purified V_o-subunits in proteoliposomes. Since bafilomycin A_1 blocks H^+-translocation of such reassembled V_0-domains lacking V_1-subunits, it has been suggested that bafilomycin A_1 might bind to the V_o-domain (Crider *et al.*, 1994; Zhang *et al.*, 1994). Subunit a is a potential candidate for bafilomycin binding.

Divalent metal-nucleotide complexes can serve as a substrate for the V-ATPase (for references, see Lüttge and Ratajczak, 1997). The nucleotide and divalent metal specificity is ATP $\gg$ GTP $>$ NTP and $Mg^{2+} \geq Mn^{2+} \gg Ca^{2+}$, Co^{2+}. V-ATPase activity is inhibited by Zn^{2+}, while $ZnATP^{2-}$ can serve as a substrate. For plant V-ATPases, apparent K_M values for $MgATP^{2-}$ range 0.2–0.8 mM (see Lüttge and Ratajczak, 1997, and references therein). Warren *et al.* (1992) determined two distinct K_M values of 0.77 mM and 2 μM for partially purified V-ATPase of *Kalanchoë daigremontiana* reconstituted in proteoliposomes, indicating either the existence of two or more catalytic centers or a cooperativity between nucleotide binding sites. This partially purified V-ATPase had a specific ATP-hydrolysis activity of 180 μmol ATP $mg_{protein}^{-1}$ h^{-1}. Taking into account the molecular mass of the V-ATPase holoenzyme, which can be calculated to be 821 kDa, and assuming the subunit stoichiometry of $A_3B_3C_1D_1E_1F_1G_1H_1a_1c/c'_6c''_1d_1$, this V-ATPase has a substrate turnover rate of 41 ATP s^{-1}. For the V-ATPase in tonoplast vesicles of *Me. crystallinum*, a specific ATP-hydrolysis activity of 300 μmol ATP $mg_{V-ATPase}^{-1}$ h^{-1} was calculated by relating the rate of ATP hydrolysis to the immunologically determined amount of V-ATPase protein (Ratajczak *et al.*, 1994). Thus, the substrate turnover rate of the *Me. crystallinum* V-ATPase present in tonoplast vesicles, a situation which might be closer to the *in vivo* situation as compared to a purified and reconstituted enzyme, is 68 ATP s^{-1}. Proton transport activities of 60 and 90 $H^+ s^{-1}$ have been determined in studies on tonoplast-enriched membrane vesicles isolated from *Av. sativa* (Ward and Sze, 1992b) and *V. radiata* (Matsuura-Endo *et al.*, 1990).

Covalent binding of N,N′-dicyclohexylcarbodiimide (DCCD) to subunit c inhibits V-ATPase activity (Manolson *et al.*, 1985; Mandala and Taiz, 1986; Randall and Sze 1986; Rea *et al.*, 1987), indicating that the integral membrane subunit c is involved in proton transport. Inhibition by binding of one single DCCD molecule per V-ATPase holoenzyme indicates a tight cooperation of c-subunits during proton translocation (Rea *et al.*, 1987; Kaestner *et al.*, 1988). The sites of DCCD-binding are negatively charged amino acid side-chains of subunit c. A good candidate for the DCCD-binding site is the carboxyl group of an essential glutamate. Replacing Glu^{137} by several other amino acids except aspartate, abolished V-ATPase activity in yeast subunit c (Noumi *et al.*, 1991). Hydropathy plots indicate that subunit c contains four membrane-spanning domains (Lai *et al.*, 1991). The critical glutamate is located in the center of membrane-spanning domain IV, which exhibits a highly conserved amino acid sequence (Lai *et al.*, 1991). According to the rotor-stator model of F- and V-ATPase function (Junge *et al.*, 1997), the critical glutamate is directly involved in proton transport. Coupling of ATP hydrolysis and proton transport seems to be achieved by the interaction of a rotor domain consisting of a ring of proteolipids coupled to the central-stalk subunit D, and a stator domain consisting of V_o subunits and the V-ATPase head connected to the peripheral stalk. ATP binding and hydrolysis leads to conformational changes in the head and to a rotation of subunit D. This rotation is transduced to the proteolipid ring, located in the membrane. Protons reach the critical glutamate residue of subunit c from the cytoplasmic side via a semi-channel, which is probably a domain of one of the V_o-subunits. By rotational movement of the proteolipid ring, the proton bound to the critical glutamate reaches another semi-channel, where it is released from subunit c and can reach the vacuolar lumen. Taking into account the threefold symmetry of the V-ATPase head, and assuming that six proteolipids make up the proteolipid ring, one may suspect that hydrolysis of the γ-phosphate bond of three ATP would lead to one entire revolution of the proteolipid ring and to the translocation of six protons from the cytoplasm to the vacuolar lumen (for a detailed description of the rotary mechanism of F- and V-ATPases and its kinetics, see Junge *et al.*, 1997; Pänke and Rumberg, 1999). This coupling ratio ($H^+_{transported}/ATP_{hydrolyzed}$) of two corresponds to the value considered in early studies of the V-ATPase of higher plants (Lüttge *et al.*, 1981; Smith *et al.*, 1982; Bennett and Spanswick, 1984).

Interestingly, by patch-clamp analysis, it was found that the coupling ratio of the V-ATPase might vary due to the pH of the medium, ranging 1.75–3.28 $H^+_{transported}/ATP_{hydrolyzed}$ (Davies *et al.*, 1994). This could be achieved by an indirect coupling mechanism, as proposed by Brauer *et al.* (1993), based on the observation that the H^+-transport and ATP-hydrolysis activity of *Zea mays* V-ATPase are inhibited by nitrate to different extents. Moriyama and Nelson (1988) have suggested an intrinsic uncoupling ('slip') of the V-ATPase holoenzyme. Varying coupling ratios might be due to changes of the 'slip' or uncoupling rates of the V-ATPase as has been reported by Müller *et al.* (1997), who observed a 2.0–2.4-fold intrinsic uncoupling of the reconstituted *Ci. limon* epicotyl V-ATPase compared to the fruit V-ATPase.

In addition to 'slip', another possible way of obtaining varying V-ATPase coupling ratios would be an incorporation of proteolipid isoforms, differing in the amount of critical negatively-charged amino acid side-chains or number of proteolipids in the proteolipid ring. The latter has been suggested to occur in *Me. crystallinum* during the shift from C_3-photosynthesis to crassulacean acid metabolism (CAM), based on the following observations: i) intramembrane particles (IMPs) corresponding to the V_o-domain appear larger in diameter when visualized on freeze-fracture replicas of tonoplast vesicles from CAM plants than from C_3-plants (Klink and Lüttge, 1992; Rockel *et al.*, 1994); and ii) during the C_3-CAM shift, the staining intensity of subunit c in polyacrylamide gels increases to a higher extent than the staining intensity of all other V-ATPase subunits (Rockel *et al.*, 1994), indicating a change of subunit stoichiometry. A change in coupling ratio by incorporation of different isoforms of subunit c was suggested to occur in leaves of tobacco plants grown under low (2 mM) and high (40 mM) nitrate supply (Fischer-Schliebs *et al.*, 2000). While ATP-hydrolysis activity of tonoplast vesicles isolated from both types of plants was not significantly different, H^+-transport activity was higher in low-nitrate plants. Two antisera were used to determine the protein amount of subunit c present in tonoplast vesicles. The fact that one of these antisera detected equal amounts of subunit c, while another antiserum exhibited a lower cross-reactivity with the c subunit of low-nitrate plants, indicates that structurally distinct c subunit isoforms are expressed due to differences in nitrate supply (Fischer-Schliebs *et al.*, 2000).

For energization to occur, it is not only the activity, but also the amount of the V-ATPase at the tonoplast that is important. For different plant species (for references, see Lüttge and Ratajczak, 1997), densities of 3.3×10^6–225.6×10^6 V-ATPases per vacuole or cell and 970–3380 V-ATPases per μm^2 tonoplast area have been determined by electron microscopy and or patch-clamp studies.

7.2.2 *Assembly, targeting and protein turnover*

V-ATPases are not found exclusively at the tonoplast. From studies of fractionated membrane vesicles, there is evidence that V-ATPases are located in various intracellular membranes, including the endoplasmic reticulum (ER) and the

Golgi apparatus, and in the plasma membrane (see Lüttge and Ratajczak, 1997, and references therein). These biochemical data are supported by immunocytochemical investigations showing that antibodies directed against V-ATPase subunits label proteins located in the ER, the Golgi apparatus, clathrin-coated vesicles and the plasma membrane (Depta *et al.*, 1991; Herman *et al.*, 1994; Oberbeck *et al.*, 1994; Ratajczak *et al.*, 1999). This wide distribution of V-ATPases within the cell raises several questions concerning the assembly, targeting and turnover of the V-ATPase holoenzyme. In yeast, two different pathways for the assembly of the V-ATPase holoenzyme seem to exist. One possibility is that the V_0- and the V_1-domain are independently synthesized (Kane and Stevens, 1992). The subunits of V_0 are co-translationally inserted into the ER and assembled to the intact V_0-domain, to which the assembled V_1-domain can bind (Kane and Stevens, 1992; Bauerle *et al.*, 1993). Another mechanism, which might be the main route for V-ATPase assembly in yeast, was proposed by Kane *et al.* (1999). Using pulse-chase experiments, these authors found that subunits of V_1 associate with the 100 kDa integral membrane subunit a within 3–5 min, followed by the binding of all other V_0-subunits to this subcomplex. Interestingly, an unknown 19 kDa polypeptide seems to bind to the subunit a/V_1-subcomplex prior to the binding of the other V_0-subunits. It was proposed that the 19 kDa protein acts as an assembly factor, appearing during early stages of V_0V_1- and V_0-assembly (Kane *et al.*, 1999). In plants, the pathway for the assembly of the V-ATPase holoenzyme at the ER seems to be different. In an elegant study by Sze and co-workers (Li *et al.*, 1998), two molecular chaperones of the ER, namely, calnexin (64–67 kDa) and BiP (binding protein; 78 kDa), have been shown to be involved in mediating the assembly of the V-ATPase in oat root cells. The authors suggested that the integral membrane chaperone, calnexin, associates with regions of V_0-subunits that are exposed to the lumen and/or to transmembrane domains. The soluble BiP, which is located in the lumen of the ER, binds to lumenal domains of the V_0-subunits. Both chaperones seem to bind sequentially in facilitating the assembly of the V_0-domain and the binding of V_1-subunits or completely assembled V_1-domains to V_0. As in yeast, a 100 kDa polypeptide (possibly subunit a) seems to play a prominent role in V-ATPase assembly (Li and Sze, 1999). The mature V-ATPase holoenzyme could be transported to the vacuole in membrane vesicles directly from the ER or via the Golgi apparatus (see Lüttge and Ratajczak, 1997, and references therein). This might explain the occurrence of V-ATPases in the ER and the Golgi apparatus due to intracellular vesicle trafficking. However, it does not explain the finding that in some plant species, V-ATPases are located in the plasma membrane, and raises the question of whether a fraction of V-ATPases is directly targeted to the plasma membrane.

Another indication of differential targeting of the V-ATPase holoenzyme within the cell comes from the observation that V-ATPases are found mainly in 'aqueous' vacuoles of *Mimosa pudica* motor cells, while almost no V-ATPase is present in the tannin vacuoles of the same cells (Fleurat-Lessard *et al.*, 1997). In plants, no factors influencing the targeting of V-ATPases are yet known. In yeast, the new small gene family, *VTC* (vacuolar transporter chaperone), was recently discovered (Cohen *et al.*, 1999). Since the deletion of the *VTC1* gene resulted in a reduced amount of V-ATPase in the vacuole, it was suggested that the members of this gene family play a role in V-ATPase targeting. Besides the *VTC* gene products, the two isoforms of subunit a, Stv1p and Vph1p, have been proposed to be involved in differential targeting of V-ATPase holoenzymes to the Golgi apparatus and vacuole, respectively (Manolson *et al.*, 1994).

Few data are available concerning *in vivo* turnover rates of the V-ATPase holoenzyme and possible mechanisms of protein degradation. After salt treatment of *Me. crystallinum* plants, the abundance of V-ATPase in tonoplast vesicles, as determined by radial immunodiffusion, increased by a factor of 2.5 (Ratajczak *et al.*, 1994). Within 2 days following removal of the salt from the soil, the amount of V-ATPase protein decreased again to levels found in control plants, indicating a degradation of V-ATPase protein. One of the first targets in degradation of the V-ATPase might be subunit B. After induction of CAM by salinity stress, the V-ATPase B-subunit is cleaved *in vivo* at the carboxyl side of Met^{192}, resulting in the novel appearance of the D_i polypeptide, exhibiting a molecular mass of 32 kDa (Zhigang *et al.*, 1996). The formation of the D_i polypeptide might have physiological significance. Interestingly, V-ATPases from plants of *Me. crystallinum* in the C_3-state and the CAM-state of photosynthesis have different properties. The CAM form of the V-ATPase shows a greater stability when incubated in solutions of high ionic strength (Lüttge *et al.*, 1995a) and detergents (Ratajczak, 1994), and seems to have a lower proton transport/ATP hydrolysis ratio (E. Berndt and R. Ratajczak, unpublished). It cannot be ruled out that the appearance of the D_i polypeptide in the V-ATPase holoenzyme is responsible, at least in part, for these differences. On the other hand, it is also possible that V-ATPases containing the D_i polypeptide represent a fraction which is in the first stage of protein degradation. Initial experiments show that the cleavage of subunit B is mediated by a cytosolic protease and/or by a reactive oxygen species (Krisch *et al.*, 2000). For the yeast V-ATPase, it has been observed that Vph1p is rapidly degraded to an 80 kDa polypeptide by endogenous proteinase A-dependent proteases during standard isolation procedures (Manolson *et al.*, 1992). It remains to be determined whether these proteases have an impact on V-ATPase turnover *in vivo*.

7.2.3 Regulation of V-ATPase expression

The regulation of V-ATPases is a highly complex process as it requires the expression of genes encoding the various subunits, assembly and targeting of the V-ATPase holoenzyme to functionally diverse subcellular compartments, and modulation of activity in response to a host of developmental, physiological and environmental stimuli. Although useful for the determination of V-ATPase subunit composition, efforts to study V-ATPase structure and function by generating gene knockouts in animals and fungi have proved unsuccessful, resulting in lethal phenotypes (Davies *et al.*, 1996; Xie *et al.*, 1996; Dow *et al.*, 1997; Inoue *et al.*, 1999). These results clearly underscore the essential nature of the V-ATPase. Disruption of yeast V-ATPase genes can, however, produce a conditional lethal phenotype in which *vma* mutants are able to grow under acidic conditions (pH 5.5) but not at pH 7.5. Much of our current understanding of V-ATPase structure, and the steps involved in the assembly of the V_0V_1 holoenzyme is, in fact, derived almost entirely from molecular genetic and biochemical analysis of the *VMA* mutants (Finbow and Harrison, 1997; Stevens and Forgac, 1997, and references therein). Functional complementation of *VMA1* and *VMA10* mutants by plant genes encoding homologs to subunits A (Kim *et al.*, 1999) and G (Rouquie *et al.*, 1998), respectively, restores the wild-type phenotype and attests to the conservation of V-ATPase structure and function in eukaryotes during the course of evolution (Stevens and Forgac, 1997).

The number of V-ATPase subunit isoforms identified in both plants and animals is steadily rising, suggesting that these multicellular organisms have adapted additional means to regulate V-ATPase activity. In animal cells, the synthesis of membrane- and tissue-specific subunit isoforms (Hemken *et al.*, 1992; Peng *et al.*, 1994; Hernando *et al.*, 1995) is the result of alternative mRNA splicing. For example, alternative splicing of subunit A transcripts generated by a single, ubiquitously expressed gene in chicken osteoclasts produces two isoforms with different catalytic properties. The A2 isoform lacks the P-loop motif and the capacity to bind ATP. Thus, the catalytic properties of V-ATPase depend on the stoichiometry of the A1/A2 isoforms in the V_1 headgroup (Hernando *et al.*, 1995). Plants are unique among the eukaryotes in that small multigene families encode closely-related isoforms of subunits A, B and c (see Lüttge and Ratajcak, 1997, and references therein). Subunits A and c are highly conserved, and have been among the first of the V-ATPase genes to be isolated and studied in plants. Determining the number of genes encoding any particular subunit, how the genes are regulated, and how differential expression of the genes may influence the subunit composition, subcellular distribution, assembly and/or activity of the V-ATPase are questions of major importance that need to be answered.

The first compelling evidence that plant V-ATPase subunit A is encoded by two genes came from studies on the gene structure and genome organization of subunit A genes in cotton (*Gossypium hirsutum*) (Wilkins *et al.*, 1994). The ancient duplication event that initially gave rise to the two subunit A genes occurred following the bifurcation of plants and fungi during the course of evolution. Although the size and nucleotide composition of the introns varies between the two genes, the number and position of the introns is highly conserved, even among plant species (Starke *et al.*, 1991; Wilkins *et al.*, 1994;

Lehr *et al.*, 1999; Kawamura *et al.*, 2000), although the intron reported by Starke *et al.* (1991) may have been lost from a barley gene (F. DuPont, personal communication). Genome analysis has identified at least two genes in the diploid wheat (J. Dubcovsky, personal communication), one of which has been mapped to the long arm of chromosome 7A in *Triticum monococcum* (Dubcovsky *et al.*, 1995). In cotton, the two ancestral V-ATPase subunit A genes have been amplified to form discrete multigene families, in a manner that is consistent with the multiple polyploidization events that have occurred in the evolutionary history of this genus (Wilkins *et al.*, 1994). The number of subunit A genes is, therefore, likely to be species-specific, depending on the role polyploidization has played in the evolution of the species. The cotton A1 and A2 isoforms are near identical polypeptides with slightly different isoelectric points, and while it is difficult to conceive that the few amino acid substitutions would significantly impact catalytic activity, the differential regulation of these genes does significantly influence the subunit composition of the tonoplast V-ATPase (see Section 7.2.5). Both isoforms, which share only ~60% amino acid identity with the yeast subunit, are able to rescue the *VMA* mutant phenotype in yeast in functional complementation experiments, although ATPase hydrolytic and proton pumping activity is only partially restored by the plant-yeast hybrid holoenzyme (Kim *et al.*, 1999). In domain substitution experiments, increased V-ATPase activity by a plant-yeast chimeric subunit A indicated that divergent N- and C-terminal domains in the plant subunit play an important functional role in V-ATPase activity.

At least four genes encoding near identical V_0 subunit c 16 kDa proteolipids in cotton (Hasenfratz *et al.*, 1995), oat (Lai *et al.*, 1991) and *Arabidopsis* (Perera *et al.*, 1995) have been shown to be differentially regulated. In cotton, although gene transcripts for two genes (*CVA16.2* and *CVA16.4*) are present in all tissue types, expression of both genes is highly induced in expanding petals, root meristems and mature anthers. Expression of genes encoding subunits A and B produces a similar pattern (T.A. Wilkins, unpublished), indicating that V_0 and V_1 genes are coordinately regulated during plant growth and development. However, the two *CVA* genes are differentially regulated in developing ovules, resulting in the preferential accumulation of *CVA16.4* transcripts (Hasenfratz *et al.*, 1995). The subunit c genes in *Arabidopsis*, however, show two discrete patterns of expression (Perera *et al.*, 1995; Sze *et al.*, 1999). *AVA-P1* and *AVA-P2* are constitutively expressed in vegetative and reproductive tissues, indicating a housekeeping role for these particular genes, whereas the tissue-specific expression of *AVA-P3* is restricted to root and shoot apexes. These results indicate that V-ATPase multigene families perform overlapping, and possibly redundant functions to some extent. Induced gene expression suggests an increased need for V-ATPase activity in highly active cell-types, although the signals responsible for regulating V-ATPase gene expression under these conditions are not yet known.

Numerous groups, employing a variety of approaches, continue the effort to target the gene isolation of V_1 and V_0 subunits not yet characterized, as well as the accessory proteins required for V-ATPase assembly and transport. Genome initiatives are aiding this research as the number of V-ATPase genes appearing in expressed sequence tag (EST) databases steadily increases. Clones encoding V_1 stalk subunits D, E and F have been reported. Not all of the plant stalk subunits (e.g. subunits C, D and E) are as well conserved relative to other species at the primary amino acid level, and it has been suggested that, in these cases, it is the secondary structure of these subunits that may be of primary importance. For instance, the amino acid identity of subunit C in plants, animals and fungi is only about 37–38% (Schumacher *et al.*, 1999; Tavakoli *et al.*, 1999). Interestingly, although the *VMA5* gene encoding subunit C of yeast was able to restore the wild-type phenotype in the subunit C-defective *det3* mutant of *Arabidopsis*, *DET3* was not successful in rescuing the yeast *vma5* mutant (Schumacher *et al.*, 1999). Subunit D encodes a 29 kDa polypeptide that shares about 48% amino acid identity with yeast Vma8p (Kluge *et al.*, 1999). The V_1 E subunits of plants, isolated from *Arabidopsis* (Accession No. AL049876), barley (Dietz *et al.*, 1995), cotton (Kim and Wilkins, 1997), lemon (Reuveni and Avi, 2000), *Me. crystallinum* (Dietz and Arbinger, 1996) and spinach (Dietz *et al.*, 1996), share ~75% amino acid identity, but show less than 30% homology to mammalian and fungal subunits (Kim and Wilkins, 1997). The molecular mass of E subunits varies considerably among plant, animal and fungal species, and several immunologically-related E isoforms ranging from 27 to 42 kDa have been detected in plants (Kawamura *et al.*, 2000), suggesting that subunit E in plants is also encoded by a multigene family. Understanding the regulation of V_1 stalk subunits is attracting heightened interest, especially as changes in subunit isoforms are becoming increasingly associated with V-ATPase isotypes exhibiting discrete biochemical properties (see Section 7.2.6).

7.2.4 Abiotic induction and downregulation

A variety of abiotic factors have been reported to influence V-ATPase activity and expression. In *Zea mays* seedlings light inhibition of cell extension growth is accompanied by a decrease in subunit A and c steady-state message levels (Viereck *et al.*, 1996). Blue light treatment of *Z. mays* coleoptiles leads to a reversible inhibition of V-ATPase activity, which might be mediated by reactive oxygen species oxidizing essential SH-groups of the enzyme (Krauss *et al.*, 1987). Treatment of *Ho. vulgare* roots with the heavy metal salt, $AlCl_3$, decreased V-ATPase H^+-pumping (Kasai *et al.*, 1993a). This decrease was correlated with an increase in the abscisic acid content of the tissue, suggesting an involvement of the phytohormone in V-ATPase regulation. Several reports have described temperature-dependent responses of the V-ATPase. ATP-dependent H^+-transport of tonoplast vesicles from hypocotyls of chilling-sensitive mung bean seedlings declined markedly when plants were grown below 10°C, while there was no response even at 0°C in chilling-insensitive pea seedlings (Yoshida and Matsuura-Endo, 1991). The decrease in activity was only observed after chilling *in vivo* and not *in vitro*, and was not due to a change in the ion permeability of the tonoplast. Moriyama and Nelson (1989) reported an irreversible *in vitro* cold-inactivation of the bovine chromaffin granule V-ATPase correlating with the dissociation of V_1-subunits from V_0, as was also found for the mung bean and oat V-ATPase (Yoshida, 1991; Ward *et al.*, 1992). An involvement of V_0V_1-dissociation in cold-inactivation is supported by the observations that recovery of V-ATPase activity did not depend on *de novo* protein synthesis (Yoshida, 1991), and that, after a 3 day exposure of mung bean hypocotyls to 0°C, the levels of several V_1-subunits in the tonoplast vesicle fraction decreased, while the proteolipid level did not change (Matsuura-Endo *et al.*, 1992). Cold-sensitivity of the V-ATPase might depend on the developmental status of the cell, since the V-ATPase of mung bean suspension cells was very cold sensitive at the early stage of exponential growth and less cold sensitive at the late state (Yoshida *et al.*, 1993). Acclimation to chilling conditions was observed in tomato cell-suspension cultures, where the temperature optimum of the V-ATPase broadened and shifted to a lower temperature range in tonoplast membranes from cells grown at 9°C as compared to 28°C (DuPont and Mudd, 1985).

The best studied environmental factor influencing V-ATPase expression and activity is salinity. Although more information is still required to understand the mechanism of the regulation of V-ATPase under saline conditions, the ability to upregulate V-ATPase seems to be a major element of the salt tolerance of halophytes. An increase of message amount for V-ATPase subunits as a response to NaCl exposure has been reported for *Arabidopsis thaliana* (Perera *et al.*, 1995), *Beta vulgaris* (Kirsch *et al.*, 1996; Lehr *et al.*, 1999), *Daucus carota* (Löw and Rausch, 1996), *Lycopersicon esculentum* (Binzel, 1994, 1995; Binzel and Dunlap, 1995), *Me. crystallinum* (Löw *et al.*, 1996; Tsiantis *et al.*, 1996; Rockel *et al.*, 1998) and *Ni. tabacum* (Narasimhan *et al.*, 1991). In *Me. crystallinum*, the expressional response to salinity seems to be regulated at tissue and cell levels, and is under developmental control (D. Golldack and K.-J. Dietz, personal communication). While transcript levels of V-ATPase subunits increased in leaves of mature plants, there was no response in juvenile plants. Moreover, in roots of adult plants there was no change in V-ATPase subunit message amount (D. Golldack and K.-J. Dietz, personal communication). So far, an increased level of V-ATPase protein after salt treatment has been reported only for *Me. crystallinum* (Ratajczak *et al.*, 1994), while in *D. carota* (Löw *et al.*, 1996; Rausch *et al.*, 1996) and *Ho. vulgare* (Hurkman *et al.*, 1988; DuPont, 1992; Mariaux *et al.*, 1997) there was no change, and in *Ni. tabacum* (Reuveni *et al.*, 1990) and *Triticum aestivum* (Wang *et al.*, 2000) the V-ATPase protein amount decreased. A salinity-induced increase in ATP-hydrolysis activity of V-ATPase has been observed in *Acer pseudoplatanus* (Zingarelli *et al.*, 1994), *Ly. esculentum* (Sanchez-Aguyo *et al.*, 1991), *Me. crystallinum* (Struve and Lüttge, 1987; Bremberger *et al.*, 1988; Vera-Estrella *et al.*, 1999), *Ni. tabacum*

(Reuveni *et al.*, 1990), *Sorghum bicolor* (Koyro *et al.*, 1992) and *V. radiata* (Nakamura *et al.*, 1992). However, this was not observed in *Helianthus annuus* (Ballesteros *et al.*, 1996, 1997), *Ho. vulgare* (Mariaux *et al.*, 1997) and *Spartina townsendii* (Koyro *et al.*, 1992). In some cases, the changes of V-ATPase amount do not correlate with the changes of V-ATPase hydrolytic activity; an explanation for this might be that, in most cases, ATP-hydrolysis activities were expressed in relation to total tonoplast protein and not to the amount of V-ATPase protein present in tonoplast vesicles. Contradictory results were obtained for cultured cells of *D. carota*: while Colombo and Cerana (1993) found no changes of ATP-hydrolysis activity due to salt treatment, Löw and Rausch (1996) reported a decrease in V-ATPase hydrolytic activity. Interestingly, in the latter study, the authors observed an increase in H^+-transport activity of the V-ATPase, indicating a change in the coupling ratio of ATP-hydrolysis and H^+-transport. Similar observations, that is, a difference in behavior of the two activities of the V-ATPase under salt-stress conditions, have been made for *Ho. vulgare* (Mariaux *et al.*, 1997), *Ho. annuus* (Ballestoros *et al.*, 1996), and *Me. crystallinum* (E. Berndt and R. Ratajczak, unpublished). An explanation for this could be a change in the 'slip' of the V-ATPase holoenzyme, or a change in the number of proteolipids in the V-ATPase holoenzyme under certain conditions, as discussed in Section 7.2.1.

7.2.5 Developmental regulation

Differential expression of V-ATPase genes in a tissue- and/or cell-specific manner is under the regulatory control of developmental signals that mediate plant growth and development, and encompasses basic processes, such as leaf ontogeny, flower opening, flower color, pollen fertility, fruit development and ripening, seed germination and seedling growth, and light-mediated responses. As often happens, there is a common thread that links these developmental events, and that thread is cell expansion. Cell expansion is a fundamental cellular process, and one that requires the activity of the V-ATPase in vacuoles to go beyond the housekeeping functions ascribed to this proton pump.

In keeping with the statement that V-ATPases are developmentally regulated during cell expansion, V-ATPase gene expression has been observed to increase significantly in tissues and cell-types preparing to undergo rapid elongation, or already actively engaged in expanding. For instance, gene transcripts of subunit A of lemon and tomato (Binzel and Dunlap, 1995; Takanokura *et al.*, 1998), and subunit E of barley (Dietz *et al.*, 1995) are more abundant in expanding leaves than in mature leaves. The expansion of the corolla during opening of the flower at anthesis is also associated with an increase in subunit c gene expression (Hasenfratz *et al.*, 1995) and V-ATPase activity (Reuveni *et al.*, 2000). The expression of subunit A is also developmentally regulated during fruit growth and development in lemon (Takanokura *et al.*, 1998) and cotton

(Smart *et al.*, 1998). In developing cotton fibers, V-ATPase activity at the tonoplast is regulated in parallel with the rate of fiber expansion (Smart *et al.*, 1998). The dramatic increase of V-ATPase transcripts, to peak levels around 10–12 days post-anthesis, is followed closely by a corresponding increase in the abundance of subunits A, B and c, and an increase in V-ATPase activity coincident with an increase in the rate of fiber expansion. As the rate of expansion decreases, so do V-ATPase RNA, protein and activity, which eventually return to basal levels. The results indicate that V-ATPase activity is regulated primarily at the transcriptional level via the coordinate regulation of V-ATPase genes for subunits A, B, and c (Smart *et al.*, 1998; T.A. Wilkins, unpublished). The inducstion of constitutively expressed V-ATPase genes is an essential step in cell expansion, as a fiber mutant that does not have the developmental increase in V-ATPase expression and activity does not undergo rapid elongation, resulting in abnormally short fibers (T.A. Wilkins, unpublished). Further evidence that expression of V-ATPase genes is coordinately regulated during expansion is the downregulation of subunit A and c RNA transcripts by light-induced inhibition of mesocotyl growth (Viereck *et al.*, 1996).

Plants in which the V-ATPase has been crippled provide the most compelling evidence that the upregulation of the V-ATPase is vital to cell expansion. Antisense suppression of subunit A expression in transgenic carrots results in a dwarf phenotype (Gogarten *et al.*, 1992). In another striking example, reduced cell expansion in the *det3* mutant of *Arabidopsis* is organ-specific, and reduced hypocotyl elongation is conditional (Cabrera y Poch *et al.*, 1993). The *det3* phenotype is the result of a splicing defect in the *DET3* gene, which encodes subunit C of the V-ATPase (Schumacher *et al.*, 1999). The *det3* mutation results in a twofold decrease in expression of a functional subunit, membrane-bound V_1 headgroups and V-ATPase activity, the impact of which is most profound in rapidly growing cell-types. In both studies, the suppression of subunits A and C gene expression in carrot and *Arabidopsis*, respectively, is presumed to inhibit V-ATPase-dependent solute uptake by the vacuoles, leading to a collapse in cell turgor, the driving force of cell expansion.

The number of examples implicating a role for hormone signal transduction pathways in the developmental regulation of the V-ATPase is steadily increasing. Abscisic acid (ABA) treatments result in an increase in V-ATPase activity that mimics salt-stress induced changes in halophytes (see Section 7.2.4). However, it was recently shown that induction of the V-ATPase is differentially regulated in response to the two stimuli. Surprisingly, the salt-stress response in *Me. crystallinum*, including upregulation of V-ATPase, apparently occurs via an ABA-independent pathway (Barkla *et al.*, 1999). Moreover, the ABA-induced increase in V-ATPase activity, which accompanies the developmentally programmed shift from C_3 metabolism to CAM, is age-dependent, whereby V-ATPase activity and CAM are induced only in adult and not in juvenile plants (Adams *et al.*, 1998; Barkla *et al.*, 1999). ABA also induces an increase in V-ATPase activity in roots of barley (Kasai *et al.*, 1993a, 1994), subunit A gene transcripts in tobacco cell suspension cultures and leaves of *Brassica* (Narasimhan *et al.*, 1991; Orr *et al.*, 1995), and subunit c transcripts in *Me. crystallinum* (Tsiantis *et al.*, 1996).

Interestingly, neither salt stress nor ABA treatments induced gene expression of subunit D in *Arabidopsis* seedlings (Kluge *et al.*, 1999). However, this apparent incongruity could be due to several overriding factors. Firstly, *Arabidopsis* is salt-sensitive relative to the other species. Secondly, the age of the *Arabidopsis* plants may be a consideration, since the ABA response was tested only in young seedlings. Lastly, there is no evidence yet available to suggest that expression of all V-ATPase subunits is necessarily induced by ABA, even though salt stress stimulates expression of V_1 subunits A, B, C and E (Narasimhan *et al.*, 1991; Ratajczak *et al.*, 1994; Binzel, 1995; Dietz *et al.*, 1995, 1996). In *Me. crystallinum*, for instance, there is no coordinate induction of V-ATPase expression by salt stress, and only subunit c has been shown to increase after a few hours of salt stress (Löw *et al.*, 1996; Tsiantis *et al.*, 1996). Furthermore, the abundance of tonoplast-associated V-ATPase subunits A, B, D and E is not altered in response to salt stress (Vera-Estrella *et al.*, 1999). The situation is actually much more complex, since ABA-dependent and ABA-independent pathways are both involved in signaling the stress response to salinity, cold and drought (Ishitani *et al.*, 1997). This is evidenced by the fact that salt stress also leads to a increase in endogenous levels of ABA in lemon seedlings (Gómez-Cadenas *et al.*, 1998).The increase of V-ATPase transcripts in response to ABA suggests that regulation is at the transcriptional level, as with many other ABA-responsive genes. Promoter activity of genes from subunits A and c from red beet increases in response to salt (Lehr *et al.*, 1999), but it is as yet undetermined if conserved G-box motifs in these promoters function as *cis*-regulatory elements in ABA signal transduction. Treatments with the phytohormone, cytokinin, did not effect V-ATPase activity in barley roots (Kasai *et al.*, 1993b). An increase in the ethylene precursor, aminocyclopropane-1-carboxylic acid, was reported in salt-treated plants, although this response has yet to be studied in sufficient detail to ascribe a role for ethylene in regulating V-ATPase activity with any degree of certainty (Gómez-Cadenas *et al.*, 1998).

Recent studies have indicated that other phytohormones, such as gibberellins (GA), methyl jasmonate and brassinosteroids (BR), may also play a key role in the differential regulation of V-ATPase activity. During germination, V-ATPase subunits A and B, and subunit c mRNA increase locally in the micropylar region of tomato seeds prior to emergence of the radicle in a GA-dependent manner (Cooley *et al.*, 1999). V-ATPase subunits noticeably absent in seeds of a GA-deficient (*gib-1*) mutant, appear in these seeds when exogenous GA is applied. These results indicate that V-ATPase genes are regulated in a developmental and tissue-specific manner in response to GA at the transcriptional level. Treatment of barley leaves with methyl jasmonate induces senescence, and has also been shown to induce changes in the subunit composition of the V-ATPase, pH optima of the enzyme and its sensitivity to inhibitors, although it has yet to be determined whether methyl jasmonate acts at the gene level (Ratajczak *et al.*, 1998). The case for BR is less clear but intriguing nonetheless. Subunit C (*DET3*) is required for BR-induced cell elongation of the hypocotyl in *Arabidopsis*, and it has been proposed that the V-ATPase is a downstream target of the BR signal transduction pathway as a means to regulate V-ATPase assembly during cell expansion (Schumacher *et al.*, 1999). An alternative explanation for the reduced sensitivity of the *det3* mutant to BR is that the BR receptor is missorted or that second messenger systems are perturbed as a result of the decrease in V-ATPase activity. It is interesting to note that in support of the latter scenario, it was recently shown that Ca^{2+} signaling may play an important role in regulating V-ATPase expression (Bressan *et al.*, 1998).

7.2.6 *Post-translational modification*

Expression studies clearly indicate that regulation of genes encoding V-ATPase subunit isoforms at the transcriptional level plays a primary role in determining the subunit composition of the V-ATPase holoenzyme (see Table 7.1). In developing cotton fibers, both A1 and A2 isoforms are found in association with the tonoplast. However, the genes encoding the two isoforms are differentially expressed, resulting in a transient shift in the A1/A2 ratio during expansion, in favor of the A1 isoform (T.A. Wilkins, unpublished). The two A isoforms predicted to exist by the cotton study (Wilkins *et al.*, 1994) have been identified at the gene or protein level in more than eight other species so far, including *Av. sativa*, *Chenopodium* sp., *Clematis* sp., *D. carota*, *Hydrastis* sp., *Ly. esculentum*, *Ni. tabacum* and *Me. crystallinum* (for review, see Lüttge and Ratajczak, 1997). Other subunits of the V_1 headgroup for which two isoforms exist include: subunit B (Berkelman *et al.*, 1994; Wan and Wilkins, 1994; Löw *et al.*, 1996); subunit E (Kawamura *et al.*, 2000; Reuveni *et al.*, 2000); and subunit G (Rouquié *et al.*, 1998). Near identical V_0 isoforms of subunit c that vary by a single amino acid are encoded by genes differentially and developmentally regulated in a tissue-specific manner (Hasenfratz *et al.*, 1995; Perera *et al.*, 1995; Tsiantis *et al.*, 1996). In fact, evidence that control of subunit composition at the gene level is associated with the genesis of functionally distinct V-ATPase isotypes exhibiting discrete biochemical properties is steadily increasing. Such isotypes characteristically possess activities and sensitivities to inhibitors that are specific to the V-ATPase.

In addition to being regulated by developmental and environmental cues in a tissue- and cell-specific manner, as described in the preceding paragraphs, organelle-specific V-ATPase isotypes are apparently tailor-made to suit the physiological needs of the endomembrane compartment. For instance, Golgi and vacuolar isotypes vary in subunit composition as well as showing differential sensitivities to inhibitors (Gogarten *et al.*, 1992; Matsuoka *et al.*, 1997; Li and Sze, 1999). The regulation of V-ATPase in vacuoles of lemon juice sacs is beginning to emerge as an interesting process. Although most vacuoles are acidic with a pH of ~5.5, juice sac vacuoles may be acidified to a pH as low as 2. Typical of most V-ATPases, the V-ATPase in lemon epicotyls, which is structurally similar to the juice sac V-ATPase, is sensitive to bafilomycin and nitrate (Müller *et al.*, 1997). In sharp contrast, the juice sac V-ATPase is insensitive to bafilomycin and nitrate, and is partially inhibited by orthovanadate (Müller *et al.*, 1996), the latter feature being a trait not normally associated with V-ATPases, but with P-type E_1E_2-ATPases. In these P-ATPases, vanadate prevents the formation of a phosphorylated intermediate, which is an important step in the catalytic cycle of the enzyme (Michelet and Boutry, 1995). Thus, vanadate inhibition of a special type of V-ATPase might indicate a phosphorylation step important for enzyme activity.

The link between V-ATPase structure and function has recently been demonstrated in two independent studies, in which characteristic properties and activity of the V-ATPase in vacuoles are associated with the differential expression of subunit E isoforms. Using petunia flowers that differ in vacuolar pH as a model system, Reuveni *et al.* (2000) determined that the decrease in vacuolar pH is accompanied by a corresponding increase in V-ATPase activity that is at least three fold higher than in flowers with a higher vacuolar pH, although there is no decrease in V-ATPase activity associated with an increase in vacuolar pH. Intriguingly, while the abundance of V-ATPase subunits A and B appears relatively unchanged in both types of vacuole, there was a considerable difference observed between the two V-ATPases with respect to subunit E. The subunit E isoform associated with low vacuolar pH is more abundant and of lower molecular mass than the subunit E isoform in vacuoles of higher pH. A similar picture is emerging for tissue-specific V-ATPase isotypes in vacuoles of leaves and epicotyls of pea (Kawamura *et al.*, 2000). The V-ATPases isolated from these two tissues display unique kinetic properties, with the leaf enzyme being more efficient in terms of proton pumping. The subunit composition of the V-ATPase isotypes appeared similar, except for subunit E. The 37 kDa E2 subunit is found in all tissues examined, but while E2 is highly enriched in leaves and cotyledons, the 40 kDa E1 isoform is barely detectable in these tissues. Although it has been suggested that immunologically-related subunit E isoforms in plants are also encoded by a multigene family, this has yet to be firmly established.

Cellular control, not only of subunit isoforms but also of subunit stoichiometry, may be a key factor in regulating V-ATPase assembly, targeting and/or activity. For instance, it has been shown that the hexameric V_1 headgroup of plant V-ATPases can also exist as a pentamer (Kramer *et al.*, 1995). The stoichiometry of subunit c may also be variable, based on the observation that the switch to CAM induced by salt stress in *Me. crystallinum* is accompanied by an increase in subunit c transcript levels (Löw *et al.*, 1996) and size of V_0 intramembrane particles (see Lüttge and Ratajczak, 1997). In yeast, the 100 kDa subunit isoforms, Stv1p and Vph1p, function in assembly and sorting (Leng *et al.*, 1996, 1998). In plants, however, the 100 kDa subunit is not always found in association with fully active V-ATPases, and this subunit is notably absent from several plant species, including the V-ATPase of oat (Ward *et al.*, 1992). Sze and co-workers discovered that, while the 100 kDa polypeptide associates with fully assembled, active V-ATPases in sub-stoichiometric levels, the subunit is part of the V_0 subcomplex (Li and Sze, 1999). These authors concluded from their study that, in plants, the 100 kDa subunit plays a primary role in facilitating assembly of the V-ATPase holoenzyme. The question then arises as to how a cell controls the spatial dynamics of V-ATPases in subcellular compartments of the endomembrane system. Are there as yet unidentified accessory proteins that transiently associate with the complex to direct V-ATPase traffic to its final destination, or is this function inherent to one of the subunit isoforms? For instance, the increase in V-ATPase density observed in pulvini motor cells of *Mi. pudica* is specific to one of two types of vacuole involved in osmoregulation (Fleurat-Lessard *et al.*, 1997), yet this does not necessarily require upregulation of all V-ATPase subunits, as only key subunits may be needed to direct assembly of functional V-ATPases from cytoplasmic subunit pools (Stevens and Forgac, 1997). Indeed, studies of the *det3* mutant indicate that the V-ATPase can assemble via alternative mechanisms under conditions that vary the stoichiometry of subunit C in the holoenzyme (Schumacher *et al.*, 1999).

Several possible mechanisms could be proposed to account for the subunit composition, stoichiometry and subcellular distribution of V-ATPases in a given cell-type, other than differential gene expression of subunit isoforms and accessory proteins. One of several possible post-translational mechanisms is the controlled assembly and disassembly of the V-ATPase holoenzyme, which may be coupled with protein synthesis and turnover. Such a mechanism would provide the cell with the means to rapidly modulate compartment acidification, as well as to modify the spatial distribution and subunit composition of V-ATPase. The disassembly and reassembly of V-ATPase has been shown to occur in yeast (Kane, 1995) and tobacco hornworm larvae (Graf *et al.*, 1996). In yeast, the V_1 subcomplex, which rapidly dissociates from V-ATPase under stress-induced conditions, readily reassembles on membranes to form an active V-ATPase upon removal of the stress stimuli. The reassembly of the active V-ATPase does not require *de novo* protein synthesis, indicating that assembly utilizes a pool of pre-existing V_1 subunits. Although this mode of regulating V-ATPases in plants remains, for the most part, speculative, the reversible dissociation of V_1 and V_0 domains may play a role in the recovery of V-ATPase activity following low-temperature stress (Yoshida, 1991; Matsuura-Endo *et al.*, 1992). In chilling-sensitive mung bean seedlings, low-temperature treatments result in a decrease in V-ATPase activity and the selective loss of membrane-associated

V_1 subunits. V-ATPase activity was gradually restored, independent of protein synthesis, in seedlings returned to normal temperatures. The physiological state of the cell can also have a direct bearing on V-ATPase activity through substrate availability (Dietz *et al.*, 1998) and the oxidation/reduction of disulfide bonds formed by essential sulfhydryl groups in subunits A and B (Hager and Lanz, 1989; Yamanishi and Kasamo, 1992; Feng and Forgac, 1994; Dschida and Bowman, 1995).

7.3 V-PPase

7.3.1 Structure, biochemistry and gene organization

In contrast to the multimeric V-ATPase, the functional form of the V-PPase appears to be a homodimer of a single polypeptide, that is, the 'substrate-binding subunit', with an apparent molecular mass ranging 64–72 kDa in extracts from different plant species when fractionated by sodium dodecyl sulphate polyacrylamide gel electrophoresis (SDS-PAGE) (for reviews, see Leigh *et al.*, 1994; Zhen *et al.*, 1997a). The term substrate-binding subunit is somehow misleading, indicating that several different polypeptides are part of the V-PPase. However, this was suggested because, in early studies, different polypeptides ranging 20–40 kDa co-purified with V-PPase activity (for review, see Leigh *et al.*, 1994). However, Sato *et al.* (1994) have shown that the 75 kDa substrate-binding subunit, purified from pumpkin seedlings, is sufficient for V-PPase activity. Further evidence that the catalytic subunit is the only polypeptide necessary for V-PPase activity came from the heterologous expression of an *Ar. thaliana* V-PPase cDNA in yeast (Kim and Rea, 1994).

An important tool for the identification and characterization of V-PPases from different species has been a variety of different antisera generated against the V-PPase, which have been produced and generously distributed by Dr Masayoshi Maeshima (Nagoya, Japan). Cloning and sequencing of the V-PPase from different plant species has shown that the enzyme contains 760–775 amino acids with a molecular mass of 79–81 kDa (for reviews, see Leigh *et al.*, 1994 and Zhen *et al.*, 1997a). Sarafian *et al.* (1992) suggested 13 membrane-spanning domains for the catalytic subunit, however re-evaluation of the sequence data with different software packages for predicting polypeptide topology has led to a revised model of the V-PPase, with 16 transmembrane domains and the N- and C-termini located in the vacuolar lumen (Zhen *et al.*, 1997a).

The substrate-binding subunit contains all the functional features necessary for enzyme activity, that is, a substrate-binding domain, a proton-translocating domain and effectors-binding domains. From sequence comparisons with soluble PPases, two amino acid sequences have been proposed to be the catalytic site of the V-PPase, one of which, (E/D)-xxxxxxxKxE, is common amongst V-PPases (Rea *et al.*, 1992; Rea and Poole, 1993). Antibodies raised against a synthetic peptide [DVGADLVGKVE], which represent the putative substrate binding site of the *V. radiata* V-PPase, cross-reacted with V-PPases from different plant species and inhibited substrate hydrolysis and proton transport activity. In contrast, antibodies directed against the C-terminal region of the *V. radiata* V-PPase, reacted only with the *V. radiata* V-PPase inhibiting enzyme activity (Takasu *et al.*, 1997). This indicates that the putative substrate binding site is a common motif of V-PPases from different species and is structurally much more conserved when compared to the C-terminus of the enzyme. The V-PPase is irreversibly inhibited by NEM (Wang *et al.*, 1986; Verstappen and Rausch, 1988; Britten *et al.*, 1989).

Zhen *et al.* (1994a) demonstrated that a 14 kDa V8-protease fragment of the V-PPase is labeled by ^{14}C-NEM. By site-directed mutagenesis the NEM-binding site was determined as Cys^{634} of the *Ar. thaliana* V-PPase (Kim *et al.*, 1995). Substitution of Cys^{634} with Ser or Ala resulted in a NEM-insensitive V-PPase active in substrate hydrolysis and proton transport activity, indicating that Cys^{634} is not directly involved in catalysis and that NEM-inhibition results from steric changes near the catalytic site. Labeling with DCCD and DCCD analogs has revealed the importance of negatively-charged amino acid side-chains for V-PPase activity, that is, Glu^{229} of the *Ar. thaliana* (Zhen *et al.*, 1997b), Glu^{749} of the pumpkin (Maruyama *et al.*, 1998) and Asp^{283} of the *V. radiata* V-PPase (Yang *et al.*, 1999). Inhibition of V-PPase activity by covalent binding of tetranitromethane suggests the involvement of at least one tyrosine residue in catalysis (Yang *et al.*, 1996). Interestingly, the stretch of amino acids 227–245 of the *Ar. thaliana* V-PPase exhibits high sequence similarity to the membrane-spanning alpha-helical domain of the F-ATPase c-subunit containing the glutamate residue essential for proton translocation (Nyren *et al.*, 1993).

V-PPases which exhibit high sequence similarity to the membrane-bound H^+-pumping inorganic pyrophosphate synthase of the photosynthetic, purple, non-sulfur bacterium, *Rhodospirillum rubrum* (see Baltscheffsky *et al.*, 1999), have been suggested to be unique to plants and some phototrophic bacteria (Rea *et al.*, 1992). However, the recent findings that V-PPase-type activities and/or polypeptides cross-reacting with V-PPase antibodies are found in promastigotes of *Leishmania donovani* (Rodriguez *et al.*, 1999a), *Plasmodium falciparum* trophozoites (Luo *et al.*, 1999), acido-calcisomes of *Trypanosoma cruzi* (Scott *et al.*, 1998) and *Trypanosoma brucei* (Rodriguez *et al.*, 1999b), and even in the thermophilic archaeon, *Pyrobaculum aerophilum* (Drozdowicz *et al.*, 1999), imply that PP_i-energized H^+-transport is phylogenetically more deeply rooted than previously thought.

In various plants, multiple isoforms of the V-PPase have been found, for instance, *BVP1* and *BVP2* of *B. vulgaris* (Kim *et al.*, 1994), *TPV5*, *TVP9* and *TVP31* of *Ni. tabacum* (Lerchl *et al.*, 1995) and *OVP1* and *OVP2* of *Oryza sativa* (Sakakibara *et al.*, 1996). Moreover, from polypeptide microsequencing there is evidence for two distinct V-PPase polypeptides in *V. radiata* (Zhen *et al.*, 1994a).

While a tissue-specific expression of V-PPase isoforms was found in tobacco (Lerchl *et al.*, 1995), virtually nothing is known about the role of V-PPase isoforms in other plants. The localization of V-PPase in the plasma membrane of cells of *Z. mays* roots (Oberbeck *et al.*, 1994), *Ricinus communis* seedlings (Long *et al.*, 1995), *Chlamydomonas reinhardtii* (Robinson *et al.*, 1998) and *Brassica oleracea* inflorescence (Ratajczak *et al.*, 1999) indicates that different V-PPase isoforms might be targeted to various cellular membranes.

For optimal activity, the V-PPase requires PP_i, Mg^{2+} and K^+, however, the enzyme shows complicated responses to changes in total PP_i and Mg^{2+} concentrations (for reviews, see Leigh *et al.*, 1994; Zhen *et al.*, 1997a). $MgPP_i$ (Wang *et al.*, 1986; Johannes and Felle, 1990; White *et al.*, 1990), and both $MgPP_i$ and Mg_2PP_i together (Walker and Leigh, 1981), or Mg_2PP_i alone (Leigh *et al.*, 1992) have been suggested to represent the V-PPase substrate. Activation through Mg^{2+} has been attributed to cooperative (White *et al.*, 1990) or non-cooperative (Leigh *et al.*, 1992) binding. V-PPase activity is insensitive to azide, orthovanadate and bafilomycin, and inhibited by potassium fluoride, an inhibitor of various PP_i hydrolases. Thus, V-PPase activity was generally characterized as a K^+-stimulated, PP_i-dependent H^+-transport activity, until Baykov *et al.* (1993) and Zhen *et al.* (1994b) discovered diphosphonates as specific inhibitors of V-PPase activity. Aminomethylene diphosphonate seems to be the most specific and potent inhibitor of this group of molecules (see Zhen *et al.*, 1997a).

V-PPase constitutes up to 5–10% of the total tonoplast protein (Maeshima and Yoshida, 1989; Ratajczak *et al.*, 1995), and is thus three to four times less abundant as compared to the V-ATPase. However, since the molecular mass of the V-PPase dimer is approximately four times lower than the molecular mass of the V-ATPase holoenzyme, almost equal numbers of active enzymes are present at the tonoplast. Therefore, it is not surprising that the V-PPase is able to generate an electrochemical proton gradient across the tonoplast that is similar to or greater than the gradient formed by the action of the V-ATPase at the same membrane (Pope and Leigh, 1987; Maeshima and Yoshida, 1989; Johannes and Felle, 1990; Rea *et al.*, 1992; Terrier *et al.*, 1998). This indicates that both the V-PPase and the V-ATPase play a major role in tonoplast energization. It is still unclear why there are two primary-active proton pumps at the tonoplast of plant cells, although a lot of work has been done to discover specific metabolic functions of the two enzymes since the question was first raised (Rea and Sanders, 1987). Different functions have been suggested for the V-PPase (for reviews, see Lüttge *et al.*, 1995a,b; Zhen *et al.*, 1997a). The first function is, of course, energization of the tonoplast for secondary-active transport events by H^+-transport driven by PP_i hydrolysis. The use of PP_i as a substrate might be of extraordinary importance in cells containing high amounts of PP_i in the cytoplasm, since the V-PPase is likely to be the only enzyme responsible for direct hydrolysis of cytoplasmic PP_i to P_i (Weiner *et al.*, 1987). In this respect, the V-PPase could be involved in the regulation of metabolic pathways depending on the cytoplasmic PP_i concentration. Secondly, the V-PPase has been suggested to synthesize PP_i using the trans-tonoplast proton gradient generated by the V-ATPase (Façanha and de Meis, 1998). Another possible function of the V-PPase is the primary active transport of K^+ from the cytoplasm to the vacuolar lumen. The first evidence for direct K^+ transport via the V-PPase came from patch-clamp studies on *Be. vulgaris* vacuoles (Davies *et al.*, 1992). In this respect, it should be mentioned that the V-PPase of pumpkin exhibits significant sequence homology to the K^+-translocating Kdp-ATPase A chain of *Escherichia coli* (Sato, 1997). However, the data from patch-clamp studies have not yet been verified by biochemical investigations (Sato *et al.*, 1994; Ros *et al.*, 1995). A third possible function of the V-PPase could be H^+-transport into the vacuole under conditions where the cytoplasmic ATP concentration is limiting, such as low temperatures and anoxia (Carystinos *et al.*, 1995; Darley *et al.*, 1995). Finally, the V-PPase could be involved in activation of the V-ATPase H^+-transport activity, as has been shown with tonoplast vesicles isolated from leaves of *K. daigremontiana* (Marquardt-Jarczyk and Lüttge, 1990) and *K. blossfeldiana* (Fischer-Schliebs *et al.*, 1997).

7.3.2 Developmental and environmental regulation

On the whole, the V-PPase has not been studied as thoroughly as the V-ATPase. Although the genes encoding the V-PPase subunit have been isolated and characterized from several plant species (for review, see Zhen *et al.*, 1997a), relatively little is known about the expression of these genes, especially in relation to their activity and cellular functions. However, it can be stated that the V-PPase is regulated in developing seeds and fruits, and responds to environmental stress.

The isolation and characterization of the V-PPase in developing cotton ovules is one of but a few studies in which the expression of the V-PPase at the RNA, protein and activity levels has been investigated (Smart *et al.*, 1998). The results from this study provide compelling evidence that V-PPase is differentially regulated at the translational and post-translational levels during development of cotton fibers, and is thus regulated via a mechanism distinct from that of the V-ATPase. Ribonuclease protection assays (RPAs) show that a cotton fiber V-PPase gene is constitutively expressed in developing ovules and fibers (Smart *et al.*, 1998). However, because of the sensitivity of RPAs, other V-PPase genes may remain undetected. The V-PPase gene expressed in cotton leaves is encoded by a second gene (T.A. Wilkins, unpublished). Interestingly, the V-PPase polypeptide does, in fact, exhibit a developmental profile but not during peak fiber expansion. The abundance of the V-PPase increases dramatically in developing ovules, peaking in the days immediately preceding the onset of fiber expansion, before rapidly decreasing to low levels that then remain relatively constant throughout development. What makes the cotton V-PPase story even more intriguing is the developmentally regulated increase in

V-PPase activity, while the RNA transcripts and protein remain unchanged at low levels. Although V-PPase activity steadily increases coincident with rapid fiber elongation, V-PPase activity peaks at around 20 days post-anthesis, when the rate of expansion has already significantly decreased. The peak of V-PPase activity occurs during the transition period between primary and secondary cell wall synthesis, and a stage of fiber development also characterized by the transient synthesis of callose. It has been proposed that the role of V-PPase in this instance may be to scavenge PP_i that may be generated as a by-product of callose biosynthesis (Wilkins and Jernstedt, 1999). The differential increase in V-PPase activity relative to V-PPase mRNA and protein during fiber expansion indicates that activity is regulated post-translationally via a mechanism that probably involves controlled phosphorylation and dephosphorylation. It is also very interesting to note that the spike in V-PPase protein observed prior to fiber expansion is not accompanied by a significant change in hydrolytic activity. This has led to the following hypothesis in terms of vacuole biogenesis (T.A. Wilkins, unpublished). Based on current evidence, it is believed that the V-PPase is resynthesized and inserted into the tonoplast of a large, prevacuolar compartment in fiber cells, but remains in an inactive state. In the earliest stages of fiber expansion, newly formed provacuoles are added to the prevacuolar compartment, which serves to nucleate the formation of the central vacuole. In so doing, the V-PPase is effectively diluted on a protein per membrane basis, but is also activated in response to developmental signals.

The observed increase in V-PPase activity during germination of pumpkin seeds is also related to the formation of the central vacuole and hydrolytic functions (Maeshima *et al.*, 1994). The developmental regulation of the V-PPase appears to be more straightforward in this case. In seeds, the V-PPase is present at very low levels in the tonoplast of storage vacuoles. Germination is accompanied by an increase in V-PPase protein and activity in association with conversion of storage vacuoles to a central vacuole, and the acquisition of lytic functions in pumpkin seedlings and castor bean endosperm.

Several other examples indicate a developmental regulation of V-PPase expression, however, it is difficult to draw a general conclusion on the mechanism of V-PPase expression from the data published so far. Nakanishi and Maeshima (1998) reported high V-PPase mRNA levels in the elongating region of 3 day old *V. radiata* hypocotyls, while the message amount in mature regions of 5 day old hypocotyls was extremely low. Both V-PPase activity and protein amount were lower in the mature region compared to the elongating region. The level of V-PPase mRNA was very low in tissues of young pear fruits and increased with maturation (Suzuki *et al.*, 1999). However, in tonoplast vesicles isolated from pear fruits at different stages of development, V-PPase activity and protein amount remained constant (Shiratake *et al.*, 1997).

Carystinos *et al.* (1995) showed that under anoxia the V-PPase message increased in rice seedlings and decreased again when the seedlings were returned to air. V-PPase activity and protein amount in tonoplast vesicles increased during anoxia treatment, while there was no response in microsomes from corn roots. In the same study, it was demonstrated that chilling of rice seedlings for 6 days at 10°C increased specific V-PPase activity by a factor of 20. Studies on *Secale cereale* plants have indicated that growth under conditions deficient in mineral nutrition could lead to an increase in V-PPase activity (Kasai *et al.*, 1998). Some information is available about the impact of salinity on V-PPase expression. In *Me. crystallinum*, V-PPase hydrolytic activity (Bremberger *et al.*, 1988; Bremberger and Lüttge, 1992) and amount (Rockel *et al.*, 1994) decreased with prolonged salt treatment. However, similar effects were observed when the plants reached the CAM-state in the absence of salt as a result of a developmental program (Rockel *et al.*, 1994). Thus, the decrease in V-PPase activity and amount might be correlated to the expression of CAM and not to salinity *per se*. On the other hand, experiments performed with the C_3-CAM intermediate plant *Kalanchoë blossfeldiana* revealed that this cannot be generalized. In *K. blossfeldiana*, CAM-inducted by a change in the length of the photoperiod increased V-PPase activity and protein amount (Fischer-Schliebs *et al.*, 1998). In cultured cells of *D. carota* (Colombo and Cerana, 1993) and *Ac. pseudoplatanus* (Zingarelli *et al.*, 1994), V-PPase activity was induced by salinity. Salt-treatment of the halophyte, *Suaeda maritima* (Leach *et al.*, 1990), and cultured cells of *Ni. tabacum* (Reuveni *et al.*, 1990), did not change V-PPase hydrolytic activity, while there was a decrease in V-PPase activity in root cells of *V. radiata* (Nakamura *et al.*, 1992). Ballesteros *et al.* (1996) found that salt-treatment of *He. annuus* led to a slight increase of root cell V-PPase substrate hydrolysis activity in tonoplast-enriched membrane vesicle fractions, while there was a more pronounced increase in H^+-transport activity. This indicates a modulation of the coupling ratio of the V-PPase under certain metabolic conditions.

7.4 Differential regulation of V-ATPase *vs* V-PPase

As mentioned previously (see Section 7.3.1), it is still not known why the plant tonoplast possesses two H^+-translocating enzymes. Much effort has been made to find a scenario in which the V-ATPase is upregulated and the V-PPase is downregulated or vice versa. However, at present, we are far from resolving this puzzle. Even antisense suppression of one of the V-PPase isoforms in *N. tabacum* has not yet resulted in phenotypic alterations (Lerchl *et al.*, 1995). For two reasons, one would expect a high V-PPase activity in young, developing tissues. Firstly, in growing tissues, RNAs, proteins and cellulose are actively synthesized for the construction of cells and, as a result, a large amount of PP_i is produced. Secondly, taking into account the fact that the V-PPase might be involved in potassium transport (Davies *et al.*, 1992), high V-PPase activity can be expected in growing tissues since growing cells have a huge demand for potassium transport across the tonoplast. Transport of osmotica across the tonoplast drives vacuolar and, thus, cellular expansion (Tomos and Pritchard, 1994). There are two observations supporting this assumption: first, in young developing pumpkin cotelydons, substrate-hydrolysis and H^+-transport activities of the V-PPase are high and decline with cotelydon aging, while the respective V-ATPase activities increase (Suzuki and Kasamo, 1993); and second, during aging and C_3-CAM transition of *Me. crystallinum*, V-PPase amount and activity decrease, while protein amount and V-ATPase activity increase (Bremberger and Lüttge, 1992; Ratajczak *et al.*, 1994).

However, there are two examples for concordant time-dependent pattern of activities and protein amount of the two proton pumps. Induction of CAM in *K. blossfeldiana* by short-day conditions led to an increase in activity and protein amount of both, V-PPase and V-ATPase, while in long-day grown plants activity and amount of both enzymes decreased (Fischer-Schliebs *et al.*, 1998). In *Ri. communis* stem-tumors induced by infection with *Agrobacterium tumefaciens*, V-PPase and V-ATPase amount and activity remained constant at a high level, while in non-infected stem tissue activity and amount of both proton pumps decreased (Fischer-Schliebs *et al.*, 1998). While V-ATPase protein amount increased with pear fruit development, V-PPase amount remained at a constant level (Shiratake *et al.*, 1997). In developing cotton fibers, the expression of V-ATPase and V-PPase showed different patterns (Smart *et al.*, 1998). While V-ATPase expression was high at the early stages of development, V-PPase expression was constitutively high. Moreover, expression of both enzymes did not correspond to the peaks of maximal enzyme activity and proton pump protein amount (see Section 7.3.2). Thus, various scenarios of the relationship of V-PPase and V-ATPase amount and activity during development of different plant species have been found, suggesting that the two proton pumps are differentially regulated. However, more detailed analyses are required to verify this conclusion.

7.5 Future directions

Although there is much information available concerning the molecular characteristics of the V-ATPase, some of the V-ATPase subunits found in yeast have not yet been cloned in plants. Efforts must be made to find these missing subunits in order to verify the structural model of the plant V-ATPase. In this respect, it is also important to unequivocally identify the second, peripheral stalk of a plant V-ATPase, which is suggested to be necessary for V-ATPase function.

There are several studies showing a differential expression of V-ATPase subunit isoforms, however, more information is required about the role of these isoforms in V-ATPase function, about their developmental and environmental control of expression, and about the importance of such isoforms for intracellular targeting of the holoenzyme.

While genes encoding the V-PPase have been cloned in several plants, detailed expression analyses in response to environmental changes and to developmental programs, which could provide information about the role of V-PPase as a second proton-translocating enzyme energizing the tonoplast, are rare. To obtain more information about the putative cooperation of V-ATPase and V-PPase in energizing the tonoplast, comparative studies of the expression of both proton pumps are required.

Few data are available about the control of gene expression of both proton pumps. Thus, one major goal for future research will be the characterization of promotors of V-ATPase and V-PPase genes and the identification of signal transduction pathways involved in proton pump gene expression.

Acknowledgments

The authors thank all colleagues for providing relevant unpublished data, which are cited in this manuscript as personal communication, and especially Dr Elke Fischer-Schliebs (Darmstadt, Germany) for helpful discussions.

References

Abrahams, J.P., Leslie, A.G.W., Lutter, R. and Walker, J.E. (1994) Structure at 2.8 A resolution of F_1-ATPase from bovine heart mitochondria. *Nature*, **370** 621-28.

Adachi, I., Puopolo, K., Marques-Sterling, N., Arai, H. and Forgac, M. (1990) Dissociation, cross-linking and glycosylation of the coated vesicle proton pump. *J. Biol. Chem.*, **265** 967-73.

Adams, P., Nelson, D.E., Yamada, S., Chmara, W., Jensen, R.G., Bohnert, H.J. and Griffiths, H. (1998) Growth and development of *Mesembryanthemum crystallinum* (Aizoaceae). *New Phytol.*, **138** 171-90.

Allen, G.J. and Sanders, D. (1997) Vacuolar ion channels of higher plants, in *The Plant Vacuole* (eds. R.A. Leigh and D. Sanders), Advances in Botanical Research, Vol. 25, Academic Press, San Diego, London, Boston, New York, Sydney, Tokyo, Toronto, pp. 217-52.

Arai, H., Terres, G., Pink, S. and Forgac, M. (1988) Topography and subunit composition of the coated vesicle proton pump. *J. Biol. Chem.*, **263** 8796-802.

Ballesteros, E., Donaire, J.P. and Belver, A. (1996) Effects of salt stress on H^+-ATPase and H^+-PPase activities of tonoplast-enriched membrane vesicles isolated from sunflower roots. *Physiol. Plant.*, **97** 259-68.

Ballesteros, E., Blumwald, E., Donaire, J.P. and Belver, A. (1997) Na^+/H^+ antiport in tonoplast vesicles isolated from sunflower roots induced by NaCl stress. *Physiol. Plant.*, **99** 328-34.

Baltscheffsky, M., Schultz, A. and Baltscheffsky, H. (1999) H^+-proton-pumping inorganic pyrophosphatase: a tightly membrane-bound family. *FEBS Lett.*, **452** 121-27.

Barkla, B.J., Vera-Estrella, R., Maldonado-Gama, M. and Pantoja, O. (1999) Abscisic acid induction of vacuolar H^+-ATPase activity in *Mesembryanthemum crystallinum* is developmentally regulated. *Plant Physiol.*, **120** 811-20.

Bauerle, C., Ho, M.N., Lindorfer, M.A. and Stevens, T.H. (1993) The *Saccharomyces cerevisiae VMA6* gene encodes the 36 kDa subunit of the vacuolar (H^+)-ATPase membrane sector. *J. Biol. Chem.*, **268** 12749-57.

Baykov, A.A., Dubnova, E.B., Bakuleva, N.P., Evtushenko, O.A., Zhen, R.-G. and Rea, P.A. (1993) Differential sensitivity of membrane-associated pyrophosphatases to inhibition by diphosphonates and fluoride delineates two classes of enzyme. *FEBS Lett.*, **327** 199-202.

Bennett, A.B. and Spanswick, R.M. (1983) Optical measurements of ΔpH and $\Delta\psi$ in corn root membrane vesicles: kinetic analysis of Cl^- effects on a proton-translocating ATPase. *J. Membr. Biol.*, **71** 95-107.

Bennett, A.B. and Spanswick, R.M. (1984) H^+-ATPase activity from storage tissue of *Beta vulgaris*. II. H^+/ATP stoichiometry of an anion-sensitive H^+-ATPase. *Plant Physiol.*, **74** 545-48.

Berkelman, T., Houtchens, K.A. and Dupont, F.M. (1994) Two cDNA clones encoding isoforms of the B subunit of the vacuolar ATPase from barley roots. *Plant Physiol.*, **104** 287-88.

Binzel, M.L. (1994) Alterations in H^+-ATPase gene expression in response to salt, in *Molecular and Cellular Mechanisms of Stress Tolerance in Plants* (ed. J.H. Cherry), NATO ASI Series, Vol. H86, Springer-Verlag, Berlin, pp. 429-42.

Binzel, M.L. (1995) NaCl-induced accumulation of tonoplast and plasma membrane H^+-ATPase message in tomato. *Physiol. Plant.*, **94** 722-28.

Binzel, M.L. and Dunlap, J.R. (1995) Abscisic acid does not mediate NaCl-induced accumulation of 70 kDa subunit tonoplast H^+-ATPase message in tomato. *Planta*, **179** 563-68.

Blumwald, E. and Gelli, A. (1997) Secondary inorganic ion transport at the tonoplast, in *The Plant Vacuole* (eds. R.A. Leigh and D. Sanders), Advances in Botanical Research, Vol. 25, Academic Press, San Diego, London, Boston, New York, Sydney, Tokyo, Toronto, pp. 401-17.

Boekema, E., Ubbink-Kok, T., Lolkema, J.S., Brisson, A. and Konings, W.N. (1997) Visualization of a peripheral stalk in V-type ATPase: evidence for the stator structure essential to rotational catalysis. *Proc. Natl. Acad. Sci. USA*, **94** 14291-93.

Böttcher, B., Schwarz, L. and Gräber, P. (1998) Direct indication for the existence of a double stalk in CF_0F_1. *J. Mol. Biol.*, **281** 757-62.

Bowman, E.J., Siebers, A. and Altendorf, K. (1988) Bafilomycins: a class of inhibitors of membrane ATPases from microorganisms, animal cells and plant cells. *Proc. Natl. Acad. Sci. USA*, **85** 7972-76.

Brauer, D., Tu, S.I., Hsu, A.F. and Patterson, D. (1993) Evidence for an indirect coupling mechanism for the nitrate-sensitive proton pump from corn root tonoplast membranes. *Physiol. Plant.*, **89** 588-91.

Bremberger, C. and Lüttge, U. (1992) Dynamics of tonoplast proton pumps and other tonoplast proteins of *Mesembryanthemum crystallinum* during the induction of crassulacean acid metabolism. *Planta*, **188** 575-80.

Bremberger, C., Haschke, H.-P. and Lüttge, U. (1988) Separation and purification of the tonoplast ATPase and pyrophosphatase from plants with constitutive and inducible crassulacean acid metabolism. *Planta*, **175** 465-70.

Bressan, R.A., Hasegawa, P.M. and Pardo, J.M. (1998) Plants use calcium to resolve salt stress. *Trends Plant Sci.*, **3** 411-12.

Britten, C.J., Turner, J.C. and Rea, P.A. (1989) Identification and purification of substrate-binding subunit of higher plant H^+-translocating inorganic pyrophosphatase. *FEBS Lett.*, **256** 200-206.

Cabrera y Poch, H., Peto, C. and Chory, J. (1993) A mutation in *Arabidopsis DET3* gene uncouples photoregulated leaf development from gene expression and chloroplast biogenesis. *Plant J.*, **4** 6671-82.

Carystinos, G.D., Heather, M.R., Monroy, A.F., Rajinder, D.S. and Poole, R.J. (1995) Vacuolar H^+-translocating pyrophosphatase is induced by anoxia and chilling in seedlings of rice. *Plant Physiol.*, **108** 641-49.

Cohen, A., Perzov, N., Nelson, H. and Nelson, N. (1999) A novel family of yeast chaperons involved in the distribution of V-ATPase and other membrane proteins. *J. Biol. Chem.*, **274** 26885-93.

Colombo, R. and Cerana, R. (1993) Enhanced activity of tonoplast pyrophosphatase in NaCl-grown cells of *Daucus carota. J. Plant Physiol.*, **142** 226-29.

Cooley, M.B., Yang, H., Dahal, P., Mella, R.A., Downie, A.B., Haigh, A.M. and Bradford, K.J. (1999) Vacuolar H^+-ATPase is expressed in response to gibberellin during tomato seed germination. *Plant Physiol.*, **121** 1339-47.

Crider, B.P., Xie, X.S. and Stone, D.K. (1994) Bafilomycin inhibits proton flow through the H^+-channel of vacuolar proton pumps. *J. Biol. Chem.*, **269** 17379-81.

Darley, C.P., Davies, J.M. and Sanders, D. (1995) Chilling-induced changes in the activity and abundance of the vacuolar proton-pumping pyrophosphatase. *Plant Physiol.*, **109** 659-65.

Davies, J.M., Poole, R.M., Rea, P.A. and Sanders, D. (1992) Potassium transport into plant vacuoles energized directly by a proton-pumping inorganic pyrophosphatase. *Proc. Natl. Acad. Sci. USA*, **89** 11701-705.

Davies, J.M., Hunt, I. and Sanders, D. (1994) Vacuolar H^+-pumping ATPase variable transport coupling ratio controlled by pH. *Proc. Natl. Acad. Sci. USA*, **91** 8547-51.

Davies, S.A., Goodwin, S.F., Kelly, D.C., Wang, Z., Sozen, M.A., Kaiser, K. and Dow, J.A.T. (1996) Analysis and inactivation of *vha55*, the gene encoding the V-ATPase B subunit in *Drosphila* reveals a larval lethal phenotype. *J. Biol. Chem.*, **271** 30677-84.

Depta, H., Holstein, S.E.H., Robinson, D.G., Lützelschwab, M. and Michalke, W. (1991) Membrane markers in highly purified clathrin-coated vesicles from *Cucurbita* hypocotyls. *Planta*, **183** 434-42.

Dietz, K.J. and Arbinger, B. (1996) cDNA sequence and expression of subunit E of the vacuolar H^+-ATPase in the inducible crassulacean acid metabolism plant, *Mesembryanthemum crystallinum. Biochim. Biophys. Acta*, **1281** 134-38.

Dietz, K.J., Rudloff, S., Ageorges, A., Eckerskorn, C., Fischer, K. and Arbinger, B. (1995) Subunit E of the vacuolar H^+-ATPase of *Hordeum vulgare* L.: cDNA cloning, expression and immunological analysis. *Plant J.*, **8** 521-29.

Dietz, K.-J., Hollenbach, J. and Arnold, J. (1996) Nucleotide sequence of subunit E of the vacuolar proton-ATPase of *Spinacia oleracea* and *Arabidopsis thaliana. Plant Physiol.*, **111** 652.

Dietz, K.-J., Heber, H. and Mimura, T. (1998) Modulation of the vacuolar H^+-ATPase by adenylates as basis for the transient CO_2-dependent acidification of the leaf vacuole upon illumination. *Biochim. Biophys. Acta*, **1373** 87-92.

Dow, J.A.T., Davies, S.A., Guo, Y., Graham, S., Finbow, M.E. and Kaiser, K. (1997) Molecular genetic analysis of V-ATPase function in *Drosophila melanogaster. J. Exp. Biol.*, **200** 237-45.

Dröse, S., Bindseil, K.U., Bowman, E.J., Siebers, A., Zeeck, A. and Altendorf, K. (1993) Inhibitory effects of modified bafilomycins and concanamycins on P- and V-type adenosine triphosphatases. *Biochemistry*, **32** 3902-906.

Drozdowicz, Y.M., Lu, Y.P., Patel, V., Fitz-Gibbon, S., Miller, S.H. and Rea, P.A. (1999) A thermostable vacuolar-type membrane pyrophosphatase from the archeon, *Pyrobaculum aerophilum*: implications for the origins of pyrophosphate-energized pumps. *FEBS Lett.*, **460** 505-12.

Dschida, W.J. and Bowman, B.J. (1992) Structure of the vacuolar ATPase from *Neurospora crassa* as determined by electron microscopy. *J. Biol. Chem.*, **267** 18783-89.

Dschida, W.J.A. and Bowman, B.J. (1995) The vacuolar ATPase: sulfite stabilization and the mechanism of nitrate inactivation. *J. Biol. Chem.*, **270** 1557-63.

Dubcovsky, J., Luo, M.-C. and Dvorak, J. (1995) Linkage relationships among stress-induced genes in wheat. *Theor. Appl. Genet.*, **91** 795-801.

DuPont, F.M. (1992) Salt-induced changes in ion transport: regulation of primary pumps and secondary transporters, in *Transport and Receptor Proteins of Plant Membranes* (eds. D.T. Cooke and D.T. Clarkson), Plenum Press, New York, pp. 91-100.

DuPont, F.M. and Mudd, J.B. (1985) Acclimation to low temperature by microsomal membranes from tomato cell cultures. *Plant Physiol.*, **77** 74-78.

DuPont, F.M. and Morrissey, P.J. (1992) Subunit composition and Ca-ATPase activity of the vacuolar ATPase from barley roots. *Arch. Biochem. Biophys.*, **294** 341-46.

Emig, I. (1995) *Elektronenmikroskopische Untersuchungen der Seitenansicht der Tonoplasten-H^+-ATPase bei der C_3-CAM intermediären Pflanze* Mesembryanthemum crystallinum *L.*, Diploma thesis, Technische Hochschule Darmstadt, Darmstadt, Germany.

Façanha, A.R. and de Meis, L. (1998) Reversibility of H^+-ATPase and H^+-pyrophosphatase in tonoplast vesicles from maize coleoptiles and seeds. *Plant Physiol.*, **116** 1487-95.

Feng, Y. and Forgac, M. (1992) Cysteine 254 of the 73 kDa A subunit is responsible for inhibition of the coated vesicle (H^+)-ATPase upon modification by sulphydryl reagents. *J. Biol. Chem.*, **267** 5817-22.

Feng, Y. and Forgac, M. (1994) Inhibition of vacuolar H^+-ATPase by disulfide bond formation between cysteine 254 and cysteine 532 in subunit A. *J. Biol. Chem.*, **269** 13224-30.

Finbow, M.E. and Harrison, M.A. (1997) The vacuolar H^+-ATPase: a universal proton pump of eukaryotes. *Biochem. J.*, **324** 697-712.

Fischer-Schliebs, E., Mariaux, J.-B. and Lüttge, U. (1997) Stimulation of H^+-transport activity of vacuolar H^+-ATPase by activation of H^+-PPase in *Kalanchoë blossfeldiana*. *Biol. Plant.*, **39** 169-77.

Fischer-Schliebs, E., Ratajczak, R., Weber, P., Tavakoli, N., Ullrich, C.I. and Lüttge, U. (1998) Concordant time-dependent patterns of activities and enzyme protein amounts of V-PPase and V-ATPase in induced (flowering and CAM or tumour) and non-induced plant tissues. *Bot. Acta*, **111** 130-36.

Fischer-Schliebs, E., Drobny, M., Ball, E., Ratajczak, R. and Lüttge, U. (2000) Variation in nitrate nutrition leads to changes in the performance of the V-ATPase and immunological differences of proteolipid subunit c in tobacco (*Nicotiana tabacum* L.) leaves. *Aust. J. Plant Physiol.*, **27** 639-48.

Fleurat-Lessard, P., Frangne, N., Maeshima, M., Ratajczak, R., Bonnemain, J.L. and Martinoia, E. (1997) Increased expression of vacuolar aquaporin and H^+-ATPase related to motor cell function in *Mimosa pudica*. *Plant Physiol.*, **114** 827-34.

Forgac, M. (1999) Structure and properties of the vacuolar (H^+)-ATPases. *J. Biol. Chem.*, **274** 12951-54.

Getz, H.P. and Klein, M. (1995) The vacuolar ATPase of red beet storage tissue: electron microscopic demonstration of the 'head-and-stalk' structure. *Bot. Acta*, **108** 14-23.

Gogarten, J.P., Fichmann, J., Braun, Y., Morgan, L., Styles, P., Taiz, S.L., Delapp, K. and Taiz, L. (1992) The use of antisense mRNA to inhibit the tonoplast H^+-ATPase in carrot. *Plant Cell*, **4** 851-64.

Gómez-Cadenas, A., Tadeo, F.R., Primo-Millo, E. and Talon, M. (1998) Involvement of abscisic acid and ethylene in the response of *Citrus* seedlings to salt shock. *Physiol. Plant.*, **103** 475-84.

Graf, R., Harvey, W.R. and Wieczorek, H. (1996) Purification and properties of a cytosolic V_1-ATPase. *J. Biol. Chem.*, **271** 20908-13.

Hager, A. and Lanz, C. (1989) Essential sulfhydryl groups in the catalytic center of the tonoplast H^+-ATPase from coleoptiles of *Zea mays* L. as demonstrated by the biotin-streptavidin-peroxidase system. *Planta*, **180** 116-22.

Hasenfratz, M.-P., Tsou, C.-L. and Wilkins, T.A. (1995) Expression of two related vacuolar H^+-ATPase proteolipid genes is differentially regulated in a tissue-specific manner. *Plant Physiol.*, **108** 1395-404.

Hedrich, R., Kurkdjian, A., Guern, J. and Flügge, U.I. (1989) Comparative studies on the electrical properties of the H^+ translocating ATPase and pyrophosphatase of the vacuolar/lysosomal compartment. *EMBO J.*, **8** 2835-41.

Hemken, P., Guo, X.L., Wang, Z.Q., Zhang, K. and Gluck, S. (1992) Immunological evidence that vacuolar H^+-ATPases with heterogeneous forms of $Mr = 31{,}000$ subunit have different membrane distributions in mammalian kidney. *J. Biol. Chem.*, **267** 9948-57.

Herman, E.M., Li, X., Su, R.T., Larsen, P., Hsu, H.-T. and Sze, H. (1994) Vacuolar-type H^+-ATPases are associated with the endoplasmic reticulum and provacuoles of root tip cells. *Plant Physiol.*, **106** 1313-24.

Hernando, N., Barkiewicz, M., Collin-Osdoby, P. and Baron, R. (1995) Alternative splicing generates a second isoform of the catalytic subunit of the vacuolar H^+-ATPase. *Proc. Natl. Acad. Sci. USA*, **92** 6087-91.

Hirata, R., Ohsumk, Y., Nakano, A., Kawasaki, H., Suzuki, K. and Anraku, Y. (1990) Molecular structure of a gene, *VMA1*, encoding the catalytic subunit of H^+-translocating adenosine triphosphatase from vacuolar membranes of *Saccharomyces cerevisiae*. *J. Biol. Chem.*, **265** 6726-33.

Holzenberg, A., Jones, P.C., Franklin, T., Pali, T., Heimburg, T., Marsh, D., Findlay, J. and Finbow, M.E. (1993) Evidence for a common structure for a class of membrane channels. *Eur. J. Biochem.*, **213** 21-30.

Hosaka, M., Kanayama, Y., Shiratake, K. and Yamaki, S. (1994) Tonoplast H^+-ATPase of mature pear fruit. *Phytochemistry*, **36** 565-67.

Hurkman, W.J., Tanaka, C.K. and DuPont, F.M. (1988) The effect of salt stress on polypeptides in membrane fractions from barley roots. *Plant Physiol.*, **88** 1263-73.

Inoue, H., Noumi, T., Nagata, M., Murakami, H. and Kanazawa, H. (1999) Targeted disruption of the gene encoding the proteolipid subunit of mouse vacuolar H^+-ATPase leads to early embryonic lethality. *Biochim. Biophys. Acta*, **1413** 130-38.

Ishitani, M., Xiong, L., Stevenson, B. and Zhu, J.-K. (1997) Genetic analysis of osmotic and cold stress signal transduction in *Arabidopsis*: interactions and convergence of abscisic acid-dependent and abscisic acid-independent pathways. *Plant Cell*, **9** 1935-49.

Johannes, E. and Felle, H. (1990) Proton gradient across the tonoplast of *Riccia fluitans* as a result of the joint action of two electroenzymes. *Plant Physiol.*, **93** 412-17.

Junge, W., Lill, H. and Engelbrecht, S. (1997) ATP synthase: an electrochemical transducer with rotary mechanics. *Trends Biol. Sci.*, **22** 420-23.

Kaestner, K.H., Randall, S.K. and Sze, H. (1988) N,N′-dicyclohexylcarbodiimide-binding proteolipid of the vacuoler H^+-ATPase from barley roots. *J. Biol. Chem.*, **263** 1388-93.

Kane, P.M. (1995) Disassembly and reassembly of the yeast vacuolar $H(^+)$-ATPase *in vivo*. *J. Biol. Chem.*, **270** 17025-32.

Kane, P.M. and Stevens, T.H. (1992) Subunit composition, biosynthesis and assembly of the yeast vacuolar proton-translocating ATPase. *J. Bioenerget. Biomembr.*, **24** 383-93.

Kane, P.M., Tarsio, M. and Liu, J. (1999) Early steps in the assembly of the yeast vacuolar H^+-ATPase. *J. Biol. Chem.*, **274** 17275-83.

Kasai, M., Nakamura, T., Kudo, N., Sato, H., Maeshima, M. and Sawada, S. (1998) The activity of the root vacuolar H^+-pyrophosphatase in rye plants grown under conditions deficient in mineral nutrition. *Plant Cell Physiol.*, **39** 890-94.

Kasai, M., Sasaki, M., Tanakamaru, S., Yamamoto, Y. and Matsumoto, H. (1993a) Possible involvement of abscisic acid in increases in activities of two vacuolar H^+-pumps in barley roots under aluminium stress. *Plant Cell Physiol.*, **34** 1335-38.

Kasai, M., Yamamoto, Y., Maeshima, M. and Matsumoto, H. (1993b) Effects of *in vivo* treatment with abscisic acid and/or cytokinin on activities of vacuolar H^+-ATPase pumps of tonoplast-enriched membrane vesicles prepared from barley roots. *Plant Cell Physiol.*, **34** 1107-15.

Kasai, M., Yamamoto, Y. and Matsumoto, H. (1994) *in vivo* treatment of barley roots with vanadate increases vacuolar H^+-translocating ATPase activity of the tonoplast-enriched membrane vesicles and the level of endogenous ABA. *Plant Cell Physiol.*, **35** 291-95.

Kawamura, Y., Arakawa, K., Maeshima, M. and Yoshida, S. (2000) Tissue-specificity of E subunit isoforms of plant vacuolar H^+-ATPase and the existence of isotype enzymes. *J. Biol. Chem.*, **275** 6515-22.

Kim, E.J. and Rea, P.A. (1994) Molecular analysis of the vacuolar H^+-pyrophosphatase by heterologous expression in *Saccharomyces*. *Plant Physiol.*, **105** 98.

Kim, E.J., Zhen, R.-G. and Rea, P.A. (1995) Site-directed mutagenesis of vacuolar H^+-pyrophosphatase: necessity of Cys[634] for inhibition by maleimides. *J. Biol. Chem.*, **270** 2630-35.

Kim, W. and Wilkins, T.A. (1997) Isolation of a vacuolar H^+-ATPase subunit E cDNA from developing cotton (*Gossypium hirsutum* L.) fibers (Accession No. AF009338) (PGR 97-126). *Plant Physiol.*, **115** 314.

Kim, W., Wan, C.-Y. and Wilkins, T.A. (1999) Functional complementation of yeast *vma1* cells by a plant subunit A homolog rescues the mutant phenotypes and partially restores vacuolar H^+-ATPase activity. *The Plant J.*, **17** 501-10.

Kim, Y., Kim, E.J. and Rea, P.A. (1994) Isolation and characterization of cDNAs encoding the vacuolar H^+-pyrophosphatase of *Beta vulgaris*. *Plant Physiol.*, **106** 375-82.

Kirsch, M., An, Z., Viereck, R., Löw, R. and Rausch, T. (1996) Salt stress induces an increased expression of V-type H^+-ATPase in mature sugar-beet leaves. *Plant Mol. Biol.*, **32** 543-47.

Klink, R. and Lüttge, U. (1991) Electron-microscopic demonstration of a 'head and stalk' structure of the leaf vacuolar ATPase in *Mesembryanthemum crystallinum*. *Bot. Acta*, **104** 122-31.

Klink, R. and Lüttge, U. (1992) Quantification of visible structural changes of the V_0V_1-ATPase in the leaf tonoplast of *Mesembryanthemum crystallinum* by freeze-fracture replicas prepared during the C_3-photosynthesis to CAM transition. *Bot. Acta*, **105** 414-20.

Klink, R., Haschke, H.-P., Kramer, D. and Lüttge, U. (1990) Membrane particles, proteins and ATPase activity of tonoplast vesicles of *Mesembryanthemum crystallinum* in the C_3- and CAM state. *Bot. Acta*, **103** 24-31.

Kluge, C., Golldack, D. and Dietz, K.-J. (1999) Subunit D of the vacuolar H^+-ATPase of *Arabidopsis thaliana*. *Biochim. Biophys. Acta*, **1419** 105-10.

Koyro, H.-W., Stelzer, R. and Huchzermeyer, B. (1992) ATPase activities and membrane fine structure of rhizodermal cells from *Sorghum* and *Spartina* roots grown under mild salt stress. *Bot. Acta*, **106** 110-19.

Kramer, D., Mangold, B., Hille, A., Emig, I., Hess, A., Ratajczak, R. and Lüttge, U. (1995) The head structure of a higher plant V-type H^+-ATPase is not always a hexamer but also a pentamer. *J. Exp. Bot.*, **46** 1633-36.

Krauss, W., Schiebel, G., Eberl, D. and Hager, A. (1987) Blue light induced, reversible inactivation of the tonoplast-type H^+-ATPase from corn coleoptiles in the presence of flavins. *Phytochem. Photobiol.*, **45** 837-44.

Krisch, R., Rakowski, K. and Ratajczak, R. (2000) Processing of V-ATPase subunit B of *Mesembryanthemum crystallinum* L. is mediated *in vitro* by a protease and/or reactive oxygen species. *Biol. Chem.*, **381** 583-92.

Lai, S., Randall, S.K. and Sze, H. (1988) Peripheral and integral subunits of the tonoplast H^+-ATPase from oat roots. *J. Biol. Chem.*, **263** 16731-37.

Lai, S., Watson, J.C., Hansen, J. and Sze, H. (1991) Molecular cloning and sequencing of cDNAs encoding the proteolipid subunit of the vacuolar H^+-ATPase from a higher plant. *J. Biol. Chem.*, **266** 16078-84.

Landoldt-Marticorena, C., Kahr, W.H., Zawarinski, P., Correa, J. and Manolson, M.F. (1999) Substrate- and inhibitor-induced conformational changes in the yeast V-ATPase provide evidence for communication between the catalytic and proton-translocating sectors. *J. Biol. Chem.*, **274** 26057-64.

Leach, R.P., Rogers, W.J., Wheeler, K.P., Flowers, T.J. and Yeo, A.R. (1990) Molecular markers for ion compartmentation in cells of higher plants. I. Isolation of vacuoles of high purity. *J. Exp. Bot.*, **41** 1079-87.

Lee Taiz, S. and Taiz, L. (1991) Ultrastructural comparison of the vacuolar and mitochondrial H^+-ATPases of *Daucus carota*. *Bot. Acta*, **104** 117-21.

Lehr, A., Kirsch, M., Viereck, R., Schiemann, J. and Rausch, T. (1999) cDNA and genomic cloning of sugar beet V-type H^+-ATPase subunit A and c isoforms: evidence for coordinate expression during plant development and coordinate induction in response to high salinity. *Plant Mol. Biol.*, **39** 463-75.

Leigh, R.A., Pope, A.J., Jennings, I.R. and Sanders, D. (1992) Kinetics of the vacuolar H^+-pyrophosphatase: the roles of magnesium, pyrophosphate and their complexes as substrates, activators and inhibitors. *Plant Physiol.*, **100** 1698-705.

Leigh, R.A., Gordon-Weeks, R., Steele, S.H. and Koren'kov, V.D. (1994) The H^+-pumping inorganic pyrophosphatase of the vacuolar membrane of higher plants, in *Membrane Transport in Plants and Fungi: Molecular Mechanisms and Control* (eds. M.R. Blatt, R.A. Leigh and D. Sanders), The Company of Biologists Ltd, Cambridge, pp. 61-75.

Leng, X.H., Manolson, M.F., Liu, Q. and Forgac, M. (1996) Site-directed mutagenesis of the 100 kD subunit, Vph1p, of the yeast vacuolar H^+-ATPase. *J. Biol. Chem.*, **271** 22487-93.

Leng, X.H., Manolson, M.F. and Forgac, M. (1998) Function of the COOH-terminal domain of Vph1p in activity and assembly of the yeast V-ATPase. *J. Biol. Chem.*, **273** 6717-23.

Lerchl, J., König, S. and Sonnewald, U. (1995) Molecular cloning and expression analysis of isoforms encoding tonoplast-bound proton-translocating inorganic pyrophosphatase in tobacco. *Plant Mol. Biol.*, **29** 833-40.

Li, X. and Sze, H. (1999) A 100 kD polypeptide associates with the V_0 membrane sector but not with the active oat vacuolar H^+-ATPase suggesting a role in assembly. *Plant J.*, **17** 19-30.

Li, X., Su, R.T.C., Hsu, H.-T. and Sze, H. (1998) The molecular chaperon, calnexin, associates with the vacuolar H^+-ATPase from oat roots. *Plant Cell*, **10** 119-30.

Liu, Q., Leng, X.H., Newman, P.R., Vasilyeva, E., Kane, P.M. and Forgac, M. (1997) Site-directed mutagenesis of the yeast V-ATPase A subunit. *J. Biol. Chem.*, **272** 11750-56.

Löw, R. and Rausch, T. (1996) In suspension-cultured *Daucus carota* cells salt stress stimulates H^+-transport but not ATP hydrolysis of the V-ATPase. *J. Exp. Bot.*, **47** 1725-32.

Löw, R., Rockel, B., Kirsch, M., Ratajczak, R., Hörtensteiner, S., Martinoia, E., Lüttge, U. and Rausch, T. (1996) Early salt stress effects on the differential expression of vacuolar H^+-ATPase genes in roots and leaves of *Mesembryanthemum crystallinum*. *Plant Physiol.*, **110** 259-65.

Long, A.R., Williams, L.E., Nelson, S.J. and Hall, J. (1995) Localization of pyrophosphatase activity in *Ricinus communis* seedlings. *J. Plant Physiol.*, **146** 629-38.

Lüttge, U. and Ratajczak, R. (1997) The physiology, biochemistry and molecular biology of the plant vacuolar ATPase, in *The Plant Vacuole* (eds. R.A. Leigh and D. Sanders), Advances in Botanical Research, Vol. 25, Academic Press, San Diego, London, Boston, New York, Sydney, Tokyo, Toronto, pp. 253-96.

Lüttge, U., Smith, J.A.C., Marigo, G. and Osmond, C.B. (1981) Energetics of malate accumulation in the vacuoles of *Kalanchoë tubiflora* cells. *FEBS Lett.*, **126** 81-84.

Lüttge, U., Fischer-Schliebs, E., Ratajczak, R., Kramer, D., Berndt, E. and Kluge, M. (1995a) Functioning of the tonoplast in vacuolar C-storage and remobilization in crassulacean acid metabolism. *J. Exp. Bot.*, **46** 1377-88.

Lüttge, U., Ratajczak, R., Rausch, T. and Rockel, B. (1995b) Stress responses of tonoplast proteins: an example for molecular ecophysiology and the search for eco-enzymes. *Acta Bot. Neerland.*, **44** 343-62.

Luo, S., Marchesini, N., Moreno, S.N. and Docampo, R. (1999) A plant-like vacuolar $H(^+)$-pyrophosphatase in *Plasmodium falciparum*. *FEBS Lett.*, **460** 217-20.

Maeshima, M. and Yoshida, S. (1989) Purification and properties of the vacuolar membrane proton-translocating inorganic pyrophosphatase from mung bean. *J. Biol. Chem.*, **264** 20068-73.

Maeshima, M., Hara-Nishimura, I., Takeuchi, Y. and Nishimura, M. (1994) Accumulation of vacuolar H^+-pyrophosphatase and H^+-ATPase during reformation of the central vacuole in germinating pumpkin seeds. *Plant Physiol.*, **106** 61-69.

Mandala, S. and Taiz, L. (1986) Characterization of the subunit structure of the maize tonoplast ATPase. *J. Biol. Chem.*, **261** 12850-55.

Manolson, M.F., Rea, P.A. and Poole, R.J. (1985) Identification of 3-O-(4-benzoyl)benzoyladenosine 5′-triphosphate and N,N′-dicyclohexylcarbodiimide-binding subunits of a higher plant H^+-translocating tonoplast ATPase. *J. Biol. Chem.*, **260** 12273-79.

Manolson, M.F., Quellette, B.F.F., Filion, M. and Poole, R.J. (1988) cDNA sequence and homologies of the '57 kDa' nucleotide-binding subunit of the vacuolar ATPase from *Arabidopsis*. *J. Biol. Chem.*, **263** 17987-94.

Manolson, M.F., Proteau, D., Preston, R.A., Stenbit, A., Roberts, B.T., Hoyt, M.A., Preuss, D., Mulholland, J., Botstein, D. and Jones, E.W. (1992) The VPH1 gene encodes a 95 kDa integral membrane polypeptide required for *in vivo* assembly and activity of the yeast vacuolar H^+-ATPase. *J. Biol. Chem.*, **267** 14292-303.

Manolson, M.F., Wu, B., Proteau, D. Taillon, B.E., Roberts, B.T., Hoyt, M.A. and Jones, E.W. (1994) *STV1* gene encodes functional homolog of 95 kDa yeast vacuolar H^+-ATPase subunit, Vph1p. *J. Biol. Chem.*, **269** 14064-74.

Mariaux, J.-B., Becker, A., Kemna, I., Ratajczak, R., Fischer-Schliebs, E., Kramer, D., Lüttge, U. and Marigo, G. (1994) Visualization by freeze-fracture electron microscopy of intramembraneous particles corresponding to the tonoplast H^+-pyrophosphatase and H^+-ATPase of *Kalanchoë daigremontiana* Hamet et Perrier de la Bâthie. *Bot. Acta*, **107** 321-27.

Mariaux, J.-B., Fischer-Schliebs, E., Lüttge, U. and Ratajczak, R. (1997) Dynamics of activity and structure of the tonoplast vacuolar-type H^+-ATPase in plants with differing CAM expression and in a C_3 plant under salt stress. *Protoplasma*, **196** 181-89.

Marquardt-Jarczyk, G. and Lüttge, U. (1990) PPase-activated ATP-dependent H^+-transport at the tonoplast of mesophyll cells of the CAM plant, *Kalanchoë daigremontiana*. *Bot. Acta*, **103** 203-13.

Martinoia, E. and Ratajczak, R. (1997) Transport of organic molecules across the tonoplast, in *The Plant Vacuole* (eds. R.A. Leigh and D. Sanders), Advances in Botanical Research, Vol. 25, Academic Press, San Diego, London, Boston, New York, Sydney, Tokyo, Toronto, pp. 365-400.

Maruyama, C., Tanaka, Y., Takeyasu, K., Yoshida, M. and Sato, M.H. (1998) Structural studies of the vacuolar $H(^+)$-pyrophosphatase: sequence analysis and identification of the residues modified by fluorescent cyclohexylcarbodiimide and maleimide. *Plant Cell Physiol.*, **39** 1045-53.

Matsuoka, K., Higuchi, T., Maeshima, M. and Nakamura, K. (1997) The vacuolar-type H^+-ATPase in a non-vacuolar organelle is required for sorting of soluble protein precursors in tobacco cells. *Plant Cell*, **9** 533-46.

Matsuura-Endo, C., Maeshima, M. and Yoshida, S. (1990) Subunit composition of the vacuolar membrane H^+-ATPase from mung bean. *Eur. J. Biochem.*, **187** 745-51.

Matsuura-Endo, C., Maeshima, M. and Yoshida, S. (1992) Mechanism of the decline in vacuolar H^+-ATPase activity in mung bean hypocotyls during chilling. *Plant Physiol.*, **100** 718-22.

Michelet, B. and Boutry, M. (1995) The plasma membrane H^+-ATPase: a highly-regulated enzyme with multiple physiological functions. *Plant Physiol.*, **108** 1-6.

Moriyama, Y. and Nelson, N. (1988) The vacuolar H^+-ATPase, a proton pump controlled by a slip, in *The Ion Pumps: Structure, Function and Regulation* (ed. W.D. Stein), Allan R. Liss, New York, pp. 387-94.

Moriyama, Y. and Nelson, N. (1989) Cold inactivation of vacuolar proton-ATPases. *J. Biol. Chem.*, **264** 3577-82.

Müller, M.L., Irkens-Kiesecker, U., Rubinstein, B. and Taiz, L. (1996) On the mechanism of hyper-acidification in lemon. *J. Biol. Chem.*, **271** 1916-24.

Müller, M.L., Irkens-Kiesecker, U., Kramer, D. and Taiz, L. (1997) Purification and reconstitution of the vacuolar H^+-ATPases from lemon fruits and epicotyls. *J. Biol. Chem.*, **272** 12762-70.

Nakamura, Y., Kasamo, K., Shimosato, N., Sakata, M. and Ohta, E. (1992) Stimulation of the extrusion of protons and H^+-ATPase activities with the decline in pyrophosphatase activity of the tonoplast in intact mung bean roots under high-NaCl stress and its relation to external levels of Ca^{2+} ions. *Plant Cell Physiol.*, **33** 139-49.

Nakanishi, Y. and Maeshima, M. (1998) Molecular cloning of vacuolar H^+-pyrophosphatase and its developmental expression in growing hypocotyl of mung bean. *Plant Physiol.*, **116** 589-97.

Narasimhan, M.L., Binzel, M.L., Perez-Prat, E., Chen, Z., Nelson, D.E., Singh, N.K., Bressan, R.B. and Hasegawa, P.M. (1991) NaCl regulation of tonoplast ATPase 70 kDa subunit mRNA in tobacco cells. *Plant Physiol.*, **97** 562-68.

Nelson, H. and Nelson, N. (1990) Disruption of genes encoding subunits of yeast vacuolar H^+-ATPase causes conditional lethality. *Proc. Natl. Acad. Sci. USA*, **87** 3503-507.

Nelson, H., Mandiyan, S. and Nelson, N. (1995) A bovine cDNA and a yeast gene (*VMA8*) encoding the subunit D of the vacuolar H^+-ATPase. *Proc. Natl. Acad. Sci. USA*, **92** 497-501.

Nelson, N. and Klionsky, D.J. (1996) Vacuolar H^+-ATPase: from mammals to yeast and back. *Experientia*, **52** 1101-10.

Noumi, T., Beltran, C., Nelson, H. and Nelson, N. (1991) Mutational analysis of the yeast vacuolar (H^+)-ATPase. *Proc. Natl. Acad. Sci. USA*, **88** 1938-42.

Nyren, P., Sakai-Nore, Y. and Strid, A. (1993) Amino acid sequence similarities between the vacuolar proton-pumping inorganic pyrophosphatase and the c-subunit of F_0F_1-ATPases. *Plant Cell Physiol.*, **34** 375-78.

Oberbeck, K., Drucker, M. and Robinson, D.G. (1994) V-ATPase and pyrophosphatase in endomembranes of maize roots. *J. Exp. Bot.*, **45** 235-44.

Orr, W., White, T.C., Iu, B., Robert, L. and Singh, J. (1995) Characterization of a low-temperature-induced cDNA from winter *Brassica naua* encoding the 70 kD subunit of tonoplast ATPase. *Plant Mol. Biol.*, **28** 943-48.

Pänke, O. and Rumberg, B. (1999) Kinetic modeling of rotary CF_0F_1-ATP synthase: storage of elastic energy during energy transduction. *Biochim. Biophys. Acta*, **1412** 118-28.

Peng, S.B., Crider, B.P., Xie, X.S. and Stone, D.K. (1994) Alternative mRNA splicing generates tissue-specific isoforms of the 116 kDa polypeptide of the vacuolar proton pump. *J. Biol. Chem.*, **269** 17262-66.

Perera, I.Y., Li, X. and Sze, H. (1995) Several distinct genes encode nearly identical 16 kDa proteolipids of the vacuolar H^+-ATPase from *Arabidopsis thaliana*. *Plant Mol. Biol.*, **29** 227-44.

Pope, A.J. and Leigh, R.A. (1987) Some characteristics of anion transport at the tonoplast of oat roots, determined from the effect of anions on pyrophosphate-dependent proton transport. *Planta*, **172** 91-100.

Radermacher, M., Ruiz, T., Harvey, W.R., Wieczorek, H. and Grüber, G. (1999) Molecular architecture of *Manduca sexta* V_1 ATPase visualized by electron microscopy. *FEBS Lett.*, **453** 383-86.

Randall, S.K. and Sze, H. (1986) Properties of the partially purified tonoplast H^+-pumping ATPase from oat roots. *J. Biol. Chem.*, **261** 1364-71.

Ratajczak, R. (1994) The non-ionic detergent, Brij 58, conserves the structure of the tonoplast H^+-ATPase of *Mesembryanthemum crystallinum* L. during solubilization and partial purification, *Bot. Acta*, **107** 201-209.

Ratajczak, R., Richter, J. and Lüttge U. (1994) Adaptation of the tonoplast V-type H^+-ATPase of *Mesembryanthemum crystallinum* to salt stress, C_3-CAM transition and plant age. *Plant Cell Environ.*, **17** 1101-12.

Ratajczak, R., Hille, A., Mariaux, J.-B. and Lüttge, U. (1995) Quantitative stress responses of the V_0V_1-ATPase of higher plants detected by immunoelectron microscopy. *Bot. Acta*, **108** 505-13.

Ratajczak, R., Feussner, I., Hause, B., Böhm, A., Parthier, B. and Wasternack, C. (1998) Alteration of V-type H^+-ATPase during methyljasmonate-induced senescence in barley (*Hordeum vulgare* cv. Salomé). *J. Plant Physiol.*, **152** 199-206.

Ratajczak, R., Hinz, G. and Robinson, D.G. (1999) Localization of pyrophosphatase in membranes of cauliflower inflorescence cells. *Planta*, **208** 205-11.

Rausch, T., Kirsch, M., Löw, R., Lehr, A., Viereck, R. and An Zhigang (1996) Salt stress responses of higher plants: the role of proton pumps and Na^+/H^+-antiporters. *J. Plant Physiol.*, **148** 425-33.

Rea, P.A. and Sanders, D. (1987) Tonoplast energization: two H^+ pumps, one membrane. *Physiol. Plant.*, **71** 131-41.

Rea, P.A. and Poole, R.J. (1993) Vacuolar H^+-translocating pyrophosphatase. *Annu. Rev. Plant Physiol. Plant Mol. Biol.*, **44** 157-80.

Rea, P.A., Griffith, C.J. and Sanders, D. (1987) Irreversible inhibition of H^+-ATPase of higher plant tonoplast by chaotropic anions: evidence for peripheral location of nucleotide binding subunits. *Biochim. Biophys. Acta*, **904** 1-12.

Rea, P.A., Kim, Y., Sarafian, V., Poole, R.J., Davies, J.M. and Sanders, D. (1992) Vacuolar H^+-translocating pyrophosphatases: a new category of ion translocase. *Trends Biochem. Sci.*, **17** 348-53.

Reuveni, M. and Avi, S. (2000) Isolation of vacuolar H^+-ATPase subunit E cDNA from juice sacs of *Citrus limon* (Accession No. AF165939). *Plant Physiol.*, **122** 291.

Reuveni, M., Bennett, A.B., Bressan, R.A. and Hasegawa, P.M. (1990) Enhanced H^+-transport capacity and ATP hydrolysis activity of the tonoplast H^+-ATPase after NaCl adaptation, *Plant Physiol.*, **94** 524-30.

Reuveni, M., Evenor, D., Artzi, B., Erner, Y. and Perl, A. (2000) Decrease in vacuolar pH during petunia flower opening is reflected in the activity of the tonoplast H^+-ATPase. *J. Plant Physiol.* (in press).

Robinson, D.G., Hoppenrath, M., Oberbeck, K., Luykx, P. and Ratajczak, R. (1998) Localization of pyrophosphatase and V-ATPase in *Chlamydomonas reinhardtii*. *Bot. Acta*, **111** 108-22.

Rockel, B., Ratajczak, R., Becker, A. and Lüttge, U. (1994) Changed densities and diameters of intramembrane tonoplast particles of *Mesembryanthemum crystallinum* in correlation with NaCl-induced CAM. *J. Plant Physiol.*, **143** 318-24.

Rockel, B., Lüttge, U. and Ratajczak, R. (1998) Changes of message amount of V-ATPase subunits during salt-stress induced C_3-CAM transition in *Mesembryanthemum crystallinum*. *Plant Physiol. Biochem.*, **36** 567-73.

Rodriguez, C.O., Scott, D.A. and Docampo, R. (1999a) Presence of a vacuolar H^+-pyrophosphatase in promastigotes of *Leishmania donovani* and its localization to a different compartment from the vacuolar H^+-ATPase. *Biochem. J.*, **340** 759-66.

Rodriguez, C.O., Scott, D.A. and Docampo, R. (1999b) Characterization of a vacuolar pyrophosphatase in *Trypanosoma brucei* and its localization to acidocalcisomes. *Mol. Cell Biol.*, **19** 7712-23.

Ros, R., Romieu, C., Gibrat, R. and Grignon, C. (1995) The plant inorganic pyrophosphatase does not transport K^+ in vacuolar membrane vesicles multilabeled with fluorescent probes for H^+, K^+ and membrane potential. *J. Biol. Chem.*, **270** 1-7.

Rouquié, D., Tournaire-Roux, C., Szeponarski, W., Rossignol, M. and Doumas, P. (1998) Cloning of the V-ATPase subunit G in plants: functional expression and subcellular localization. *FEBS Lett.*, **437** 287-92.

Sakakibara, Y., Kobayashi, H. and Kasamo, K. (1996) Isolation and characterization of cDNAs encoding vacuolar H^+-pyrophosphatase isoforms from rice (*Oryza sativa* L.), *Plant Mol. Biol.*, **31** 1029-38.

Sanchez-Aguyo, I., Gonzalez-Utor, A. and Medina, A. (1991) Cytochemical localization of ATPase activity in salt-treated and salt-free grown *Lycopersicon esculentum* roots. *Plant Physiol.*, **96** 153-58.

Sarafian, V., Kim, Y., Poole, R.J. and Rea, P.A. (1992) Molecular cloning and sequencing of cDNA encoding the pyrophosphatase-energized vacuolar membrane pump of *Arabidopsis thaliana*. *Proc. Natl. Acad. Sci. USA*, **89** 1775-79.

Sato, M.H. (1997) H^+-translocating inorganic pyrophosphatase of pumpkin shows significant sequence similarity to K^+-translocating Kdp-ATPase A chain of *E. coli*. *Plant Physiol.*, **114** 196.

Sato, M.H., Kasahara, M., Ishii, N., Homareda, H., Matsui, H. and Yoshida, M. (1994) Purified vacuolar inorganic pyrophosphatase consisting of a 75 kDa polypeptide can pump H^+ into reconstituted proteoliposomes. *J. Biol. Chem.*, **269** 6725-28.

Schumacher, K., Vafeados, D., McCarthy, M., Sze, H., Wilkins, T.A. and Chory, J. (1999) The *Arabidopsis det3* mutant reveals a central role for the vacuolar H^+-ATPase in plant growth and development. *Genes Dev.*, **13** 3259-70.

Scott, D.A., de Souza, W., Benchimol, M., Zhong, L., Lu, H.G., Moreno, S.N. and Docampo, R. (1998) Presence of a plant-like proton-pumping pyrophosphatase in acidocalcisomes of *Trypanosoma cruzi*. *J. Biol. Chem.*, **273** 22151-58.

Shiratake, K., Kanayama, Y., Maeshima, M. and Yamaki, S. (1997) Changes in $H(^+)$ pumps and a tonoplast intrinsic protein of vacuolar membranes during the development of pear fruits. *Plant Cell Physiol.*, **38** 1039-45.

Smart, L.B., Vojdani, F., Maeshima, M. and Wilkins, T.A. (1998) Genes involved in osmoregulation during turgor-driven cell expansion of developing cotton fibers are differentially regulated. *Plant Physiol.*, **116** 1539-49.

Smith, J.A.C., Marigo, G., Lüttge, U. and Ball, E. (1982) Adenine nucleotide levels during crassulacean acid metabolism and the energetics of malate accumulation in *Kalanchoë tubiflora*. *Plant Sci. Lett.*, **26** 13-21.

Starke, T., Linkkila, T.P. and Gogarten, J.P. (1991) Two separate genes encode the catalytic 70 kDa V-ATPase subunit in *Psilotum* and *Equisetum*. *Z. Naturforsch.*, **46c** 613-20.

Stevens, T.H. and Forgac, M. (1997) Structure, function and regulation of the vacuolar (H^+)-ATPase. *Annu. Rev. Cell Dev. Biol.*, **13** 779-808.

Struve, I. and Lüttge, U. (1987) Characteristics of $MgATP^{2-}$-dependent electrogenic proton transport in tonoplast vesicles of the facultative crassulacean acid metabolism plant, *Mesembryanthemum crystallinum* L. *Planta*, **170** 111-20.

Supekova, L., Sbia, M., Supek, F., Ma, Y.-M. and Nelson, N. (1996) A novel subunit of vacuolar H^+-ATPase related to the b subunit of F-ATPases. *J. Exp. Biol.*, **199** 1147-56.

Suzuki, K. and Kasamo, K. (1993) Effects of aging on the ATP- and pyrophosphate-dependent pumping of protons across the tonoplast isolated from pumpkin cotelydons. *Plant Cell Physiol.*, **34** 613-19.

Suzuki, Y., Maeshima, M. and Yamaki, S. (1999) Molecular cloning of vacuolar $H(^+)$-pyrophosphatase and its expression during the development of pear fruit. *Plant Cell Physiol.*, **40** 900-904.

Sze, H. (1985) H^+-translocating ATPases: advances using membrane vesicles. *Annu. Rev. Plant Physiol.*, **36** 175-208.

Sze, H., Li, X. and Palmgren, M.G. (1999) Energization of plant cell membranes by H^+-pumping ATPases: regulation and biosynthesis. *Plant Cell*, **11** 677-90.

Takanokura, Y., Komatsu, A., Omura, M and Akihama, T. (1998) Cloning and expression analysis of vacuolar H^+-ATPase 69 kDa catalytic subunit cDNA in citrus (*Citrus unshiu* Marc.). *Biochim. Biophys. Acta*, **1414** 265-72.

Takasu, A., Nakanishi, Y., Yamauchi, T. and Maeshima, M. (1997) Analysis of the substrate binding site and carboxyl terminal region of vacuolar H^+-pyrophosphatase of mung bean with peptide antibodies. *J. Biochem.*, **122** 883-89.

Tavakoli, N., Eckerskorn, C., Golldack, D. and Dietz, K.-J. (1999) Subunit C of the vacuolar H^+-ATPase of *Hordeum vulgare*. *FEBS Lett.*, **456** 68-72.

Terrier, N., Deguilloux, C., Sauvage, F.-X., Martinoia, E. and Romieu, C. (1998) Proton pumps and anion transport in *Vitis vinifera*: the inorganic pyrophosphatase plays a predominant role in the energization of the tonoplast. *Plant Physiol. Biochem.*, **36** 367-77.

Tomashek, J.J., Graham, L.A., Hutchins, M.U., Stevens, T.H. and Klionsky, D.J. (1997) V_1-situated stalk subunits of the yeast vacuolar proton-translocating ATPase. *J. Biol. Chem.*, **272** 26787-93.

Tomos, D. and Pritchard, J. (1994) Biophysical and biochemical control of cell expansion in roots and leaves. *J. Exp. Bot.*, **45** 1721-31.

Tsiantis, M., Bartholomew, D.M. and Smith, J.A.C. (1996) Salt regulation of transcript levels for the c subunit of a leaf vacuolar H^+-ATPase in the halophyte, *Mesembryanthemum crystallinum*. *Plant J.*, **9** 729-36.

Tzeng, C.M., Hsu, L.H. and Pan, R.L. (1992) Inhibition of tonoplast ATPase from etiolated mung bean seedlings by fluorescein 5′-isothiocyanate. *Biochem. J.*, **285** 737-43.

Vera-Estrella, R., Barkla, B.J., Bohnert, H.J. and Pantoja, O. (1999) Salt stress in *Mesembryanthemum crystallinum* L. cell suspensions activates adaptive mechanisms similar to those observed in the whole plant. *Planta*, **207** 426-35.

Verstappen, R. and Rausch, T. (1988) Different SH groups involved in H^+-translocation and PP_i hydrolysis of higher plants Mg^{2+}-PP_iase. *FEBS Lett.*, **236** 420-24.

Viereck, R., Kirsch, M., Löw, R. and Rausch, T. (1996) Downregulation of plant V-type H^+-ATPase genes after light-induced inhibition of growth. *FEBS Lett.*, **384** 285-88.

Walker, R.R. and Leigh, R.A. (1981) Mg^{2+}-dependent, cation-stimulated inorganic pyrophosphatase associated with vacuoles of red beet (*Beta vulgaris* L.). *Planta*, **153** 150-55.

Wan, C.-Y. and Wilkins, T.A. (1994) Isolation of multiple clones encoding the vacuolar H^+-ATPase subunit B from developing cotton (*Gossypium hirsutum* L.) ovules. *Plant Physiol.*, **106** 393-94.

Wang, B.S., Ratajczak, R. and Zhang, J.H. (2000) Activity, amount and subunit composition of vacuolar-type H^+-ATPase and H^+-PPase in wheat roots under severe NaCl stress. *J. Plant Physiol.*, **157** 109-16.

Wang, Y. and Sze, H. (1985) Similarities and differences between the tonoplast-type and the mitochondrial H^+-ATPases of oat roots. *J. Biol. Chem.*, **260** 10434-43.

Wang, Y., Leigh, R.A., Kaestner, K.H. and Sze, H. (1986) Electrogenic H^+-pumping pyrophosphatase in tonoplast vesicles of oat roots. *Plant Physiol.*, **81** 497-502.

Ward, J.M. and Sze, H. (1992a) Subunit composition and organization of the vacuolar H^+-ATPase from oat roots. *Plant Physiol.*, **99** 170-79.

Ward, J.M. and Sze, H. (1992b) Proton transport activity of the purified vacuolar H^+-ATPase from oats: direct stimulation by Cl^-. *Plant Physiol.*, **99** 925-31.

Ward, J.M., Reinders, A., Hsu, H.-T. and Sze, H. (1992) Dissociation and reassembly of the vacuolar H^+-ATPase complex from oat roots. *Plant Physiol.*, **99** 161-69.

Warren, M., Smith, J.A.C. and Apps, D.K. (1992) Rapid purification and reconstitution of a plant vacuolar ATPase using Triton X-114 fractionation: subunit composition and substrate kinetics of the H^+-ATPase from the tonoplast of *Kalanchoë daigremontiana*. *Biochim. Biophys. Acta*, **1106** 117-25.

Weiner, H., Stitt, M. and Heldt, H.W. (1987) Subcellular compartmentation of pyrophosphatase and alkaline pyrophosphatase in leaves. *Biochim. Biophys. Acta*, **893** 13-21.

White, P.J., Marshall, J. and Smith, J.A.C. (1990) Substrate kinetics of the H^+-translocating inorganic pyrophosphatase and its activation by free $Mg2^+$. *Plant Physiol.*, **93** 1063-70.

Wilkens, S., Vasilyeva, E. and Forgac, M. (1999) Structure of the vacuolar ATPase by electron microscopy. *J. Biol. Chem.*, **274** 31804-10.

Wilkins, T.A. and Jernstedt, J.A. (1999) Molecular genetics of developing cotton fibers, in *Cotton Fibers* (ed. A.M. Basra), chp. 9, Food Products Press, New York, pp. 231-67.

Wilkins, T.A., Wan, C.-Y. and Lu, C.-C. (1994) Ancient origin of the vacuolar H^+-ATPase 69 kilodalton catalytic subunit superfamily. *Theor. Appl. Genet.*, **89** 514-24.

Wink, M. (1997) Compartmentation of secondary metabolites and xenobiotics in plant vacuoles, in *The Plant Vacuole* (eds. R.A. Leigh and D. Sanders), Advances in Botanical Research, Vol. 25, Academic Press, San Diego, London, Boston, New York, Sydney, Tokyo, Toronto, pp. 141-69.

Xie, Y.-Y., Coukell, M.B. and Gombos, Z. (1996) Antisense RNA inhibition of the putative vacuolar H^+-ATPase proteolipid of *Dictyostelium* reduces intravcellular Ca^{2+} transport and cell viability. *J. Cell Sci.*, **109** 489-97.

Yamanishi, H. and Kasamo, K. (1992) Binding of 7-chloro-4-nitrobenzo-2oxa-1,3-diazole to an essential cysteine residue(s) in the tonoplast H^+-ATPase from mung bean (*Vigna radiata* L.) hypocotyls. *Plant Physiol.*, **99** 652-58.

Yang, S.J., Jiang, S.S., Tzeng, C.M., Kuo, S.Y., Hung, S.H. and Pan, R.L. (1996) Involvement of tyrosine residue in the inhibition of plant vacuolar $H(^+)$-pyrophosphatase by tetranitromethane. *Biochim. Biophys. Acta*, **1294** 89-97.

Yang, S.J., Jiang, S.S., Kuo, S.Y., Hung, S.H., Tam, M.F. and Pan, R.L. (1999) Localization of a carboxylic residue possibly involved in the inhibition of vacuolar H^+-pyrophosphatase by N,N′-dicyclohexylcarbodiimide. *Biochem. J.*, **342** 641-46.

Yoshida, S. (1991) Chilling-induced inactivation and its recovery of tonoplast H^+-ATPase in mung bean cell suspension cultures. *Plant Physiol.*, **95** 456-60.

Yoshida, S. and Matsuura-Endo, C. (1991) Comparison of temperature dependency of tonoplast proton translocation between plants sensitive and insensitive to chilling. *Plant Physiol.*, **95** 504-508.

Yoshida, S., Hattanda, Y. and Suyama, T. (1993) Variations in chilling sensitivity of suspension-cultured cells of mung bean (*Vigna radiata* L., Wilczek) during the growth cycle. *Plant Cell Physiol.*, **34** 673-79.

Zhang, J.M., Myers, M. and Forgac, M. (1992) Characterization of the V_0 domain of the coated vesicle (H^+)-ATPase. *J. Biol. Chem.*, **267** 9773-78.

Zhang, J.M., Feng, Y. and Forgac, M. (1994) Proton conduction and bafilomycin binding by the V_0 domain of the coated vesicle V-ATPase. *J. Biol. Chem.*, **269** 23518-23.

Zhang, J.M., Vasilyeva, E., Feng, Y. and Forgac, M. (1995) Inhibition and labeling of the coated vesicle V-ATPase by 2-azido-[^{32}P]ATP. *J. Biol. Chem.*, **270** 15494-500.

Zhen, R.-G., Kim, E.J. and Rea, P.A. (1994a) Localization of cytosolically-oriented, maleimide-reactive domain of vacuolar H($^+$)-pyrophosphatase. *J. Biol. Chem.*, **269** 23342-50.

Zhen, R.-G., Baykov, A.A., Bakuleva, N.P. and Rea, P.A. (1994b) Aminomethylene diphosphonate as a potent type-specific inhibitor of both plant and phototrophic bacterial H^+-pyrophosphatases. *Plant Physiol.*, **104** 153-59.

Zhen, R.-G., Kim, E.J. and Rea, P.A. (1997a) The molecular and biochemical basis of pyrophosphate-energized proton translocation at the vacuolar membrane, in *The Plant Vacuole* (eds. R.A. Leigh and D. Sanders), Advances in Botanical Research, Vol. 25, Academic Press, San Diego, London, Boston, New York, Sydney, Tokyo, Toronto, pp. 298-337.

Zhen, R.-G., Kim, E.J. and Rea, P.A. (1997b) Acidic residues necessary for pyrophosphate-energized pumping and inhibition of the vacuolar H^+-pyrophosphatase by N,N′-dicyclohexylcarbodiimide. *J. Biol. Chem.*, **272** 22340-48.

Zhigang, A., Löw, R., Rausch, T., Lüttge, U. and Ratajczak, R. (1996) The 32 kDa tonoplast polypeptide D_i associated with the V-ATPase of *Mesembryanthemum crystallinum* L. in the CAM state: a proteolytically processed subunit B? *FEBS Lett.*, **389**, 314-18.

Zingarelli, L., Anzani, P. and Lado, P. (1994) Enhanced K^+-stimulated pyrophosphatase in NaCl-adapted cells of *Acer pseudoplatanus*. *Physiol. Plant.*, **91** 510-16.

8 Spatial and temporal variation in vacuolar contents

A.D. Tomos, R.A. Leigh and O.A. Koroleva

8.1 Introduction

Much of our understanding of vacuolar function has come from the application of averaging techniques, for example whole tissue analysis, where results are dominated by the contents of the most abundant vacuoles. Such measurements on populations of vacuoles mask any cell-to-cell variation in activity. Alternatively, the properties of a single cell type, such as leaf mesophyll, guard cell or storage root parenchyma, have been studied without reference to the adjacent cells. The evidence that vacuolar composition varies significantly between different cells in the same tissue or organ is increasing. This spatial variability adds a degree of complexity to understanding vacuolar function and is additional to the more widely-appreciated temporal variability in vacuolar composition. While major steps have been made in determining intracellular solute distribution (e.g. Walker *et al.*, 1996), quantitative mapping of intercellular variations in concentrations lags behind. Single-cell-based techniques are needed to detect such detail.

Some functions assigned to the central vacuole are probably common to all cells (space filling, flexibility in solute storage, osmotic pressure generation and cytosolic ion homeostasis), while others may be unique to individual cell types. This chapter reviews the evidence for intercellular variation in the composition of the large central vacuoles of individual mature cells within tissues. The discovery that distinct types of vacuole may coexist within the same cell (e.g. Paris *et al.*, 1996) raises the possibility that there might also be variation in solute distributions between these. However, studies at this level lie beyond the scope of many techniques and are not addressed here.

8.2 Functional roles of vacuoles

Vacuoles fulfil at least four generic roles that affect their solute contents.

1. They generate osmotic pressure, which gives them the capacity for water storage, a feature that is important in maintaining and changing turgor for support and movement.

2. They act as a temporary store for solutes for later use elsewhere in the cell or plant, for example, when needed for cytosolic ion homeostasis.
3. They constitute a permanent store for those materials taken up by the plant which are present in excess, are not required or are toxic, and which must be removed from the cytosol.
4. They accumulate certain solutes that have a specific function within the vacuole. These include secondary metabolites, such as those responsible for petal colour or for discouraging herbivores.

The relative importance of each role varies in different cell types and this will influence the composition of the vacuolar sap in particular cells. Moreover, the functions are not independent of each other. For example, all solutes influence water transport, and this will become physiologically significant when the concentration of an individual solute exceeds 10–20 mM. Conversely, some mixtures of solutes are incompatible due to solubility problems or mutual reactivity, and so there will be limits on their co-occurrence in the same vacuole.

To begin to understand the extent to which the vacuoles of different cells vary, and the factors that determine this variation, measurements are needed at the resolution of individual vacuoles within functional organs under appropriate physiological conditions. Fortunately, appropriate methods are now available. Thus, while Tomos and Wyn Jones (1988) were unable to draw any significant conclusions about the extent of the heterogeneity of vacuolar contents, there is now a large amount of information from which some principles are emerging.

8.3 Generic factors affecting vacuolar solute composition

8.3.1 Mechanical limitations

It is likely that plant vacuoles evolved as a low-cost, volume-filling structure to increase the effective light- or nutrient-absorbing surface area of the cell (Wiebe, 1978; Raven, 1997). On land, this imposes mechanical requirements. The tissue must be able to support its own weight and that of any structure built upon it. In non-woody tissues, this involves turgor pressure, which results from differences in solute concentrations, and hence osmotic pressure, across the plasma membrane. Within the protoplast, the vacuolar osmotic pressure will be identical or close to that of the cytosol; although the concentrations of the individual solutes in each compartment may be very different.

Vacuolar osmotic pressure must always exceed the minimum required turgor pressure. Few calculations of this mechanical minimum have been made, but those for the bean leaf pulvinus indicate that the cell turgor pressure (0.5 MPa) (Irving *et al.*, 1997) is some 5–10 times greater than that required to support the leaf in a horizontal position (Irving and Tomos, unpublished). This would suggest that these mechanical requirements are not the main determinants of vacuolar turgor pressure and that other factors are more important.

Although cell wall strength imposes a maximum on turgor pressure, vacuolar osmotic pressure is not so limited by mechanical considerations. Thus, it is possible to have much higher total osmotic pressure in the vacuole than expected from turgor measurements because large total solute concentrations in the vacuole can be balanced by the accumulation of extracellular solutes that limit the rise in turgor (Leigh, 1997). A good example of this occurs in sugar beet, where turgor is maintained constant despite large increases in total sucrose concentration, and hence vacuolar osmotic pressure, as the storage root grows (Tomos *et al.*, 1992).

8.3.2 Osmotic requirements of the cytoplasm

As no significant hydrostatic pressure can be sustained across the tonoplast, there must be a balance of the forces of water flow. In simplest terms, this means equality of osmotic pressures in the vacuole and cytosol, although, in principle, low values of tonoplast reflection coefficient (Maurel, 1997) or even water pumping (Zeuthen *et al.*, 1997) could achieve this. Potassium is a major determinant of cytosolic osmotic pressure (Wyn Jones *et al.*, 1977), and measurements of cytosolic K^+ activity indicate that it is maintained more or less constant at about 80 mM over a wide range of external K^+ supplies, which decrease both tissue and vacuolar K^+ concentrations substantially (Walker *et al.*, 1996). This constancy of cytosolic K^+ probably reflects the high K^+ selectivity of protein synthesis (Leigh and Wyn Jones, 1984). The osmotic pressure generated by cytosolic K^+, and its counter anion, probably dictates the minimum value for the balancing vacuolar osmotic pressure.

As with mechanical considerations, cytoplasmic osmotica do not impose a maximum on vacuolar solute concentrations as the cytoplasm can also accumulate balancing solutes (Leigh, 1997). However, these extra solutes are restricted to a small number of molecules, such as polyols and quaternary ammonium compounds, that are compatible with cytoplasmic enzyme activity (Wyn Jones *et al.*, 1977; Flowers *et al.*, 1986). Very high vacuolar ionic strengths might limit enzyme functions in the vacuole but, for halophytes at least, vacuolar enzymes appear to be relatively insensitive to ionic strength (Flowers *et al.*, 1986).

8.3.3 Energy considerations

The ability of individual transporters to generate and sustain large solute or ion gradients across the tonoplast may be another potential limit to vacuolar sap composition. This was highlighted for Ca^{2+} by Blackford *et al.* (1990), who showed that measured gradients of this ion across the tonoplast could only be maintained if there was a H^+/Ca^{2+} antiport with a minimum stoichiometry of

$3H^+:1Ca^{2+}$. Smaller numbers of H^+ transported for each Ca^{2+} would result in net Ca^{2+} transport into the cytosol. A similar situation exists in lemon fruits, where vacuolar pH can be as low as 2 and, with a cytosolic pH of 7.0–7.5, is at the thermodynamic limits of the vacuolar type-H^+-transporting adenosine triphosphatase (V-ATPase), if the pump's coupling ratio is $2H^+$ per ATP hydrolysed (Müller *et al.*, 1996). A comparison of the enzyme from lemon fruit and epicotyl tissues indicated that the fruit ATPase appears to be a different type of V-ATPase (e.g. unlike the epicotyl enzyme, it is vanadate sensitive), and that a larger pH gradient can be generated across the fruit sac tonoplast because: the pump has a variable coupling ratio; its slip-rate (ATP hydrolysis not coupled to H^+ transport) shows low pH dependence; and the tonoplast has a low passive permeability to H^+ (Müller *et al.*, 1999). Without these properties, low pH could not be generated within the fruit vacuoles. Therefore, where large differences exist in ion or solute gradients in vacuoles of different cells, the properties of the transporters in the tonoplast may be different. Hence, in barley leaves, where Ca^{2+} is preferentially accumulated in epidermal cell vacuoles, and can be present at concentrations of up to 150 mM (Fricke *et al.*, 1995), the Ca^{2+} transport systems in the tonoplast of these cells might be better adapted to maintaining a large Ca^{2+} transport gradient than in mesophyll cells, where much lower vacuolar Ca^{2+} concentrations occur in the vacuole.

8.3.4 Solubility

Many calcium and magnesium salts have low solubility products. Crystals of various kinds are well known in plant vacuoles (Esau, 1965) and they provided some of the earliest information about heterogeneity of vacuolar contents. Not only do oxalate, phosphate and sulphate limit Ca^{2+} solubility, but the common vacuolar anion, malate, is also incompatible with high Ca^{2+} concentrations (calcium malate has a solubility of about 30 mM; Weast and Astle, 1980). Thus, the presence of Ca^{2+} would be expected to limit the accumulation of malate and vice versa. It is therefore surprising that in sink-cooled barley leaves, osmotically active concentrations of Ca^{2+} and malate, well in excess of 150 mM each, are found in epidermal vacuolar samples (Koroleva *et al.*, 2000). Calcium and malate are also found at high concentrations in leaf extracts of crassulacean acid metabolism (CAM) plants (Meyer and Popp, 1997).

When analysed by microdroplet X-ray microanalysis (Bates *et al.*, 1997), epidermal vacuolar samples from wheat grown in low-nitrate and high-sulphate media contained Ca^{2+} concentrations well in excess of 200 mM, and the accompanying S signal suggested SO_4^{2-} concentrations of about 180 mM. Yet, no precipitate (suggesting the presence of insoluble $CaSO_4$), could be seen. However, the sum of all the analysed solutes considerably exceeded the independently measured osmotic pressure, suggesting that the $CaSO_4$ was not in free solution, or that the Ca^{2+} was chelated. The possibility of supersaturation in vacuoles has not been considered and may reward future study. Solubility of solutes can be affected by the presence of anions, as is indicated by studies of vacuolar berberine in *Coptis japonica*, which has a solubility of 3 mM in the presence of nitrate but of 640 mM in the presence of 168 mM malate (Sato *et al.*, 1992).

8.4 Evidence for heterogeneous distribution of solutes

8.4.1 Histochemical staining

The earliest descriptions of differential chemical distribution in plant cells are those of insoluble crystalline inclusions in 'idioblast' and other cells, where identification was based on simple wet-chemistry and crystal structure. These techniques still have value, especially when used alongside more powerful modern analytical tools. Sugimura *et al.* (1999) used variable pressure scanning electron microscopy and silver-amidol staining (Gahan, 1984) to measure the deposition of Ca^{2+} and Si in the idioblasts of living sections of mulberry leaves. Microscope-based analysis is still the most flexible method for detecting metabolites, especially secondary compounds that are restricted to the vacuoles of very few cells, but when soluble components are to be measured they need to be rendered insoluble (e.g. precipitation of K^+ with sodium cobalt (III) nitrate; Itai *et al.*, 1982).

Hand dissection of tissues is still useful, but cannot detect the presence of gradients within the tissue. It was recently used to demonstrate the specific accumulation of Cd^{2+} and Zn^{2+} in the lower epidermis of heavy metal tolerant and sensitive varieties of *Silene vulgaris* (Chardonnens *et al.*, 1998, 1999). Similarly, hand-dissected material was used to demonstrate the migration of solutes from the vacuoles of one side of the pulvinus of *Vicia faba* during phototropism (Irving *et al.*, 1997).

8.4.2 Vacuole and protoplast isolation

Vacuolar contents can be quantified by the analysis of intact vacuoles, following their physical isolation from tissues (Wagner and Siegelman, 1975; Leigh and Branton, 1976). In order to provide information about heterogeneity, however, this approach needs to be preceded by some form of isolation of the parental cell types. Deitz and co-workers (1992a) separated epidermal and mesophyll protoplasts from barley leaves, using density gradient centrifugation. Qualitative differences in the solute concentrations and proteins in the two cell types were demonstrated (Dietz *et al.*, 1992a,b; Hollenbach and Dietz, 1995). Under control conditions, K^+ and NO_3^- dominated the solutes of both. Chloride and Ca^{2+} were preferentially found in the epidermis, while PO_4^{3-} was predominantly found in the mesophyll vacuoles (Dietz *et al.*, 1992b; Karley *et al.*, 2000). The sugar and amino acid content of the epidermis was very low, even in illuminated excised leaves in which photoassimilates were accumulated to high levels in the rest of the leaf (Dietz *et al.*, 1994). Heavy metals also showed specific segregation—with Cd^{2+} and Zn^{2+} accumulating preferentially in the epidermis (see Chardonnens *et al.*, 1998, 1999) and molybdenum in the mesophyll vacuoles (Brune *et al.*, 1995). The nitrate content of the epidermis dropped during nitrate starvation of the plant (Dietz *et al.*, 1994).

Attempts to explain some of these distributions by differences in the intrinsic transport properties of the cells have not been very successful. Although Dietz and co-workers (1992b) found differences in $^{36}Cl^-$ uptake between epidermal and mesophyll protoplasts when results were expressed relative to the number of protoplasts, this difference disappeared when related to membrane area. Similarly, Karley *et al.* (2000), using tracers and patch-clamping, also found the transport properties of the two cell types to be largely similar.

A considerable technical advance has been the identification of protoplasts from specific cell types following the introduction of cell specific molecular markers. Maathuis *et al.* (1998) were able to isolate and patch-clamp green fluorescent protein (GFP)-expressing protoplasts from the stele of *Arabidopsis thaliana* and compare the results with cells from non-fluorescent parts of the root. More recently, Kiegle *et al.* (2000) used *Arabidopsis* lines in which specific root cells were expressing GFP. This allowed the protoplasts of four root cell types to be isolated and identified and patch-clamping was used to characterize their Ca^{2+} channels. It is only a matter of time before this approach is extended to the analysis of vacuoles derived from protoplasts isolated from such cells.

The basis for an alternative isolation technique for specific protoplasts, and thus vacuoles, was reported by Struck *et al.* (1994), who demonstrated that epidermal and mesophyll protoplasts of barley leaves exhibit differential agglutination in the presence of concanavalin A, tetragonolobus lectin and β-D-glucosyl Yariv artificial antigen. This approach appears not to have been exploited to isolate specific vacuoles.

8.4.3 X-ray microanalysis

The longest established technique for vacuolar analysis, albeit largely qualitative, involves the use of X-ray microanalysis of suitably fixed tissue, which gives information on the elemental composition of cells and their compartments (van Steveninck and van Steveninck, 1991). Generally, it cannot be used for organic solutes because the light atoms in these emit very low energy radiation, which cannot be easily distinguished from the organic matrix, although Lichtenberger and Neumann (1997) were able to use it to measure the accumulation of phenolic chelates of copper in *Armeria* leaf idioblasts.

Pallaghy (1973) demonstrated that in freeze-substituted maize leaves there was a higher relative abundance of Cl^- in epidermal, guard cell and subsidiary cell vacuoles compared with bundle sheath and mesophyll cells. He also demonstrated an apparent light-induced shuttle of K^+ between the mesophyll and bundle sheath vacuoles. van Steveninck *et al.* (1982) demonstrated a consistent decrease of Cl^- content and an increase in phosphorus and sulphur in a centripetal direction in frozen, hydrated roots of salt-stressed *Lupinus luteus*.

In the leaves of the same species, Treeby and van Steveninck (1988) demonstrated that phosphorus was present in epidermal vacuoles at a higher concentration than in mesophyll vacuoles. This contrasts with barley, where X-ray microanalysis has shown the opposite distribution of phosphorus, but where Ca^{2+} and Cl^-, however, were preferentially concentrated in epidermal vacuoles (Leigh and Storey, 1993; Williams *et al.*, 1993). Williams *et al.* (1993) found sulphur and Mg^{2+} to be accumulated in bundle sheath vacuoles. Leigh and Storey (1993) found K^+ and Na^+ were evenly spread between the different cells but, following salt-stress, they showed that Cl^- was excluded from mesophyll until external NaCl reached 170 mM. Sodium, on the other hand, was not excluded from mesophyll. They also found cell-to-cell variation in epidermal solute concentrations.

Fitzgerald *et al.* (1992) showed that in the salt glands of mangrove most Na^+ and Cl^- was found in the water parenchyma cells, while K^+ and phosphorus were predominant in the palisade cell vacuoles. The peltate and abaxial hair cells excluded NaCl completely. The basal cells contained both Na^+ and K^+. In the secreting trichome of chickpea, Lazzaro and Thomson (1992) demonstrated dense Ca^{2+} deposits in the vacuoles of both stalk and head cells.

Rare examples of spatially-defined time courses of vacuolar content are those of Stelzer and co-workers (1990, 1993). These involve the vacuoles of three cell types (mesophyll, endodermis and transfusion parenchyma) from leaves of the conifers, *Picea abies* and *Larix decidua*, measured regularly over periods of many months. Their first paper showed that the mature endodermis of the *Picea abies* needle is characterized by high Mg^{2+}, with K^+ declining to levels lower than those in adjacent tissues. The counter anions are dominated by solutes containing sulphur and phosphorus. The authors suggested that the needle endodermis is involved in collection and buffering the supply of Mg^{2+}, sulphur and phosphorus, but not of K^+, to the leaf. The subsequent report (Stelzer *et al.*, 1993) followed the same three cell types in *L. decidua* needles. Here, the endodermal vacuoles accumulate sulphur, Cl^- and Ca^{2+}, while, in contrast to *P. abies*, it is the transfusional parenchyma vacuoles that act as the buffer store of Mg^{2+}, K^+ and sulphur. During the summer months, the inorganic content of the mesophyll vacuoles was lower than that of the other cell types and may well be linked to the expected accumulation of photoassimilates, which cannot be measured with X-ray microanalysis. Balancing the ions and sugars in the vacuole may regulate the overall osmotic pressure. In the autumn, Mg^{2+}, sulphur, Cl^- and Ca^{2+} increase in mesophyll vacuoles, corresponding to the time course of chlorophyll degradation. The Mg^{2+} may subsequently be translocated to winter-storage structures. Gierth *et al.* (1998) confirmed that the leaf endodermal vacuoles of *L. decidua* preferentially accumulate Ca^{2+} compared with the mesophyll, and that the parenchyma of the vascular bundle accumulates Ca^{2+}, in the form of calcium oxalate crystals. The endodermal vacuoles of these cells had an affinity for strontium even higher than that for Ca^{2+}.

The distribution of Ca^{2+} is also different between deciduous and long-lived leaves. In the former, accompanied by Cl^- or SO_4^{2-} (measured as sulphur), it accumulates in the endodermal vacuoles and is discarded at abscission. In *Picea*, it accumulates in the vacuoles and apoplastic spaces of the mesophyll (Stelzer *et al.*, 1990).

The distribution of silicon has been described for a number of tissues (e.g. wheat leaves, Hodson and Sangster, 1988; and sorghum roots, Hodson and Sangster, 1989). It appears likely, however, that in wheat leaves silicon is generally restricted to the abaxial epidermis, while in sorghum roots—apart from cell walls—silicon is accumulated preferentially in the endodermal and pericycle protoplasts. These workers were again able to demonstrate that Cl^- is entirely confined to the epidermal cells of wheat leaves, as is Ca^{2+} in older cells.

Recently, Ashford *et al.* (1999) showed that in the root cells of *Eucalyptus pilularis* infected by the ectomycorhiza, *Pisolithus tinctorius*, a much higher signal for K^+ was obtained from the exodermal than from the epidermial cells. Very little phosphorus was found in the host cell vacuoles, and the authors concluded that the polyphosphate measured in the entire infected roots is derived from the fungal vacuoles. In a similar work, Bucking and Heyser (2000) showed that phosphorus accumulates in the cortical vacuoles of *Pinus sylvestris* roots under conditions of high external phosphorus supply, but is restricted to the cell cytoplasm under deficient conditions, as in most phosphorus-depleted tissues (Lee and Ratcliffe, 1983; Leigh and Wyn Jones, 1986).

X-ray microanalysis is well suited for the study of heavy metal distribution between tissue vacuoles. The distribution of strontium in larch needles has been mentioned previously (Gierth *et al.*, 1998). van Steveninck and co-workers (1994) measured zinc-phytate distribution in roots of a range of crop species, and concluded that it accumulates preferentially in the endodermal vacuoles of dicotyledonous plants and in the pericycle of monocotyledonous plants. Vazquez *et al.* (1994) demonstrated a gradual increase of Zn^{2+} from vascular cylinder to epidermal cells in leaves of the Zn^{2+} hyper-accumulating plant, *Thlaspi caerulescens*, consistent with its distribution being determined by the transpiration stream. Indeed, the concentrations were so high in the vacuoles of epidermal and subepidermal cells, that they contained zinc-rich crystals.

Finally, McCully (1994) used X-ray analysis of frozen sections of a range of dicotyledonous roots to study the behaviour of K^+ in the vacuoles of developing xylem vessels. Developing secondary xylem have the highest levels (190 mM), with less in the early (100 mM) and late (110 mM) metaxylem. The lowest K^+

concentrations in the roots were found in the cortical vacuoles, with those of the stele parenchyma intermediate. During development, the K^+ concentration of each vacuole increased up to the point of cell death, when there was a precipitous fall to 4–12% of its highest value.

8.4.4 *Mass spectrometry*

Two other microbeam methods, laser microprobe mass analysis (LAMMA) and secondary ion mass spectrometry (SIMS), have been used to analyse vacuolar heterogeneity. These techniques are especially powerful when isotopes of the ions are available, and can be fed to the intact organ. Grignon *et al.* (1999) used SIMS and ^{15}N-nitrate to show that in soybean leaves nitrate was first transported to the epidermis and subsequently to other compartments. Mangabeira *et al.* (1999) followed the distribution of Al^{3+} in *Vicia faba* roots. The metal is generally excluded totally from the protoplasts but, following tissue damage, it appeared as a phosphate complex in the root cortex vacuoles. Kaldorf *et al.* (1999) compared X-ray microanalysis, LAMMA and SIMS while preparing tissue maps of a range of nutrient ions in maize roots colonized by mycorrhiza, while Kuhn *et al.* (2000) used LAMMA to follow the entry, via the apoplast, of ^{25}Mg and ^{44}Ca into mycorrhizal spruce roots.

8.4.5 *Ion-sensitive microelectrodes*

Ion-selective microelectrodes offer a precise, quantitative method for measuring the ion activities in vacuoles and other cell compartments in intact tissues. They have great potential in mapping heterogeneity of vacuolar contents but, surprisingly, have been used relatively little for this purpose. Radial profiles of vacuolar K^+ activity were measured with K^+-sensitive electrodes in roots of maize (Dunlop and Bowling, 1971), sunflower (Bowling, 1972) and rye-grass (Dunlop, 1973), and no heterogeneity was found. Similarly, Kelday and Bowling (1980) using a combination of microsampling with a microcapillary and electrometric titration found no radial gradients of Cl^- across *Commelina communis* roots. In contrast, Zhen *et al.* (1991) used nitrate-selective electrodes to show that the nitrate concentration in the vacuoles of root epidermal cells (53 mM) was lower than that in the cells immediately beneath (101 mM). van der Leij *et al.* (1998) demonstrated that epidermal vacuoles remobilized nitrate much faster than those of the cortex in nitrate-deprived barley roots.

By incorporating a barrel to measure pH in addition to the ion of interest, it is also possible to use microelectrodes to discriminate measurements arising from the cytosol and vacuole (Walker *et al.*, 1995). Using such an electrode on barley roots, Walker and co-workers (1996) showed that, while the vacuolar K^+ activities in root epidermal and cortical cells responded similarly to K^+-deficiency, the response of K^+ in the cytosol of each cell was different.

A colourful use of pH microelectrodes is the demonstration that the pH of the vacuole of morning glory flower petal rises from 6.6 in the bud to as high as 7.7 in the open flower, and causes the pigment anthocyanin to change colour (Yoshida *et al.*, 1995).

8.4.6 Single cell sampling and analysis (SiCSA)

The cell pressure probe (Hüsken *et al.*, 1978) permits the measurement of the hydrostatic pressure of individual cells in higher plants. The instrument has been adapted to extract picolitre-sized samples of sap from cells that lie within a few hundred micrometres of a tissue surface (Malone *et al.*, 1991; Tomos and Leigh, 1999). For highly-vacuolated cells the sample will be predominantly vacuolar. However, Fricke and co-workers (1994a) demonstrated the presence of cytoplasmic malate dehydrogenase from mesophyll cell samples, showing that samples from these cells are mixtures of the two compartments. This ability to sample individual cells can be coupled with appropriate microanalytical techniques to give a general method that has been termed 'single cell sampling and analysis' (Tomos *et al.*, 1994; Tomos and Leigh, 1999). Having isolated a sample from an individual cell, it can then be exposed to a wide range of analytical techniques, including: osmometry (Malone *et al.*, 1989); X-ray microanalysis (Malone *et al.*, 1991); enzyme-linked fluorescence microscope fluorimetry (Zhen *et al.*, 1991; Koroleva *et al.*, 1997, 1998); capillary electrophoresis (Bazzanella *et al.*, 1998); measurements of enzyme activity (Koroleva *et al.*, 1998); immunoassay of proteins (Koroleva *et al.*, 2000); and reverse transcriptase-polymerase chain reaction (Karrer *et al.*, 1995; Brandt *et al.*, 1999). Solute analyses can be made fully quantitative by addition of an internal standard (Tomos *et al.*, 1994). A major advantage of this method is that it permits the simultaneous and quantitative measurement of several vacuolar parameters, being limited mainly by the size of the sample and the number of separate assays that can be applied to it.

The SiCSA technique has been applied extensively to the barley leaf in an attempt to build up three dimensional maps of vacuolar solutes and to determine how these change with time and treatments. The overall conclusion is that there is considerable cell heterogeneity both in space and time. However, this heterogeneity is not random but follows distinct patterns, the basis of which is beginning to be understood.

8.4.6.1 Leaf epidermis

Use of SiCSA on barley epidermis has, in several reports (e.g. Fricke *et al.*, 1994a,b), accounted for all of the major osmotically active components, providing confidence that it is now possible to integrate the behaviour of all major solutes. As noted using other techniques, K^+ and Ca^{2+} are usually the only osmotic cations. In young leaves, K^+ dominates all cell types (Fricke *et al.*,

1994a,b). As the leaf ages, Ca^{2+} is continuously brought to the epidermis by the transpiration stream and, as it does not appear to be retranslocated, it is deposited there. With development, the Ca^{2+} displaces the K^+, which can drop to undetectable levels in some cases (Hinde, 1994). The final concentration of Ca^{2+} can exceed 150 mM. Potassium is totally redistributable, indicating circulation within the leaf. Moreover, the displacement behaviour varies considerably between cells within the epidermis. Calcium accumulates preferentially in the cell vacuoles adjacent to the stomatal pores. This is consistent with its being deposited close to the point of water evaporation and subsequent removal from the apoplast by the nearest vacuoles in order to avoid injurious effects on stomatal function. A similar protective role for vacuolar sequestration of Ca^{2+} has been proposed for leaf trichomes (De Silva *et al.*, 1996).

The effect of Ca^{2+} on epidermal K^+ concentration is dependent on leaf illumination. At low light intensity (120 μE), there is a direct displacement of K^+ (Fricke *et al.*, 1995). Under these conditions, there are negligible concentrations of organic or other multivalent anions. The stoichiometry of 2:3 (Ca:K) (Hinde, 1994) is that expected for minimum osmotic disturbance if transported K^+ and Ca^{2+} were accompanied by one and two monovalent anions, respectively. At higher light intensity, malate is found in the epidermal cells that accumulate Ca^{2+}, and no displacement of K^+ is seen (Fricke *et al.*, 1995).

Calcium can, therefore, displace K^+ from epidermal vacuoles. However, if the plants are grown at low K^+ (0.2 mM), which results in lowered vacuolar K^+, Ca^{2+} does not make up the osmotic shortfall and both epidermal osmotic and turgor pressures are depressed (Hinde, 1994). It would appear that Ca^{2+} can only displace K^+, it cannot fully replace it.

Epidermal vacuolar anions also follow distinct patterns. As indicated by other techniques, Cl^- is a major osmoticum. However, at lower light intensities and when present in the growth medium, nitrate is present at similar concentrations. Solute mapping results indicate that there are reciprocal intercell gradients of each, suggesting that each can be replaced by the other in a system in which the total osmotic balance is regulated (Fricke *et al.*, 1995). There is even the suggestion that vacuoles overlying adjacent intermediate and lateral veins have different anion profiles (Hinde, 1994; Fricke *et al.*, 1995). Predictably, nitrate limitation leads to higher Cl^- and Cl^- limitation to higher nitrate. When both are replaced by sulphate in the medium, some sulphate accumulates, but chloride, nitrate and even significant phosphate levels make up a more diverse cocktail (Richardson, 1993; Bell *et al.*, 1995).

Higher light intensities increase epidermal malate (Koroleva *et al.*, 2000) and decrease epidermal nitrate. When supplied under these conditions, with luxury levels of sulphate and little nitrate, wheat epidermal vacuolar sulphur levels, measured using microdroplet X-ray microanalysis (presumably, but not yet confirmed as, SO_4^{2-}) can rise to very high levels (>150 mM) (Bates, 1998). Ozone fumigation of barley also considerably increases the sulphur content of the epidermal vacuoles from older parts of the barley leaf (Cuin *et al.*, 1995). The identity of the sulphur-containing compound(s) is yet to be confirmed, although it seems unlikely that such a high concentration could result from glutathione or sulphur-containing amino acids.

A characteristic of barley epidermis is an almost complete lack of sugars; their total concentration does not exceed 10 mM (Dietz *et al.*, 1994, Fricke *et al.*, 1994a, Koroleva *et al.*, 1997). This is true even under conditions in which the carbohydrate load of the leaf has been artificially raised. That was achieved by high intensity illumination (1000 μE) or while reducing export of photoassimilate from the leaves by cooling of the sinks (the roots and stem apical meristem) of the barley plant (Koroleva *et al.*, 1998). Low sugar concentration is, however, not a universal feature of epidermis in general, as significant quantities of sucrose, glucose and fructose (20–60 mM of total hexoses) have recently been measured in the epidermis of the dicotyledonous species, *Alonsoa meridionalis* and *Asarina barclaiana* (Voitsekhovskaja, Koroleva and Tomos, unpublished). In potato epidermis, sugars were not only measured by SiCSA (Kehr *et al.*, 1999), but it was shown that their concentrations were increased by the transgenic insertion of sucrose transporter antisense genes.

The major organic component of the sap of the barley leaf epidermal vacuole is malate. As noted above, the intercellular distribution of several epidermal solutes is a function of the distance of a cell from the stomatal pore and underlying vein. For example, 6 days after full leaf expansion at 400 μE, malate is found predominantly in the cells closest to the stomatal guard cells (at present the cereal guard cells themselves are too small for SiCSA techniques).

The time course of epidermal malate concentration provides a further example of the dynamics of vacuolar behaviour. Koroleva and co-workers (2000) found no evidence of diurnal fluctuations of malate concentration in epidermal trough cells, in contrast to mesophyll and bundle sheath cells; it stayed in the range 120–180 mM. However, long-term accumulation from 120 to 230 mM was seen when photoassimilate export from the leaf was slowed by cooling the tissue sinks. Only some 30% of epidermal malate could be remobilized. When plants were kept in darkness for 6 days, epidermal malate dropped from the initial value (140 mM) by some 20% over the first 24 h, and then slowly decreased (to 95 mM) by the end of the experiment. Finally, epidermal malate concentrations increase in response to infection of the leaf by the powdery mildew (*Blumeria [Erisiphe] graminis*) (Richardson *et al.*, 1997).

In contrast to barley epidermal vacuoles, malate is almost absent from the vacuoles of wheat epidermis. Results from capillary electrophoresis of individual cell samples (Bazzanella *et al.*, 1998), however, indicate that an organic anion of electrical mobility similar to citrate is the dominant anion (Tomos, unpublished).

The response of the epidermal vacuoles to stressing the barley seedlings with NaCl, KCl and KNO_3 was investigated by Fricke and co-workers (1996). Under

NaCl stress, Na^+ and Cl^- were preferentially accumulated in the epidermal vacuoles. Following 50 mM external NaCl for 6–8 days, K^+, Ca^{2+}, nitrate and malate levels were greatly reduced, and Na^+ and Cl^- concentrations rose to 300 mM. Na^+ accumulated to the highest levels in cells adjacent to the stomatal pores, while K^+ levels remained highest in those vacuoles furthest from the stomata. At 150 mM external NaCl, the effects were even more extreme. Salinisation with 100 mM KCl resulted in the very high epidermal loads of K^+ (490 mM) and Cl^- (456 mM), whilst salinisation with 100 mM KNO_3, resulted in epidermal vacuoles accumulating K^+ (363 mM) and NO_3^- (325 mM).

8.4.6.2 Leaf mesophyll and bundle sheath

Mesophyll SiCSA samples, unlike those of the epidermis, are not free of cytoplasmic contamination and vacuolar concentrations can only be inferred. For the major osmotica, such as sugars, K^+ and nitrate, however, the high concentrations must be due to the vacuolar contribution (Fricke *et al.*, 1994a,b, 1995, 1996). Diurnal patterns of accumulation have been demonstrated for sucrose (Koroleva *et al.*, 1997, 1998) and malate (Koroleva *et al.*, 2000) in both mesophyll and bundle sheath cells. Fructan synthesis and accumulation were induced in both the mesophyll and the bundle sheath cells under conditions when photosynthetic production of sugars exceeded the capacity of their export from the source leaf (high light intensity or by cooling root and shoot apical meristems; Koroleva *et al.*, 1998). Sugars and fructan also accumulated in the mesophyll and bundle sheath cells when initially carbohydrate-depleted, excised leaves were fed sucrose through the transpiration stream, in darkness (Koroleva *et al.*, 2000). Koroleva and co-workers suggest that the bundle sheath cells have a lower threshold of sucrose concentration (a substrate for fructan synthesis) for the initiation of the fructan synthesis than do mesophyll cells. Therefore, as a result of conversion of sucrose into fructan, a downhill gradient of sucrose concentration is maintained between mesophyll and bundle sheath cells (Koroleva *et al.*, 1997).

The first measurement of ^{14}C assimilation from SiCSA samples (Koroleva *et al.*, 2000) indicated that the mesophyll sucrose reached isotopic equilibrium with the external CO_2 within the 30 min of the experiment. Although close to the current limit of the technique, this would also appear to be true for the epidermal malate. If this is indeed the case, it would mean that the apparently immobile malate pool of the epidermal vacuoles is in dynamic equilibrium with that of the mesophyll.

The concentrations of inorganic solutes in the mesophyll cells and their response to external ion supply was described by Fricke and co-workers (1994a, 1996). Potassium (230 mM) and nitrate (208 mM) were the predominant inorganic solutes under control conditions, with lower concentrations of Ca^{2+} (11 mM), phosphorus (29 mM) and Cl^- (6 mM). Following NaCl treatment, Na^+ and Cl^- were preferentially accumulated in the epidermal vacuoles, but also accumulated in the mesophyll. In contrast to the report by Leigh and Storey (1993), there was no apparent external threshold for Cl^- accumulation. At 50 mM external NaCl, 50 mM Cl^- was found in the mesophyll. At 150 mM external NaCl, this rose to 167 mM, with no apparent influence on the net rate of photosynthesis—contradicting the earlier suggestion by Leigh and Storey (1993) that Cl^- is strongly excluded from the mesophyll in order to protect carbon fixation. Treatment with 100 mM KCl led to K^+ accumulation by all cell types, but also to a preferential increase in the mesophyll Cl^- (216 mM) in excess of that seen with 100 mM NaCl (117 mM).

8.4.6.3 Roots

As with mature leaf epidermal cells, the composition of SiCSA samples from mature root cells will be dominated by that of the vacuole. The osmotic pressure of a cell is a direct measure of the total solute concentration of the vacuole. Pritchard *et al.* (1989) demonstrated that both turgor pressure and osmotic pressure were uniform across the expanding tips of wheat roots, although in more mature regions, tissue corresponding to the stele had a turgor pressure (and presumably osmotic pressure) twice that of the cortex and epidermis. Subsequently, Rygol and co-workers (1993) showed that transpiration effects could impose a gradient (higher innermost) in both turgor and osmotic pressure across the cortex of both wheat and maize. This had also been demonstrated for the halophytes, *Mesembryanthemum crystallinum* (Rygol and Zimmermann, 1990) and *Aster tripolium* (Zimmermann *et al.*, 1992). These transpiration-related gradients are not easy to explain. Rygol and co-workers (1993) proposed that they were due to a sweep away effect of solutes across the tissue. If this were the case, it would need to involve transport across the tonoplast into consecutive vacuoles. The speed at which the gradients dissipated, however, is not consistent with the expected permeability of the vacuolar solutes. This observation serves to emphasize the importance of the physiological state of the plant when mapping vacuolar composition.

In the case of maize, the identity of most of the vacuolar solutes was reported by Pritchard *et al.* (1996), who used SiCSA techniques to assay the contents of cells 12 mm from the root tip (just outside the expansion zone). Cells at the surface of the root, in the inner cortical cells and from the stelar parenchyma were measured. The cell compositions were measured in plants growing under three distinct hydroponic media: 0.5 mM $CaCl_2$; Cl-free nutrient solution; and $CaCl_2$ solution with 400 mM mannitol. Steps were found for all solutes between the different cell types, implying gradients across the tissue. This is in contrast to the early observations of Bowling and co-workers (e.g. Kelday and Bowling, 1980). Although there were qualitative differences between the different treatments, the qualitative distribution of osmotic pressure, potassium, chloride, hexoses and amino acids was unchanged. This was particularly noteworthy in the case of K^+ and Cl^-. The $CaCl_2$ nutrient medium is a large potential source of Cl^- but not of K^+, while the opposite is true in the Cl^--free/K^+ containing medium. The solute distribution was clearly not a passive response to the external availability of solutes. The distribution of amino acids was also not uniform, with maximum concentrations in the inner cortex.

As noted above, Zhen *et al.* (1991) also demonstrated a step in vacuolar nitrate concentration between the epidermis (53 mM) and cortical cells (101 mM), using both a nitrate sensitive electrode and SiCSA. Pritchard *et al.* (1996) also demonstrated longitudinal vacuolar solute gradients and the influence of osmotic stress on them.

Taproot solutes follow distinct patterns in both time and space. Hole (1996) showed that the carrot taproot fills in phases. Recently, Korolev *et al.* (2000) (see also Tomos *et al.*, 2000) have refined these results to the resolution of individual vacuoles within the carrot. The initial swelling of the vacuoles is achieved using amino acids as the major vacuolar osmoticum (225 mOsm at 10 days). By 20 days, these are replaced by other electrolytes (K^+, nitrate and malate). As the cells continue to expand, the concentration of the electrolytes decreases, to be replaced by hexoses (glucose and fructose) and finally sucrose. Osmotic pressure is maintained more or less constant during the hexose and sucrose accumulation periods, exemplifying polymerisation (or at least dimerisation) as a mode of osmotic regulation. There are, however, distinctive tissue patterns of the solutes. At 50 and 90 days, vacuoles farthest from active xylem and phloem maintain high levels of K^+ and malate, while those nearest have the highest sugar concentration. The exception are the cells of the cambium. These have low K^+ and sugars. Although not directly measured, the authors assume that the osmotica of these cells are amino acids, as in the root of the young seedling.

Wyse (R.E. Wyse, personal communication) observed a similar situation in sugarbeet tissue. By careful dissection of parenchyma between the concentric rings of cambia, parenchymatous cells were shown to be larger at greater distance from their cambia, but their sugar concentration was lower and electrolyte content higher. This pattern suggests that a series of 'alternative osmotica' are involved, culminating in sucrose becoming dominant on becoming available at full canopy development.

8.4.6.4 Other tissues

Arabidopsis stem. The SiCSA technique is particularly useful in the identification and measurement of vacuolar organic solutes and has wide potential in the quantification of secondary plant products. One recent development has been the identification of specific large cells in the flower stalk of *Arabidopsis thaliana*, which accumulate a sulphur-containing compound (O. Koroleva, unpublished). The distribution of five elements (K^+, Ca^{2+}, Cl^-, sulphur and phosphorus) in different highly-vacuolated cell types was analysed in freeze-dried sections of *A. thaliana* flower stalk by X-ray imaging. Their concentrations were subsequently quantified by microanalysis of single-cell samples and phloem exudates. Potassium concentration was found to be highest in the parenchymatous tissue around vascular bundles. Calcium and chloride were present mainly in the central part of the flower stalk. Phosphorus was largely located in the bundles and in the parenchyma surrounding them. The sulphur signal was extraordinarily high in groups of cells (S-cells) situated between the phloem of every vascular bundle and the endodermis. Enzymatic hydrolysis by thioglucosidase of cell sap collected from S-cells resulted in the release of glucose, indicating that these cells contain glucosinolates at high (>100 mM) concentration, which is consistent with the concentration of sulphur (>200 mM) estimated by X-ray analysis of cell sap samples. The likely identity of the sulphur-containing solute as a glucosinolate was further supported by the estimation of charge and osmotic pressure (from total charge and osmotic pressure balance of the cell sap) of the monovalent, two-sulphur-containing molecule.

Thlaspi caerulescens. Kupper *et al.* (1999) have used SiCSA to show that soluble Zn^{2+} is preferentially accumulated in epidermal cells (vacuoles) of *Thlaspi caerulescens* leaves, rather than mesophyll cells. Using X-ray analysis of microdroplets extracted using microcapillaries, they found that the former contained levels 5–6.5 times higher than the latter, with an epidermal average of 385 mM for plants growing in 4 mg Zn^{2+} per g compost.

Ricinus communis hypocotyl. Meshcheryakov *et al.* (1992) employed cell turgor and osmotic pressure measurements to determine total cell (vacuolar) solute concentrations across the radius of the growing castor bean hypocotyl. They demonstrated a large gradient in solute concentration (innermost highest) and suggested that this was due to the supply of osmotic solutes from the phloem.

8.4.7 *Optical and nuclear magnetic resonance (NMR) imaging*

The immense potential of *in vivo* imaging (Fricker and Oparka, 1999) in generating quantitative maps of vacuolar contents is yet to be tapped. Interest has centred on the cytoplasmic compartments, especially on the role of Ca^{2+} as a signalling molecule (Bush, 1995). However, imaging a vacuolar fluorescent pH-probe, 1′,7′-bis-(2-carboxyethyl)-5-(and-6)-carboxy-fluorescein (BCECF), enabled Brauer *et al.* (1995) to estimate maize root-hair cell vacuolar pH *in situ*. More recently, Roos *et al.* (1998) used confocal imaging of carboxy-seminaphthorhodafluor-1-acetoxymethylester (SNARF-1) to estimate pH of the vacuoles and cytoplasm of cultured Californian poppy cells. The vacuolar pH showed marked intercellular variation from 5.5 (the lower detection limit) to 6.3. The introduction of solute-sensitive, compartment-directed gene products, such as the calcium-sensitive GFP construct of Miyawaki *et al.* (1997), may provide an alternative, fully non-invasive approach.

Magnetic resonance imaging of plant solutes is also in its infancy. Metzler *et al.* (1995) showed that sucrose was preferentially accumulated in the vascular phloem area and in the cortex parenchyma of castor-bean hypocotyls. Glucose was higher in the pith. A heterogeneity in the distribution of some amino acids was also shown. Heidenreich *et al.* (1998) obtained some evidence of sucrose distribution in the same tissue, using ^{13}C-labelled compounds.

8.5 Storage of water

The most important reversible storage molecule of the vacuole is water, which plays key roles in the life of the plant. A vivid example of this occurs when stored corms, tubers and bulbs sprout in the total absence of added water. The water storage capacity of a vacuole is determined largely by the mechanical properties of the cell wall of the cell in which the vacuole is found (Steudle *et al.*, 1980). In general, however, the water stored in a vacuole is there for the purpose of the cell itself. In all fleshy organs it is the support medium of the plant, its low compressibility passively maintaining the tissue volume. In the scase of expansion growth, most of the volume expansion is due to vacuolar water storage. In motor cells of various types, the amount of water stored by a cell is under dynamic control (Irving *et al.*, 1997). By regulating the water stored, the volume of the cell can be made to perform work.

8.6 Conclusions

The evidence that vacuoles in different cells can have quite different compositions is now overwhelming, but the significance of this and the mechanisms that generate the patterns and maintain them remain to be identified. Nonetheless, it is clear that this heterogeneity is far from random and that within tissues the same differential distributions of solutes can be consistently measured by different techniques. This is exemplified by barley leaves, where differential distribution of Ca^{2+}, Cl^{-}, and phosphorus between the mesophyll and epidermis have been shown by X-ray microanalysis (Leigh and Storey, 1993; Williams *et al.*, 1993), protoplast isolation (Dietz *et al.*, 1992b; Karley *et al.*, 2000) and SiCSA (Fricke *et al.*, 1994a), and by the distribution of Zn in *Thlaspi* (X-ray microanalysis, Vazquez *et al.*, 1994; SiCSA, Kupper *et al.*, 1999).

However, there is not necessarily interspecies consistency in the patterns of distribution of major solutes. Thus, in lupin, phosphorus is preferentially accumulated in the epidermis not the mesophyll (Treeby and van Steveninck, 1988); whilst, in barley, the reverse is found (Dietz *et al.*, 1992b; Fricke *et al.*, 1994a; Karley *et al.*, 2000). In barley and wheat epidermis, total carbohydrates are below 10 mM (Deitz *et al.*, 1994; Fricke *et al.*, 1994a; Koroleva *et al.*, 1997), whilst in some dicotyledonous species more appreciable concentrations are found (Kehr *et al.*, 1999; Voitsekhovskaja, Koroleva and Tomos, unpublished). The differential deposition of Zn^{2+} between the endodermal and pericycle cells of roots is species dependent (van Steveninck *et al.*, 1994).

In some cases, the reasons for the differences can be inferred. For instance, the difference between the locations of vacuolar Ca^{2+} deposits in deciduous and evergreen conifers may be explicable by the way in which each disposes of excess ions. If the leaves are to be disposed of, it may not be so important to free the protoplasts of unwanted material (Stelzer *et al.*, 1990). In other situations, closely-related species may use different solutes for the same cell specific function, e.g. malate in barley epidermis but another carboxylic acid in wheat. In some cases, there may be some underlying principles that govern the separation of certain solutes. For instance, if phosphorus is to be stored to high concentration and remain available for remobilization, it must be separated from Ca^{2+} to avoid precipitation. This is thought to be the reason why Ca^{2+} is accumulated to high concentrations in barley leaf epidermis, while phosphorus is in the mesophyll (Leigh and Storey, 1993; Leigh and Tomos, 1993).

Similarly, the ability of the tonoplast to maintain high concentration gradients of particular ions may also determine its accumulation within particular cell types. Thus Cl^- is preferentially accumulated in the epidermis of barley leaves. In salt-grown plants, the concentrations of Cl^- increase in all cells, but the differential between the epidermis and mesophyll is maintained (Fricke *et al.*, 1996); in plants grown in 150 mM NaCl, the concentration in epidermal cells was 563 mM, whereas the mesophyll contained only 167 mM. The differences in Na^+ concentrations were less extreme, being 480 mM in the epidermis and 302 mM in the mesophyll (Fricke *et al.*, 1996). A possible explanation for this is that the tonoplast of the mesophyll is unable to maintain large Cl^- concentration gradients, and that, to avoid possible damage to these cells from large increases in cytosolic Cl^-, the ion is accumulated in the epidermis.

How differential distributions arise remains unclear. For solutes produced by biosynthesis, it is easy to see that differential distribution of the metabolic pathways between different cells could give rise to the intercellular distributions, but for ions that are ultimately derived from the soil, there must be differential uptake and/or retention. Explaining differences in leaves is particularly difficult because transport of ions from the xylem to the leaf cells is generally thought to involve passive movement in the transpiration stream. Leigh and Tomos (1993) suggested that the distributions mean either that cells have different transporters that allow them to select particular ions, leaving the remainder for cells later in the pathway, or there is separation at some point with ions destined for the epidermis following a different pathway to those for the mesophyll and other cells.

In some cases, distribution does appear to follow a pattern predicted by the transpiration stream. Calcium and Na^+ accumulation patterns in barley epidermis (Fricke *et al.*, 1995, 1996), or Zn^{2+} in *Thlaspi* leaves (Vazquez *et al.*,

1994) might be the result of this. The patterns of malate in barley epidermis (Fricke *et al.*, 1995) may then follow as an adaptive response. The distribution of glucosinolates in *Arabidopsis* flower stalks (O. Koroleva unpublished), of Ca^{2+} and Mg^{2+} in conifer leaves (Stelzer *et al.*, 1990) or of Zn^{2+} in roots (van Steveninck *et al.*, 1994), however, illustrates the targeted and active nature of other distributions.

Attempts to measure the transport properties of different cell types in barley leaves have not indicated any major differences between epidermal and mesophyll cells (Dietz *et al.*, 1992b; Karley *et al.*, 2000), so the basis for the solute distributions in this tissue remains unresolved. In contrast, Kiegle *et al.* (2000) have shown different electrophysiological activity from protoplasts derived from different types of root cells. The use of protoplasts or vacuoles made from specific cell types may provide a fruitful way forward. Quantitative measurements following the modification of cell or tissue targeted membrane transporters (Kehr *et al.*, 1999) is another approach.

Whatever the explanation, the existence of major differences in the solute compositions of vacuoles in different cell types raises the possibility that manipulation of tissues to accumulate high concentrations of solutes may require an approach that targets particular cells. For instance, attempts to manipulate salt tolerance in barley by increasing Cl^- transport into vacuoles in all leaf cells might fail if the properties of the mesophyll tonoplast prevent it from maintaining high Cl^- concentration gradients. The epidermal tonoplast seems better adapted for this and attempting to increase uptake and retention of Cl^- in these cells might be better.

Finally, when mapping vacuolar contents, it must be remembered that many components are highly dynamic and that measurements are of a steady state. The extreme example is water, for which the cell membranes are barely a barrier (Tomos, 1988), but the recent observations of rapid equilibration of the epidermal nitrate (Grignon *et al.*, 1999) and mesophyll sugars (Koroleva *et al.*, 2000) indicate that this also needs to be described. With so many techniques in place, it is a very exciting phase in our description and understanding of the variation of vacuolar contents with time and space.

References

Ashford, A.E., Vesk, P.A., Orlovich, D.A., Markovina, A.L. and Allaway, W.G. (1999) Dispersed polyphosphate in fungal vacuoles in *Eucalyptus pilularis/Pisolithus tinctorius* ectomycorrhizas. *Fungal Genet. Biol.*, **28** 21-33.

Bates, N.J. (1998) The distribution and mobilisation of sulphur and nitrogen in leaves of *Triticum aestivum* approaching senescence. PhD thesis, University of Wales.

Bates, N.J., Cram, W.J. and Tomos, A.D. (1997) Sulphate concentrations in wheat epidermis measured at single cell resolution: influence of N and S supply, in *Sulphur Metabolism in Higher Plants* (eds. W.J. Cram, L.J. de Kok, I. Stulen, C. Brunold and H. Rennenberg), Backhuys Publ., Leiden, pp. 315-17.

Bazzanella, A., Lochmann, H., Tomos, A.D. and Bächmann, K. (1998) Determination of inorganic cations and anions in single plant cells by capillary zone electrophoresis. *J. Chromatogr. A.*, **809** 231-39.

Bell, C., Fordham, M., Richardson, P., Cram, J. and Tomos, A.D. (1995) Cellular and subcellular compartmentation of sulphate in leaves in relation to low sulphur mobility. *Z. Pflanzen. Bodenkund.*, **158** 63-65.

Blackford, S., Rea, P.A. and Sanders, D. (1990) Voltage sensitivity of H^+/Ca^{2+} antiport in higher plant tonoplast suggests a role in vacuolar calcium accumulation. *J. Biol. Chem.*, **265** 9617-29.

Bowling, D.J.F. (1972) Measurements of profiles of potassium activity and electrical potential in the intact root. *Planta*, **108** 147-51.

Brandt, S., Kehr, J., Walz, C., Imlau, A., Willmitzer, L. and Fisahn, J. (1999) Technical advance: a rapid method for detection of plant gene transcripts from single epidermal, mesophyll and companion cells of intact leaves. *Plant J.*, **20** 245-50.

Brauer, D., Otto, J. and Tu, S.I. (1995) Selective accumulation of the fluorescent pH indicator, BCECF, in vacuoles of maize root-hair cells. *J. Plant Physiol.*, **145** 57-61.

Brune, A., Urbach, W. and Dietz, K.J. (1995) Differential toxicity of heavy metals is partly related to a loss of preferential extraplasmic compartmentation: a comparison of Cd-stress, Mo-stress, Ni-stress and Zn-stress. *New Phytol.*, **129** 403-409.

Bucking, H. and Heyser, W. (2000) Subcellular compartmentation of elements in non-mycorrhizal and mycorrhizal roots of *Pinus sylvestris*: an X-ray microanalytical study. I. The distribution of phosphate. *New Phytol.*, **145** 311-20.

Bush, D.S. (1995) Calcium regulation in plant cells and its role in signalling. *Annu. Rev. Plant Physiol. Plant Mol. Biol.*, **46** 95-122.

Chardonnens, A.N., ten Bookum, W.M., Kuijper, L.D.J., Verkleij, J.A.C. and Ernst, W.H.O. (1998) Distribution of cadmium in leaves of cadmium tolerant and sensitive ecotypes of *Silene vulgaris*. *Physiol. Plant.*, **104** 75-80.

Chardonnens, A.N., ten Bookum, W.M., Vellinga, S., Schat, H., Verkleij, J.A.C. and Ernst, W.H.O. (1999) Allocation patterns of zinc and cadmium in heavy metal tolerant and sensitive *Silene vulgaris*. *J. Plant Physiol.*, **155** 778-87.

Cuin, T., Ashenden, T. and Tomos, A.D. (1995) The influence of ozone on the distribution of solutes between individual cells of barley leaves. *J. Exp. Bot.*, **46** (Suppl.), 55-56.

De Silva, D.L.R., Hetherington, A.M. and Mansfield, T.A. (1996) Where does all the calcium go? Evidence of an important regulatory role for trichomes in two calcicoles. *Plant Cell Environ.*, **19** 880-86.

Dietz, K.J., Schramm, M., Betz, M., Busch, H., Dürr, C. and Martinoia, E. (1992a) Characterisation of the epidermis from barley primary leaves. I. Isolation of epidermal protoplasts. *Planta*, **187** 425-30.

Dietz, K.-J., Schramm, M., Lang, B., Lanzl-Schramm, A.H., Dürr, C. and Martinoia, E. (1992b) Characterization of the epidermis from barley primary leaves. II. The role of the epidermis in ion compartmentation. *Planta*, **187** 431-37.

Dietz, K.J., Hollenbach, B. and Hellwege, E. (1994) The epidermis of barley leaves is a dynamic intermediate storage compartment of carbohydrate, amino acids and nitrate. *Physiol. Plant.*, **92** 31-36.

Dunlop, J. (1973) The transport of potassium to the xylem exudate of ryegrass. *J. Exp. Bot.*, **24** 995-1002.

Dunlop, J. and Bowling, D.J.F. (1971) The movement of ions to the xylem exudate of maize roots. I. Profiles of membrane potential and vacuolar potassium activity across the root. *J. Exp. Bot.*, **22** 434-44.

Esau, K. (1965) *Plant Anatomy*, 2nd edn, John Wiley, London.

Fitzgerald, M.A., Orlovich, D.A. and Allaway, W.G. (1992) Evidence that abaxial leaf glands are the sites of salt secretion in leaves of the mangrove, *Avicennia marina* (forsk) vierh. *New Phytol.*, **120** 1-7.

Flowers, T.J., Hajibagheri, M.A. and Clipson, N.J.W. (1986) Halophytes. *Q. Rev. Biol.*, **61** 313-37.

Fricke, W., Leigh, R.A. and Tomos, A.D. (1994a) Concentrations of inorganic and organic solutes in extracts from individual epidermal, mesophyll and bundle-sheath cells of barley leaves. *Planta*, **192** 310-16.

Fricke, W., Leigh, R.A. and Tomos, A.D. (1994b) Epidermal solute concentrations and osmolality in barley leaves studied at the single-cell level: changes along the leaf blade during leaf ageing and NaCl stress. *Planta*, **192** 317-23.

Fricke, W., Hinde, P.S., Leigh, R.A. and Tomos, A.D. (1995) Vacuolar solutes in the upper epidermis of barley leaves: intercellular differences follow patterns. *Planta*, **196** 40-49.

Fricke, W., Leigh, R.A. and Tomos, A.D. (1996) The intercellular distribution of vacuolar solutes in the epidermis and mesophyll of barley leaves changes in response to NaCl. *J. Exp. Bot.*, **47** 1413-26.

Fricker, M.D. and Oparka, K.J. (1999) Imaging techniques in plant transport: meeting review. *J. Exp. Bot.*, **50** 1089-100.

Gahan, P.B. (1984) *Plant Histochemistry and Cytochemistry: An Introduction*. Academic Press, London.

Gierth, M., Stelzer, R. and Lehmann, H. (1998) Endodermal Ca and Sr partitioning in needles of the European larch (*Larix decidua* L., Mill.) *J. Plant Physiol.*, **152** 25-30.

Grignon, N., Jeusset, J., Lebeau, E., Moro, C., Gojon, A. and Fragu, P. (1999) SIMS localization of nitrogen in the leaf of soybean: basis of quantitative procedures by localized measurements of isotopic ratios. *J. Trace Microprobe Tech.*, **17** 477-90.

Heidenreich, M., Kockenberger, W., Kimmich, R., Chandrakumar, N. and Bowtell, R. (1998) Investigation of carbohydrate metabolism and transport in castor bean seedlings by cyclic J cross polarization imaging and spectroscopy. *J. Magn. Reson.*, **132** 109-24.

Hinde, P.S. (1994) The role of potassium as an osmoticum in barley leaf cells. PhD thesis, University of Wales.

Hodson, M.J. and Sangster, A.G. (1988) Observations on the distribution of mineral elements in the leaf of wheat (*Triticum aestivum* L.) with particular reference to silicon. *Ann. Bot.*, **62** 463-71.

Hodson, M.J. and Sangster, A.G. (1989) X-ray-microanalysis of the seminal root of *Sorghum bicolor* with particular reference to silicon. *Ann. Bot.*, **64** 659-67.

Hole, C.C. (1996) Carrots, in *Photoassimilate Distribution in Plant and Crops: Source-sink Relationships* (eds. E. Zamski and A.A. Schaffer), Marcel Dekker, New York, pp. 671-90.

Hollenbach, B. and Dietz, K.J. (1995) Molecular-cloning of EMIP, a member of the major intrinsic protein (MIP) gene family, preferentially expressed in epidermal cells of barley leaves. *Bot. Acta*, **108** 425-31.

Hüsken, D., Steudle, E. and Zimmermann, U. (1978) Pressure probe technique for measuring water relations of cells in higher plants. *Plant Physiol.*, **61** 158-63.

Irving, M.S., Ritter, S., Tomos, A.D. and Koller, D. (1997) Phototropic response of the bean pulvinus: movement of water and ions. *Bot. Acta*, **110** 118-26.

Itai, C., Rothbejerano, N. and Zvilich, M. (1982) Involvement of epidermal cells in stomata movement: the effect of ABA. *Physiol. Plant.*, **55** 35-38.

Kaldorf, M., Kuhn, A.J., Schroder, W.H., Hildebrandt, U. and Bothe, H. (1999) Selective element deposits in maize colonised by a heavy metal tolerance conferring arbuscular mycorrhizal fungus. *J. Plant Physiol.*, **154** 718-28.

Karley, A.J., Leigh, R.A. and Sanders, D. (2000) Differential ion accumulation and ion fluxes in the mesophyll and epidermis of barley. *Plant Physiol.*, **122** 835-39.

Karrer, E.E., Lincoln, J.E., Hogenhout, S., Bennett, A.B., Bostock, R.M., Martineau, B., Lucas, W.J., Gilchrist, D.G. and Alexander, D. (1995) *In situ* isolation of messenger rRNA from individual plant cells: creation of cell-specific cDNA libraries. *Proc. Natl. Acad. Sci. USA*, **92** 3814-18.

Kehr, J., Wagner, C., Willmitzer, L. and Fisahn, J. (1999) Effect of modified carbon allocation on turgor, osmolality, sugar and potassium content, and membrane potential in the epidermis of transgenic potato (*Solanum tuberosum* L.) plants. *J. Exp. Bot.*, **50** 565-71.

Kelday, L.S. and Bowling, D.J.F. (1980) Profiles of chloride concentration and PD in the root of *Commelina communis* L. *J. Exp. Bot.*, **31** 1347-56.

Kiegle, E., Gilliham, M., Haseloff, J. and Tester, M. (2000) Hyperpolarisation-activated calcium currents found only in cells from the elongation zone of *Arabidopsis thaliana* roots. *Plant J.*, **21** 225-29.

Korolev, A.V., Tomos, A.D., Bowtell, R. and Farrar, J.F. (2000) Spatial and temporal distribution of solutes in the developing carrot taproot measured at single-cell resolution. *J. Exp. Bot.*, **51** 567-77.

Koroleva, O.A., Farrar, J.F., Tomos, A.D. and Pollock, C.J. (1997) Solute patterns in individual mesophyll, bundle sheath and epidermal cells of barley leaves induced to accumulate carbohydrate. *New Phytol.*, **136** 97-104.

Koroleva, O.A., Farrar, J.F., Tomos, A.D. and Pollock, C.J. (1998) Carbohydrates in individual cells of epidermis, mesophyll and bundle sheath in barley leaves with changed export or photosynthetic rate. *Plant Physiol.*, **118** 1525-32.

Koroleva, O.A., Tomos, A.D., Farrar, J.F., Roberts, P. and Pollock, C.J. (2000) Tissue distribution of primary metabolism between epidermal, mesophyll and parenchymatous bundle sheath cells in barley leaves. *Aust. J. Plant Physiol.*, **27** 747-55.

Kuhn, A.J., Schroder, W.H. and Bauch, J. (2000) The kinetics of calcium and magnesium entry into mycorrhizal spruce roots. *Planta*, **210** 488-96.

Kupper, H., Zhao, F.J. and McGrath, S.P. (1999) Cellular compartmentation of zinc in leaves of the hyperaccumulator, *Thlaspi caerulescens*. *Plant Physiol.*, **119** 305-11.

Lazzaro, M.D. and Thomson, W.W. (1992) Ultrastructural localization of calcium in the organic acid secreting trichomes of chickpea (*Cicer arietinum*). *Can. J. Bot.*, **70** 2319-25.

Lee, R.B. and Ratcliffe, R.G. (1983) Phosphorus nutrition and the intracellular distribution of inorganic phosphate in pea root tips: a quantitative study using ^{31}P NMR. *J. Exp. Bot.*, **34** 1222-44.

Leigh, R.A. (1997) Solute composition of vacuoles. *Adv. Bot. Res.*, **25** 171-94.

Leigh, R.A. and Branton, D. (1976) Isolation of vacuoles from root storage tissue of *Beta vulgaris* L. *Plant Physiol.*, **58** 656-62.

Leigh, R.A. and Storey, R. (1993) Distribution of ions between epidermal and mesophyll cells in barley leaves: an assessment of the role of intercellular compartmentation in nutrition and salinity. *J. Exp. Bot.*, **44** 755-62.

Leigh, R.A. and Tomos, A.D. (1993) Ion distribution in cereal leaves: pathways and mechanisms. *Philos. Trans. R. Soc., Lond. B Biol. Sci.*, **341** 75-86.

Leigh, R.A. and Wyn Jones, R.G. (1984) A hypothesis relating critical potassium concentrations for growth to the distribution and functions of this ion in the plant cell. *New Phytol.*, **97** 1-13.

Leigh, R.A. and Wyn Jones, R.G. (1986) Cellular compartmentation in plant nutrition: the selective cytoplasm and the promiscuous vacuole. *Adv. Plant Nutr.*, **2** 249-79.

Lichtenberger, O. and Neumann, D. (1997) Analytical electron microscopy as a powerful tool in plant cell biology: examples using electron energy loss spectroscopy and X-ray microanalysis. *Eur. J. Cell Biol.*, **73** 378-86.

Maathuis, F.J.M., May, S.T., Graham, N.S., Bowen, H.C., Jelitto, T.C., Trimmer, P., Bennett, M.J., Sanders, D. and White, P.J. (1998) Cell marking in *Arabidopsis thaliana* and its application to patch-clamp studies. *Plant J.*, **15** 843-51.

Malone, M., Leigh, R.A. and Tomos, A.D. (1989) Extraction and analysis of sap from individual wheat leaf cells: the effect of sampling speed on the osmotic pressure of extracted sap. *Plant Cell Environ.*, **12** 919-26.

Malone, M., Leigh, R.A. and Tomos, A.D. (1991) Concentrations of vacuolar inorganic ions in individual cells of intact wheat leaf epidermis. *J. Exp. Bot.*, **42** 385-89.

Mangabeira, P., Mushrifah, I., Escaig, F., Laffray, D., France, M.G.C. and Galle, P. (1999) Use of SIMS microscopy and electron probe X-ray microanalysis to study the subcellular localisation of aluminium in *Vicia faba* roots cells. *Cell. Mol. Biol.*, **45** 413-22.

Maurel, C. (1997) Aquaporins and water permeability of plant membranes. *Annu. Rev. Plant Physiol. Plant Mol. Biol.*, **48** 399-429.

McCully, M.E. (1994) Accumulation of high levels of potassium in the developing xylem elements in roots of soybean and some other dicotyledons. *Protoplasma*, **183** 116-25.

Meshcheryakov, A., Steudle, E. and Komor, E. (1992) Gradients of turgor, osmotic pressure and water potential in the cortex of the hypocotyl of growing *Ricinus* seedlings: effects of the supply of water from the xylem and of solutes from the phloem. *Plant Physiol.*, **98** 840-52.

Metzler, A., Izquierdo, M., Ziegler, A., Kockenberger, W., Komor, E., Vonkienlin, M., Haase, A. and Decorps, M. (1995) Plant histochemistry by correlation peak imaging. *Proc. Natl. Acad. Sci. USA*, **92** 11912-15.

Meyer, A.J. and Popp, M. (1997) Free Ca^{2+} in tissue saps of calciotrophic CAM plants as determined with Ca^{2+}-selective electrodes. *J. Exp. Bot.*, **48** 337-44.

Miyawaki, A., Llopis, J., Heim, R., McCaffery, J.M., Adams, J.A., Ikura, M. and Tsien, R.Y. (1997) Fluorescent indicators for calcium based on green fluorescent protein and calmodulin. *Nature*, **388** 882-87.

Müller, M.L., Irkens-Kiesecker, U., Rubinstein, B. and Taiz, L. (1996) On the mechanism of hyper-acidification in lemon: comparison of vacuolar H^+-ATPase activities of fruits and epicotyls. *J. Biol. Chem.*, **271** 1916-24.

Müller, M.L., Jensen, M. and Taiz, L. (1999) The vacuolar H^+-ATPase of lemon fruits is regulated by variable H^+/ATP coupling and slip. *J. Biol. Chem.*, **274** 10706-16.

Pallaghy, C. (1973) Electron probe microanalysis of potassium and chloride in freeze-substituted leaf sections of *Zea mays*. *Aust. J. Biol. Sci.*, **26** 1015-34.

Paris, N., Stanley, C.M., Jones, R.J. and Rogers, J.C. (1996) Plant cells contain functionally distinct vacuolar compartments. *Cell*, **85** 563-72.

Pritchard, J., Williams, G., Wyn Jones, R.G. and Tomos, A.D. (1989) Radial turgor pressure profiles in growing and mature zones of wheat roots: a modification of the pressure probe technique. *J. Exp. Bot.*, **40** 567-71.

Pritchard, J., Fricke, W. and Tomos, A.D. (1996) Turgor regulation during extension growth and osmotic stress of maize roots: an example of single-cell mapping. *Plant Soil*, **187** 11-21.

Raven, J.A. (1997) The vacuole: a cost benefit analysis. *Adv. Bot. Res.*, **25** 59-86.

Richardson, P.S. (1993) Intercellular distribution of nitrate in wheat leaves. PhD thesis, University of Wales.

Richardson, P., Tomos, A.D. and Carver, T. (1997) Effects of *Erysiphe graminis* on the water and solute relations of host leaf epidermal cells. *J. Exp. Bot.*, **48** (Suppl.), p. 40.

Roos, W., Evers, S., Hieke, M., Tschope, M. and Schumann, B. (1998) Shifts of intracellular pH distribution as a part of the signal mechanism leading to the elicitation of benzophenanthridine alkaloids: phytoalexin biosynthesis in cultured cells of *Escholtzia californica*. *Plant Physiol.*, **118** 349-64.

Rygol, J. and Zimmermann, U. (1990) Radial and axial turgor pressure measurements in individual root cells of *Mesembryanthemum crystallinum* grown under various saline conditions. *Plant Cell Environ.*, **13** 15-26.

Rygol, J., Pritchard, J., Zhu, J.J., Tomos, A.D. and Zimmermann, U. (1993) Transpiration induces radial turgor pressure gradients in wheat and maize roots. *Plant Physiol.*, **103** 493-500.

Sato, H., Taguchi, G., Fukui, H. and Tabata, M. (1992) Role of malic acid in solubilizing excess berberine accumulating in vacuoles of *Coptis japonica*. *Phytochemistry*, **31** 3451-54.

Stelzer, R., Lehmann, H., Kramer, D. and Lüttge, U. (1990) X-ray microprobe analyses of vacuoles of spruce needle mesophyll, endodermis and transfusion parenchyma cells at different seasons of the year. *Bot. Acta*, **103** 415-23.

Stelzer, R., Holste, R., Groth, M. and Schmidt, A. (1993) X-ray microanalytical studies on mineral concentrations in vacuoles of needle tissues from *Larix decidua* (L.) mill. *Bot. Acta*, **106** 325-30.

Steudle, E., Smith, J.A.C. and Lüttge, U. (1980) Water-relation parameters of individual mesophyll cells of the crassulacean acid metabolism plant, *Kalanchöe daigremontiana*. *Plant Physiol.*, **66** 1155-63.

Struck, C., Rohringer, R. and Heitefus, R. (1994) Isolation and lectin-binding properties of barley epidermal and mesophyll protoplasts. *Can. J. Bot.*, **72** 1688-91.

Sugimura, Y., Mori, T., Nitta, I., Kotani, E., Furusawa, T., Tatsume, M., Kusakari, S.I., Wada, M. and Morita, Y. (1999) Calcium deposition in idioblasts of mulberry leaves. *Ann. Bot.*, **83** 543-50.

Tomos, A.D. (1988) Cellular water relations in plants, in *Water Science Reviews 3* (ed. F. Franks), Cambridge University Press, pp. 186-277.

Tomos, A.D. and Leigh, R.A. (1999) The pressure probe: a versatile tool in plant cell physiology. *Annu. Rev. Plant Physiol. Plant Mol. Biol.*, **50** 447-72.

Tomos, A.D. and Wyn Jones, R.G. (1988) Some transport properties of cells within tissues, in *Solute Transport in Plant Cells and Tissues* (eds. D.A. Baker and J.L. Hall), Longman, pp. 220-50.

Tomos, A.D., Leigh, R.A., Palta, J.A. and Williams, J.H.H. (1992) Sucrose and cell water relations, in *Carbon Partitioning Between and Within Organisms* (eds. C.J. Pollok and J.F. Farrar), Biosis Scientific Publishers, Oxford, pp. 71-89.

Tomos, A.D., Hinde, P., Richardson, P., Pritchard, J. and Fricke, W. (1994) Microsampling and measurements of solutes in single cells, in *Plant Cell Biology; A Practical Approach* (eds. N. Harris and K.J. Oparka), IRL Press, Oxford, pp. 297-314.

Tomos, A.D., Korolev, A.V., Farrar, J.F., Nicolay, K., Bowtell, R. and Koeckenberger, W. (2000) Water and solute relations of the carrot cambium studied at single-cell resolution, in *Cell and Molecular Biology of Wood Formation* (eds. R. Savidge, J. Barnett and R. Napier) Biosis Scientific Publishers, Oxford, pp. 101-112.

Treeby, M.T. and van Steveninck, R.F.M. (1988) Effects of salinity and phosphate on ion distribution in lupin leaflets. *Physiol. Plant.*, **73** 317-22.

van der Leij, M., Smith, S.J. and Miller, A.J. (1998) Remobilisation of vacuolar stored nitrate in barley root cells. *Planta*, **205** 64-72.

van Steveninck, R.F.M. and van Steveninck, M.E. (1991) Microanalysis, in *Electron Microscopy of Plant Cells* (eds. J.M. Hall and C.R. Hawes), Academic Press, London, pp. 415-55.

van Steveninck, R.F.M., van Steveninck, M.E., Stelzer, R. and Läuchli, A. (1982) Studies on the distribution of Na and Cl in two species of lupin (*Lupinus luteus* and *Lupinus angustifolius*) differing in salt tolerance. *Physiol. Plant.*, **56** 465-73.

van Steveninck, R.F.M., Babare, A., Fernando, D.R. and Vansteveninck, M.E. (1994) The binding of zinc, but not cadmium, by phytic acid in roots of crop plants. *Plant Soil*, **167** 157-64.

Vazquez, M.D., Poschenrieder, C., Barcelo, J., Baker, A.J.M., Hatton, P. and Cope, G.H. (1994) Compartmentation of zinc in roots and leaves of the zinc hyperaccumulator, *Thlaspi caerulescens* j and c presl. *Bot. Acta*, **107** 243-50.

Wagner, G.J. and Siegelman, H.W. (1975) Large scale isolation of intact vacuoles and isolation of chloroplasts from protoplasts of mature plant tissues. *Science*, **190** 1298-99.

Walker, D.J., Smith, S.J. and Miller, A.J. (1995) Simultaneous measurement of intracellular pH, K^+ or NO_3^- in barley roots using triple-barrelled ion-sensitive microelectrodes. *Plant Physiol.*, **108** 743-51.

Walker, D.J., Leigh, R.A. and Miller, A.J. (1996) Potassium homeostasis in vacuolate plant cells. *Proc. Natl. Acad. Sci. USA*, **93** 10510-14.

Weast, R.C. and Astle, M.J. (1980) *CRC Handbook of Chemistry and Physics*, 61st edn, CRC Press, Boca Raton, FL.

Wiebe, H.H. (1978) The significance of plant vacuoles. *Bioscience*, **25** 327-31.

Williams, M.L., Thomas, B.J., Farrar, J.F. and Pollock, C.J. (1993) Visualizing the distribution of elements within barley leaves by energy dispersive X-ray image maps (EDX maps). *New Phytol.*, **125** 367-72.

Wyn Jones, R.G., Storey, R., Leigh, R.A., Ahmad, N. and Pollard, A. (1977) A hypothesis on cytoplasmic osmoregulation, in *Regulation of Cell Membrane Activities in Plants* (eds. E. Marré and O. Ciferri), Elsevier/North Holland Biomedical Press, Amsterdam, pp. 121-36.

Yoshida, K., Kondo, T., Okazaki, Y. and Katou, K. (1995) Cause of blue petal colour. *Nature*, **373** 291.

Zeuthen, T., Meinild, A.K., Klaerke, D.A., Loo, D.D.F., Wright, E.M., Belhage, B. and Litman, T. (1997) Water transport by the Na^+/glucose co-transporter under isotonic conditions. *Biol. Cell*, **89** 307-12.

Zhen, R.-G., Koyro, H.W., Leigh, R.A., Tomos, A.D. and Miller, A.J. (1991) Compartmental nitrate concentrations in barley root cells measured with nitrate-selective microelectrodes and by single-cell sap sampling. *Planta*, **185** 356-61.

Zimmermann, U., Rygol, J., Balling, A., Klock, G., Metzler, A. and Haase, A. (1992) Radial turgor and osmotic pressure profiles in intact and excised roots of *Aster tripolium*: pressure probe measurements and nuclear magnetic resonance-imaging analysis. *Plant Physiol.*, **99** 186-96.

9 Ion channels and ion co-transporters in the tonoplast

Omar Pantoja, Bronwyn J. Barkla
and Rosario Vera-Estrella

9.1 Introduction

Our knowledge of the transport properties of the tonoplast from higher plant cells has increased considerably during the past 15 yrs, as a result of the application of the patch-clamp technique to isolated vacuoles and the design of methodology to purify the vacuolar membrane of the plant cell. More recently, advances in the fields of yeast and animal membrane biology have augmented the research in this area, by the rescuing of yeast transport mutants with plant cDNAs and the identification of novel secondary messengers involved in the release of vacuolar Ca^{2+}, as specific examples.

Since the first study of vacuoles from red beet (Hedrich *et al.*, 1986), we have learned that the tonoplast of most species present at least two types of cation channel, the so-called SV (slowly-activating vacuolar) and FV (fast-activating vacuolar) channels (see Barkla and Pantoja, 1996). In particular cases, such as the vacuole from guard cells, a third cation channel has been described, the VK channel (vacuolar K^+ channel; Ward and Schroeder, 1994). It has also been demonstrated that an anion channel, uniquely selective to an organic anion (malate; VMAL) rather than to inorganic anions, may be ubiquitous in plant vacuoles (Iwasaki *et al.*, 1992; Pantoja *et al.*, 1992b; Pei *et al.*, 1998).

The presence of ion/H^+ co-transporters for Na^+, Ca^{2+}, Cd^{2+} and Mg^{2+} on the tonoplast has also been described (see Barkla and Pantoja, 1996). However, a large advancement in this field has been achieved only recently, with the molecular characterization of the putative transporters for all but Cd^{2+} (Hirschi *et al.*, 1996; Apse *et al.*, 1999; Gaxiola *et al.*, 1999; Shaul *et al.*, 1999).

This chapter reviews the most recent progress towards understanding the biochemistry, physiology, biophysics and molecular biology of ion channels and ion co-transporters present in the tonoplast of higher plants. The review focuses on papers published after 1995, since an earlier report (Barkla and Pantoja, 1996) reviewed the progress achieved between 1986 and 1995. When necessary, earlier works are used to present a clearer discussion of the papers published during the last 5 yrs.

9.2 Ion channels

9.2.1 Voltage gated cation and anion channels

Voltage gated channels are ion channels activated by changes in the membrane potential. Variations in membrane potential may induce two responses, changes in the current magnitude and/or changes in the open-probability of the single channels. Because of the possibility of these two different responses, several situations can occur and these can easily be understood analysing events at the single channel level. If the magnitude of the single channel currents show a linear relationship with respect to voltage, the channel is denominated as voltage-independent. In contrast, if the current–voltage relationship is nonlinear, then the channel is referred to as voltage-dependent. If the open-probability of a channel showing a linear voltage-current relationship varies with voltage, the channel is then classified as voltage-dependent. When the open-probability is constant with respect to voltage, the channel remains as a voltage-independent channel. At the whole cell level, the voltage dependency or independency of the currents depends on the product of these two parameters, since the whole cell current is given by:

$$I_{wc}(V) = i_{sc}(V) \cdot P_o(V) \cdot n$$

where, I_{wc} is the whole vacuole current at voltage V, i_{sc} and P_o are the single channel current and the open-probability at the holding potential V, and n is the total number of channels present in the membrane. As will be discussed below for the SV channels, the properties at the single channel level can be completely different to those observed at the whole vacuole, as a result of the product of P_o and i_{sc}. Therefore, in order to have an unequivocal view of the behavior of a particular channel, it is essential that the results and analysis at both levels lead to the same conclusion.

9.2.1.1 SV cation channels

SV channels have been found in the tonoplast of all species studied so far. At the whole vacuole level, these channels are characterized by a strong outward rectification (large currents at positive tonoplast potentials), and by their unphysiological activation at μM concentrations of cytoplasmic Ca^{2+} (Ca^{2+}_{cyt}). Because of these properties, a physiological role for these channels has been difficult to ascribe.

The SV outward rectification observed at the whole vacuole level is in contrast with the linear (Gambale *et al.*, 1996) or inward rectification (Pottosin *et al.*, 1997) observed at the single channel level. These results indicate that the whole vacuole outward rectification is due to the voltage-dependent open-probability of the SV channel, as demonstrated by Gambale *et al.* (1996) and Pottosin *et al.* (1997). The open-probability of the SV channels at negative physiological tonoplast potentials is very low, requiring the application of tail current protocols to be able to observe single channel openings at these potentials (Gambale *et al.*, 1996; Pottosin *et al.*, 1997). The unitary conductance of SV channels with symmetrical 100 and 200 mM KCl solutions varies between 26–102 pS and 120–280 pS, respectively (see Schulz-Lessdorf and Hedrich, 1995).

Some controversy exists regarding the selectivity of SV channels. Based on results from tail current experiments in asymmetrical KCl solutions, several authors have reported a measurable permeability of the SV channels to Cl^- (Coyaud *et al.*, 1987; Schulz-Lessdorf and Hedrich, 1995). Deviations of E_{rev} from E_K led these groups to report relatively low P_K/P_{Cl} values of 6–8. However, ion substitution experiments, replacing Cl^- with large organic (presumably impermeable) anions, suggested that this was not the case. Substitution of Cl^- by gluconate on the cytoplasmic side did not affect the reversal potential of single channel currents from onion guard cell vacuoles (Amodeo *et al.*, 1994), or from sugar beet vacuoles (Pantoja *et al.*, 1992a). These early indications of the impermeability of the SV channels to Cl^- were confirmed by Ward *et al.* (1995). When working under conditions where E_K, E_{Ca}, E_H and E_{Mg} were set close to +30 mV, Ward *et al.* (1995) recorded an E_{rev} close to this value but far from E_{Cl}, clearly indicating that SV channels are impermeable to anions and that the indiscriminate use of the Goldman-Hodgkin-Katz equation to determine permeability ratio values may lead to incorrect conclusions.

Since the first report by Hedrich and Neher (1987), there have been several detailed papers addressing the dependence of SV channels not only on Ca^{2+}_{cyt} but also on vacuolar Ca^{2+} (Ca^{2+}_{vac}) (Ward and Schroeder, 1994; Schulz-Lessdorf and Hedrich, 1995; Pottosin *et al.*, 1997). At physiological concentrations of Ca^{2+}_{vac} (0.5–5.0 mM) and in a background of 100–200 mM symmetrical KCl, increases in Ca^{2+}_{cyt} shifted the SV channel activation potential (E_{ac}) to less positive tonoplast potentials (Schulz-Lessdorf, 1995), without affecting the gating charge (Pottosin *et al.*, 1997), resulting in the stimulation of the SV current. In contrast, at high Ca^{2+}_{cyt} (0.5–2 mM), increases in Ca^{2+}_{vac} inhibited the SV channels by decreasing the open probability of single channels, causing a change in E_{ac} to more depolarized potentials, with a concomitant increase in the gating charge from 0.9 to 1.7 (Pottosin *et al.*, 1997). Activation of SV channels by Mg^{2+}_{cyt} has been shown by Allen and Sanders (1996) and Davies and Sanders (1995), who assigned a physiological role to this effect in Ca^{2+}-induced Ca^{2+} release. However, these effects of Mg^{2+} on the SV channels may be the result of Ca^{2+} contamination, an issue that was addressed by Pottosin *et al.* (1997). Working with solutions containing ethylene glycol tetraacetic acid (EGTA), these authors clearly demonstrated that the Ca^{2+} present as contaminant in the chemicals used for the preparation of experimental solutions, was sufficient to stimulate the SV currents, thus invalidating the possible role of Mg^{2+}_{cyt} in regulating these currents. By the same token, the recent report by Pei *et al.* (1999), who proposed a synergistic effect of Mg^{2+}_{cyt} and Ca^{2+}_{cyt} on the activation of SV channels, may also be explained by Ca^{2+} contamination, since the required precautions were not taken. Possible contamination of the Ca^{2+}-free solutions by Ca^{2+} is supported by additional results in the same paper, but for the FV channels (Pei *et al.*, 1999). FV channel activity was inhibited by increasing concentrations of Mg^{2+} at both sides of the tonoplast, an effect due to the presence of Ca^{2+}, as clearly shown by Tikhonova *et al.* (1997) (see Section 9.2.1.2). These results highlight the need to eliminate possible Ca^{2+} contamination of solutions by including the specific Ca^{2+} chelators EGTA or bis-amino phenoxyethane tetraacetic acid (BAPTA).

Another cytoplasmic factor that has been studied as a possible regulator of SV channels is pH, due mainly to its assigned role as a second messenger (Felle, 1989b). Schulz-Lessdorf and Hedrich (1995), working with 1 mM Ca^{2+} at both sides of the tonoplast, reported that alkalinization of the cytoplasmic side stimulated SV currents as a result of a shift in E_{ac} towards less positive potentials, without affecting the single channel conductance. Similar results were reported by Pottosin *et al.* (1997) but only when Ca^{2+}_{vac} was below 0.5 μM; at physiological Ca^{2+}_{vac}, however, acidification of pH_{cyt} failed to affect SV currents. These contrasting results are difficult to reconcile but it may be that they are due to the different cell type or species employed in these studies. Schulz-Lessdorf and Hedrich (1995) also reported the complete inhibition of SV channels by acidification of the vacuolar lumen to pH_{vac} of 3.8, although this inhibition was 100-fold less sensitive than that caused by changes in pH_{cyt}. Regulation of SV channels by pH_{cyt} has been proposed as an important factor in guard cell physiology, specifically during abscisic acid (ABA)-induced stomatal closing, which causes cytoplasmic alkalinization (Irving *et al.*, 1992), however, because of the low open-probability of the SV channels at physiological tonoplast potentials (Gambale *et al.*, 1996; Pottosin *et al.*, 1997), the role of SV channels in K^+ release into the cytoplasm would be unlikely.

SV channels have been proposed to participate in the phenomenon of Ca^{2+}-induced Ca^{2+} release in plant cells, primarily because of their permeability to Ca^{2+} (Ward and Schroeder, 1994). Initial indications that SV channels were permeable to Ca^{2+} were presented by Pantoja *et al.* (1992c), who demonstrated the activity of SV-like currents in the tonoplast of sugar beet cell cultures carried by Ba^{2+}, a Ca^{2+} analog. Since then, the permeability of the SV channels to Ca^{2+} has been clearly demonstrated in several studies (Ward and Schroeder, 1994; Schulz-Lessdorf and Hedrich, 1995; Allen and Sanders, 1996; Pottosin *et al.*, 1997). The high selectivity of SV channels for Ca^{2+} was indicated by the almost identical values of E_{rev} and E_{Ca} obtained in the presence of a $CaCl_2$ gradient (Ward and Schroeder, 1994; Schulz-Lessdorf and Hedrich, 1995). A shift in E_{rev} by 5 mV with 50 μM Ca^{2+}_{cyt} and 2 mM Ca^{2+}_{vac}, under a background of symmetrical 100 mM KCl, also indicated that the SV channels from barley mesophyll vacuoles were highly selective to Ca^{2+} (Pottosin *et al.*, 1997). Under conditions where $Ca^{2+}(Ba^{2+})$ and K^+ were present at opposite sides of the tonoplast, permeability ratios $P_{Ca(Ba)}/P_K$ between 5 and 7 were obtained (Pantoja *et al.*, 1992c; Ward and Schroeder, 1994). These results led Allen and Sanders (1996) to carefully study the variation of E_{rev} with changes in the ionic composition of the solutions at both sides of the tonoplast. These authors reported that the permeability of SV channels to Ca^{2+} was dependent on the Ca^{2+} mole fraction on either side of the tonoplast. Very high P_{Ca}/P_K values (240) were calculated when the Ca^{2+} mole fraction on the vacuolar side was small. When the Ca^{2+} mole fraction was set low at the cytoplasmic side, Ca^{2+} permeability values decreased 16-fold. However, if the Ca^{2+} mole fraction was set high at either side of the tonoplast, P_{Ca}/P_K decreased significantly, with values close to 0.5 (Allen and Sanders, 1996). All these results indicated that the SV channels were multi ion pore channels, where ions do not move through the pore independently of each other, restricting the application of the Goldman-Hodgkin-Katz equation for the determination of P_{Ca}/P_K values (Allen and Sanders, 1996).

As mentioned previously, the physiological role ascribed to the SV channels has been in mediating Ca^{2+}-induced Ca^{2+}-release (Ward and Schroeder, 1994; Ward *et al.*, 1995; Allen and Sanders, 1996), however this view has recently been challenged (Pottosin *et al.*, 1997). Based on the results obtained from vacuoles of barley mesophyll cells (Pottosin *et al.*, 1997), and those reported for vacuoles of *Vicia faba* guard cells (Schulz-Lessdorf and Hedrich, 1995), Pottosin *et al.* (1997) analysed the dependence of the open-probability on the Ca^{2+} gradients across the tonoplast. They calculated that under physiological Ca^{2+}_{vac} (50 μM to 2 mM), the concentration of Ca^{2+}_{cyt} and the tonoplast potential required to maintain the SV channels open with a probability as low as 1%, would establish a net efflux of Ca^{2+} from the cytoplasm into the vacuole, instead of an influx of Ca^{2+}. Pottosin *et al.* (1997) also concluded that the parameter controlling the voltage-dependent gating of the SV channels is the Ca^{2+} gradient across the tonoplast rather than the absolute concentration of Ca^{2+}_{cyt}. The analysis presented by Pottosin *et al.* (1997) clearly indicates that under physiological conditions where the electrochemical potential for Ca^{2+} is poised for its release into the cytoplasm, the SV channels do not open, thus the role of these channels in Ca^{2+}-induced Ca^{2+} release (Ward and Schroeder, 1994; Allen and Sanders, 1996) has to be re-evaluated.

9.2.1.2 FV K^+ channels

During the characterization of the SV channels, Hedrich and Neher (1987) identified a second class of ion channels activated instantaneously at submicromolar Ca^{2+}_{cyt}, the so-called FV channels. These channels have also been recorded in vacuoles from *V. faba* guard cells (Ward and Schroeder, 1994; Allen and Sanders, 1996), sugar beet cell cultures (Pantoja *et al.*, 1992b) and mesophyll cells from barley (Tikhonova *et al.*, 1997).

FV channels show a slight outward rectification in symmetrical 100 mM KCl solutions, with a reduced conductance at physiological potentials (Hedrich and Neher, 1987; Pantoja *et al.*, 1992b; Tikhonova *et al.*, 1997). FV channels are inhibited by increases in Ca^{2+} at both sides of the tonoplast (Tikhonova *et al.*, 1997). At physiological concentrations of Ca^{2+}_{vac}, FV currents decreased proportionally with increasing concentrations of Ca^{2+}_{cyt} (Tikhonova *et al.*, 1997). With vacuoles exposed to zero Ca^{2+}_{vac}, FV currents showed similar amplitudes at positive and negative potentials, with the low conductance observed at tonoplast potentials close to -40 mV, and were equally blocked by μM concentrations of Ca^{2+}_{cyt} (Tikhonova *et al.*, 1997). Inhibition of FV currents by Ca^{2+}_{cyt} at high and low concentrations of Ca^{2+}_{vac} showed similar affinities, with a K_d of 6 μM and a Hill coefficient close to one. Inhibition of FV currents by Ca^{2+}_{vac} occurred with a lower affinity, showing a K_d of 14 μM and a lower Hill coefficient of 0.5 (Tikhonova *et al.*, 1997). Vacuolar Ca^{2+} inhibited inward FV currents more strongly than outward. The mechanism of inhibition by Ca^{2+} was different: while Ca^{2+}_{cyt} caused a decrease in the open-probability, Ca^{2+}_{vac} reduced the single channel currents, again with a greater effect on the inward single channel currents (80% inhibition at -100 mV). Vacuolar Ca^{2+} also diminished the rate of gating of the single channels (Tikhonova *et al.*, 1997). FV channels are K^+ selective, with a P_K/P_{Cl} of 30. They show Michaelis–Menten kinetics for K^+, with a K_m of 33.5 mM and a maximal conductance V_{max} of 23.8 Sm^{-2} (Tikhonova *et al.*, 1997).

The slight outward rectification observed for the FV channels at the whole vacuole level can be explained by the different conductance shown by the single channel currents. A positive potentials, FV outward currents displayed a conductance of 25 pS compared to 17.5 pS determined for the inward channels (Tikhonova *et al.*, 1997). Moreover, the low conductance recorded at physiological potentials can be explained by a change in the open-probability of the single channels, which showed a smaller value at these potentials.

As observed for SV channels, pH also plays an important role as a regulator of FV channels. Acidification of pH_{cyt} from 7.5 to 5.1 caused a decrease in magnitude of almost 50% in the FV currents from the green alga, *Eremosphaera viridis*; alkalinization of the vacuolar lumen also inhibited FV currents, with a larger inhibition observed at positive potentials ($\sim$40%) as compared to that recorded for the inward current ($\sim$5%) (Linz and Köhler, 1994). FV channels also seem to be regulated by nucleotides but not by direct phosphorylation, since addition of ATP in the absence of Mg^{2+} or the non-hydrolyzable analog ATPγS, caused a stimulation of FV-like currents in red beet vacuoles (Davies and Sanders, 1995). Several physiological roles have been proposed for the FV channels, including: i) as the shunt conductance for the tonoplast H^+ pumps, which by dissipating the electrical potential would allow the acidification of the vacuolar lumen (Davies and Sanders, 1995), and thus maintain the ΔpH across the membrane necessary for the activation of the secondary active transporters; ii) as the pathway for vacuolar K^+ release during stomatal closing induced by ABA, a response that could be the result of FV channel activation caused by the cytoplasmic alkalinization induced by this hormone (Allen and Sanders, 1996).

9.2.1.3 *VK channels*

A third type of voltage gated cation channel is present in the tonoplast of higher plants, the VK channels, which have so far only been described in guard cell vacuoles from *V. faba* (Ward and Schroeder, 1994; Allen and Sanders, 1996). VK channels are activated at intermediate concentrations of Ca^{2+}_{cyt} (1–5 μM), and are time- and voltage-independent, with a single channel conductance of 70 pS in 100 mM KCl symmetrical solutions (Ward and Schroeder, 1994). In contrast to FV channels, VK channels are more selective for K^+ than for Cl^-, are impermeable to Ca^{2+} and show the following selectivity sequence among monovalent cations, $K^+ > Rb^+ > NH_4^+ \gg Cs^+ \sim Na^+ \sim Li^+$ (Ward and Schroeder, 1994). Studies on the regulation of VK channels by pH_{cyt} have demonstrated that maximal activity is observed at pH 6.5, with cytoplasmic alkalinization causing a strong inhibition (~85%) (Ward and Schroeder, 1994).

It has been suggested that VK channels, like FV channels, are also involved in the release of vacuolar K^+ during stomatal closing, upon their activation by increases in Ca^{2+}_{cyt} induced by ABA (Ward and Schroeder, 1994; Allen and Sanders, 1996). However, this upregulation would be diminished by the cytoplasmic alkalinization that would also occur in response to ABA addition (Irving *et al.*, 1992), indicating that VK channels may not be the only mechanism mediating K^+ release from the vacuole. An alternative mechanism for long-term K^+ release could be the SV channels (Ward and Schroeder, 1994); however, in view of the low open-probability of these channels at negative tonoplast potentials (Gambale *et al.*, 1996; Pottosin *et al.*, 1997), the involvement of an additional mechanism has to be postulated.

9.2.1.4 *Vacuolar malate ion channels (VMAL)*

Among the voltage-dependent channels so far identified in the tonoplast of higher plant cells are those selective for chloride and malate, the two most abundant anions in plant cell vacuoles. Because of the central role of the organic anion, malate, in plant cell metabolism (Martinoia and Rentsch, 1994), the mechanism for vacuolar accumulation in plant cells has been studied for many years. Early results led some authors to postulate that malate uptake was mediated by a carrier, with a K_m for malate in the low mM range (Martinoia *et al.*, 1985; Rentsch and Martinoia, 1991). Malate uptake into barley mesophyll cell vacuoles is ATP-dependent and, according to uptake experiments at different pHs, it was suggested that the divalent malate anion was the transported species (Martinoia *et al.*, 1985; Rentsch and Martinoia, 1991). Similar results were reported for malate transport across the tonoplast of *Kalanchoë daigremontiana* and *Catharanthus roseus* (Marigo *et al.*, 1988; White and Smith, 1989). Although a putative malate transporter was partially purified and functionally reconstituted into liposomes (Martinoia *et al.*, 1991; Ratajczak *et al.*, 1994), the actual protein responsible for malate transport was never identified.

Clearer evidence on the mechanism responsible for the accumulation of malate into the vacuole has been obtained employing the patch-clamp technique. Using isolated vacuoles from *Graptopetalum paraguayense*, a crassulacean acid metabolism (CAM) plant, and sugar beet cell cultures, Iwasaki *et al.* (1992) and Pantoja *et al.* (1992b) were able to identify inward currents that, according to the experimental conditions, were carried by malate^{2-} (mal^{2-}). These mal^{2-} currents were only activated at physiological levels of both Ca^{2+}_{cyt} (between 10^{-6} and 10^{-8} M) and tonoplast potential (more negative than −20 mV). Whole vacuole recordings in sugar beet showed strong inward rectification, and from tail current experiments, a P_{mal}/P_K of 6–10 was derived (Pantoja *et al.*, 1992b). The conductance at the single channel level was 120–125 pS in *G. paraguayense*, with an open-probability slightly voltage-dependent (Iwasaki *et al.*, 1992). The physiological role of this ion channel in mediating vacuolar mal^{2-} accumulation is thermodynamically viable when considering a tonoplast potential close to −50 mV; and, in the particular case of CAM plants, because of the steep mal^{2-} concentration gradient that is established by the protonation to Hmal^{1-} in the vacuolar lumen. Similar whole vacuolar mal^{2-} currents have been reported in *Arabidopsis thaliana* (Cerana *et al.*, 1995) and *K. daigremontiana* (Cheffings *et al.*, 1997), although in these species the mal^{2-} currents were independent of Ca^{2+}_{cyt}. In contrast to all previous reports, mal^{2-} currents in guard cells from *V. faba* required the presence of a constitutively active calcium-dependent protein kinase (CDPK) (Pei *et al.*, 1998), an observation that may be of particular importance in stomatal physiology in view of the postulated role that increases in Ca^{2+}_{cyt} play during the light-stimulated opening of guard cells (Irving *et al.*, 1992).

9.2.1.5 Vacuolar chloride channels (VCl)

So far, only a single report by Pei *et al.* (1998) has demonstrated the presence of Cl^- currents in the tonoplast of *V. faba* guard cells. The activity of these currents showed inward rectification, indicating the movement of Cl^- from the cytosol into the vacuolar lumen, since the activity of the FV and VK channels was prevented by the presence of Mg^{2+} and the use of low KCl concentrations in the vacuolar lumen, respectively. From this report, it is clear that the presence of a CDPK on the cytoplasmic side of the tonoplast is a prerequisite to observe the activity of the Cl^- channels and may also explain why its identification has been difficult to ascertain. Activation of the Cl^- current in the tonoplast of guard cells was clearly a result of channel phosphorylation mediated by the CDPK, since Cl^- currents were not observed in the absence of either Ca^{2+}_{cyt} or ATP. At the single channel level, the CDPK-induced currents flickered and a conductance of 34 pS was calculated with 90 mM Cl^- at both sides of the membrane. Tonoplast inward currents carried by chloride and regulated by a constitutively active

CDPK were also reported for red beet, where a simultaneous activation of outward currents in the presence of Cl^- was also observed (Pei *et al.*, 1998). If the presence of a CDPK-regulated chloride conductance in the central vacuole of higher plant cells is a common feature, this could suggest a primary role for Cl^- channels in volume regulation.

9.2.2 Ligand gated channels

A different category of ion channels present in the tonoplast of higher plants comprises those activated by the binding of a specific ligand. In contrast to the voltage gated channels, opening of ligand gated channels is induced upon the binding of a particular effector; in plant cells this has been shown to be mediated by the second messengers, inositol 1,4,5-trisphosphate (IP_3) and cyclic adenosine 5′-diphosphoribose (cADPR) (Alexandre *et al.*, 1990; Allen *et al.*, 1995). In the past 5 yrs, no further reports have been published on IP_3-gated channels and the reader is refered to an earlier review for details of the works published on this subject (Barkla and Pantoja, 1996).

Several advances, however, have been made not only on cADPR-gated channels but on the role of cADPR as a signaling molecule involved in ABA perception. The NAD^+ metabolite, cADPR, is synthesized by the activity of ADP-ribosyl cyclase (Glick *et al.*, 1991), and its role in the release of Ca^{2+} from the endoplasmic reticulum (ER) has been well characterized in animal cells (Rusinko and Lee, 1989), where its participation in Ca^{2+}-induced Ca^{2+} release has also been demonstrated (Galione *et al.*, 1993). Initial results on the role of cADPR in plants were presented by Allen *et al.* (1995), who demonstrated the release of Ca^{2+} from tonoplast vesicles upon the addition of 100 nM cADPR and the inhibition of this effect by ruthenium red, a specific inhibitor of this process in animal cells (Galione, 1994). The possible intracellular target for cADPR and mediator of Ca^{2+} release action was demonstrated by patch-clamp studies in isolated red beet vacuoles. Addition of 100 nM cADPR to isolated vacuoles under the whole vacuole configuration induced the activation of instantaneous currents, not only at negative (Allen *et al.*, 1995) but also at positive tonoplast potentials (Leckie *et al.*, 1998). Specificity for cADPR was demonstrated by the lack of stimulation of the vacuolar currents upon addition of the non-cyclic analog, adenosine 5′-diphosphoribose (ADPR) (Allen *et al.*, 1995). The effect of cADPR on vacuolar currents was voltage independent, with a K_m for cADPR close to 25 nM. The cADPR-induced currents showed a selectivity for Ca^{2+}, with a P_{Ca}/P_K of 15 (Leckie *et al.*, 1998). However, in marked contrast to the stimulation by Ca^{2+}_{cyt} observed in animal cells (Galione *et al.*, 1991), the vacuolar currents induced by cADPR were completely inhibited by Ca^{2+}_{cyt} at 600 nM (Leckie *et al.*, 1998). The effects of cADPR on tonoplast cytoplasmic side out patches were difficult to assess because of ill-defined single channel events (Allen *et al.*, 1995), thus preventing the confirmation that cADPR exerts its effect directly on ion channels. Allen *et al.* (1995) have proposed that the additive stimulation of vacuolar currents by the subsequent addition of IP_3 and cADPR indicated that these two messengers activated the release of Ca^{2+}_{vac} through separate pathways. They noted the importance of the latter observation in Ca^{2+} signaling and postulated that the presence of two distinct ligand-gated Ca^{2+} channels in the tonoplast may be significant in generating specific Ca^{2+} responses depending on the stimulus involved (Allen *et al.*, 1995).

Further direct support for the involvement of cADPR in the physiology of the plant cell has come from studies of ABA-induced responses (Wu *et al.*, 1997; Leckie *et al.*, 1998). Employing two ABA-regulated *A. thaliana* genes, *rd29A* and *kin2*, Wu *et al.* (1997) generated chimeric gene constructs from the promoters of these two genes with a β-glucuronidase (GUS) reporter gene, and demonstrated that the expression of either of these genes could be induced indistinctively by the co-injection of the gene together with cADPR or Ca^{2+}. The injection of the specific inhibitor of cADPR activity, 8-amino-cADPR or EGTA, prevented the induction of *rd29A* and *kin2* by ABA, supporting the role of these second messengers in ABA signaling. Further experiments demonstrated that injection of EGTA blocked the expression of *rd29A* and *kin2* induced by cADPR (Wu *et al.*, 1997), associating cADPR with Ca^{2+} release from intracellular stores. Conclusive evidence on the role of cADPR in ABA signaling was obtained by evaluating changes in the *in vivo* levels of cADPR in response to ABA. Extracts of plants treated with 50 μM ABA for different time periods induced the release of Ca^{2+} from sea urchin microsomes, with peaks after 1 and 4 h of ABA treatment and previous to the induction of the *kin2* gene (Wu *et al.*, 1997), suggesting the ABA-induced presence of cADPR in the plant extracts.

Direct involvement of cADPR in another ABA-mediated response with the participation of increases in Ca^{2+}_{cyt} has been demonstrated for stomatal closing (Leckie *et al.*, 1998). Microinjection of cADPR into a single stomatal cell caused an increase in Ca^{2+}_{cyt} that lasted several minutes, followed by a decrease in cell turgor. This response was not observed with the injection of the non-cyclic isomer, ADPR, which induced only a transient increase in Ca^{2+}_{cyt} and no change in cell turgor (Leckie *et al.*, 1998). As with the GUS reporter gene assays, injection of 8-amino-cADPR prevented the ABA-induced stomatal closing, although with some degree of variance (Leckie *et al.*, 1998). These results provide further and stronger evidence for the participation of cADPR as a second messenger in plants in general, and in ABA-mediated responses in particular.

9.3 Ion co-transporters

9.3.1 Na^+/H^+ antiporter

In plants, vacuolar Na^+ accumulation, via a Na^+/H^+ antiport or exchanger, utilizes the proton motive force generated by the activity of the two tonoplast primary proton pumps, the vacuolar-type H^+-transporting adenosine triphosphatase (V-ATPase), and the vacuolar-type H^+-transporting inorganic pyrophosphatase (V-PPase) (reviewed in this volume). Active Na^+ uptake, against a concentration gradient, is therefore driven by the downhill movement of H^+ out of the vacuole. As observed for the mammalian Na^+/H^+ exchangers (NHE isoforms) (reviewed in Counillon and Pouysségur, 2000), the stoichiometry of the plant vacuolar Na^+/H^+ exchanger is assumed to be one cytoplasmic Na^+ for one vacuolar H^+, resulting in electroneutral transport across the tonoplast (Blumwald *et al.*, 1987; Barkla *et al.*, 1995). Whereas, in animals, Na^+/H^+ exchange functions to stabilize intracellular pH close to neutrality, thus avoiding cellular acidification, in plants, the tonoplast Na^+/H^+ antiport serves to maintain low cytoplasmic Na^+ concentrations, protecting against the toxic effects of this ion on metabolic processes, while allowing the sequestered Na^+ to be utilized as an intracellular osmoticum for the maintenance of cellular water homeostasis. Thus, in plants, the antiport is thought to play a key role in salt-tolerance and adaptation.

Tonoplast Na^+/H^+ antiport activity has been documented in roots or root storage tissue of a diverse number of plant species, ranging in salt-tolerance from glycophytes that are salt-sensitive, encompassing cultivars of such crop plants as rice, tomato and barley (Fan *et al.*, 1989; Mennen *et al.*, 1990; Fukada *et al.*, 1998), to those glycophytes that are classified as considerably salt-tolerant, including sunflower, *B. vulgaris*, *Plantago maritima* and specific barley cultivars (Blumwald and Poole, 1985; Niemietz and Willenbrink, 1985; Garbarino and Dupont, 1988; Staal *et al.*, 1991; Ballesteros *et al.*, 1997). Antiport activity has also been recorded in cell suspension cultures of *B. vulgaris* and *C. roseus* (Blumwald *et al.*, 1987; Guern *et al.*, 1989). In contrast, activity of a tonoplast Na^+/H^+ antiport in leaf tissue has only been observed in the halophytic plant species, *Mesembryanthemum crystallinum* and *Atriplex gmelini* (Matoh *et al.*, 1989; Barkla *et al.*, 1995), suggesting that it may not be simply the presence of an antiport that confers salt-tolerance, but, rather more importantly, the degree of salt-tolerance exhibited by a given species depends on where in the plant the antiport protein is expressed. This is supported by comparative studies of halophyte species, which show that salt-tolerance in these plants can be positively correlated with the capacity of the plants to transport and sequester Na^+ into the shoot tissue, where it is then used for osmotic adjustment (Glenn *et al.*, 1999). On average 90% of the Na^+ in halophytes is located in the shoot system, whereas, net shoot transfer rates for Na^+ in glycophytes, which are typically considered shoot 'excluders' of Na^+, relying on extrusion of Na^+ across the plasma membrane of root cortical cells and retention of Na^+ in vacuoles of root cells, have been shown to be extremely low (Flowers *et al.*, 1977).

Evidence for a fundamental role of the tonoplast Na^+/H^+ antiport in salt-tolerance comes from studies demonstrating the induction or upregulation of activity of this transporter upon the imposition of a salt-stress. In barley, *P. maritima* and sunflower, antiport activity is absent in untreated plants, whereas in salt-treated plants activity is induced (Garbarino and Dupont, 1988; Staal *et al.*, 1991; Ballesteros *et al.*, 1997). *B. vulgaris* and *M. crystallinum*, which show constitutive antiport activity when grown in the absence of Na^+, show significant increases in activity upon Na^+-treatment of the cell suspension cultures or plants, respectively (Blumwald *et al.*, 1987; Barkla *et al.*, 1995).

The mechanism(s) by which Na^+ regulates the activity of the plant tonoplast Na^+/H^+ antiport is (largely) unknown. The rapid Na^+-induction of antiport activity in barley, with maximal levels observed within 50 min of NaCl treatment, would indicate post-translational regulation of the enzyme, rather than control of protein amount at the level of transcription or translation (Garbarino and Dupont, 1989). The plant growth regulator, ABA, which has been shown to mediate a wide range of salt- and drought-stress responses (Bray, 1997), does not appear to play a role in the Na^+-induced upregulation of Na^+/H^+ antiport activity in *M. crystallinum*, as exogenous foliar application of ABA did not affect antiport activity (Barkla *et al.*, 1999). More over, the Na^+ induction of antiport activity in *M. crystallinum* does not appear to be coordinated with the developmentally-programmed salt- and drought-induced switch from C_3 metabolism to CAM observed in adult plants (Cushman *et al.*, 1990). Rather, the salt-induced increase in activity is independent of plant age and can be observed in both juvenile and adult plants (Barkla *et al.*, 1999). A greater understanding of the regulation of Na^+/H^+ antiports comes from the more widely-studied animal NHE isoforms. Extracellular stimuli, including hormones, integrins, growth factors and mechanical stress, activate the exchanger by increasing the affinity of the transporter for protons. Downstream intracellular messengers of these stimuli include Ca^+/calmodulin, cAMP and protein kinase C, which converge on the C-terminal region of the protein (Wakabayashi *et al.*, 1997; Counillon and Pouysségur, 2000).

The diuretic drug, amiloride (3,5-diamino-6-chloro-N-[diaminomethylene] pyrazine carboxamide), and its alkyl 5-*N*-substituted derivatives have been shown to be potent, competitive inhibitors of Na^+/H^+ exchangers (Harris and Fliegel, 1999). Amiloride has been shown to inhibit the activity of several plant tonoplast Na^+/H^+ antiporters, with K_is for inhibition in the range 100–200 μM (Blumwald *et al.*, 1987; Barkla *et al.*, 1995; Ballesteros *et al.*, 1997), while the more specific amiloride analogs, including MIA (5-[N-methyl-N-isobutyl] amiloride) and EIPA (5-[N-ethyl-N-isopropyl] amiloride), inhibit activity, with K_is for inhibition in the range 2–6 μM (Blumwald *et al.*, 1987; Barkla and Blumwald, 1991; Fukada *et al.*, 1998). Interestingly, while barley antiport activity was shown to be insensitive to amiloride concentrations up to 500 μM (Garbarino and Dupont, 1988), analogs of amiloride were shown to inhibit activity, with a high specificity (Fukada *et al.*, 1998).

The recent cloning of a plant endomembrane Na^+/H^+ antiport in *A. thaliana* (*AtNHX1*) (Apse *et al.*, 1999; Gaxiola *et al.*, 1999) and rice (*OsNHX1*) (Fukada *et al.*, 1999), was aided by the earlier cloning of its homolog in the yeast,

Saccharomyces cerevisiae (*NHX1*) (Nass *et al.*, 1997; Numata *et al.*, 1998). NHX1 was shown to be involved in yeast Na^+ tolerance and presented homology to mammalian amiloride sensitive NHEs but appeared to cluster with a new family of antiporters with members in bacteria, animals and yeast, that reside exclusively on internal membranes rather than on the plasma membrane (Nass *et al.*, 1997). NHX1 shows highest sequence identity to NHE6, a mammalian mitochondria-located exchanger, and arguments have been presented for NHX1 localization in yeast mitochondria (Numata *et al.*, 1998). However, a contrasting view for the internal site of NHX1 localization has been forwarded by Nass and Rao (1998), who conclude that NHX1 resides on membranes derived from prevacuolar compartments.

Overexpression of the plant *AtNHX1* in yeast partially suppressed phenotypes of the yeast *nhx1* mutant, which suggested some similarity in function between yeast NHX1 and *A. thaliana* AtNHX1 (Gaxiola *et al.*, 1999). More definitive results for the involvement of AtNHX1 in Na^+/H^+ exchange and a role in plant salt-tolerance have come from expression studies in *A. thaliana* (Apse *et al.*, 1999). Overexpressing *AtNHX1* plants were significantly more salt-tolerant than wild-type plants, being able to grow and reproduce in Na^+ concentrations up to 200 mM NaCl (Apse *et al.*, 1999). Vacuoles isolated from leaves of these transgenic plants also showed high levels of Na^+/H^+ exchange, which was absent in vacuoles isolated from leaves of wild-type plants (Apse *et al.*, 1999). Surprisingly, although no specific Na^+/H^+ exchange is recorded in leaves of wild-type plants, both the AtNHX1 transcript and protein are abundant, and the relatively small increases in mRNA and protein observed in transgenic overexpressing plants as compared to wild-type plants did not correlate with the large differences in exchange activity observed. The authors argue that this could be attributed to repression of Na^+/H^+ exchange activity under normal growth conditions (Apse *et al.*, 1999). However, Frommer *et al.* (1999) suggest that such contradictions could be attributed to inherent problems in protein engineering, leading to missorting of the overexpressed AtNHX1 protein to either the tonoplast and/or plasma membrane from its native endomembrane location. Although the AtNHX1 protein was shown to co-migrate with the tonoplast marker enzymes, V-ATPase and V-PPase, on sucrose density gradients (Apse *et al.*, 1999), the presence of contaminating membranes within this fraction was not addressed. Further localization studies are required before a clear answer to the intracellular site of expression of AtNHX1 can be reached.

Tissue expression studies of both *AtNHX1* and *OsNHX1* in control and salt-treated wild-type *A. thaliana* and rice plants have uncovered some conflicting results. Gaxiola *et al.* (1999) presented evidence for a greater than four-fold increase in *AtNHX1* mRNA in plants treated for 6 h with 250 mM NaCl or KCl. In contrast, Apse *et al.* (1999), analyzing mRNA from different growth stages and tissues, in *A. thaliana* plants treated for 6–36 h with 50–250 mM

NaCl, observed no such increases, nor did they observe increases in AtNHX1 protein levels or Na^+/H^+ exchange activity under these conditions. Moreover, the induction of *AtNHX1* mRNA by either NaCl or KCl, as reported by Gaxiola *et al.* (1999), does not agree with the high specificity for Na^+ of antiport activity (Barkla *et al.*, 1995; Apse *et al.*, 1999). In rice, upregulation of transcript levels for *OsNHX1* upon salt-stress were observed in both root and shoot tissue (Fukada *et al.*, 1999). However, in tonoplast vesicles isolated from rice roots, no Na^+-induced increase in Na^+/H^+ antiport activity was observed (Fukada *et al.*, 1998). Clearly, although preliminary, these pioneering molecular studies present important new findings on plant Na^+/H^+ transporters and should enhance and expedite our knowledge in this field in the very near future.

9.3.2 Ca^{2+}/H^+ antiporter

In plant cells under 'resting' conditions, concentrations of Ca^{2+}_{cyt} are strictly maintained in the nM range. However, levels of calcium can be rapidly increased by more than 100-fold upon the perception of a wide range of external stimuli through the action of plasma membrane and intracellular calcium permeable ion 'influx' or 'release' channels, respectively. Biotic and abiotic factors capable of inducing such transient increases in Ca^{2+}_{cyt} include, sugars, cold-, heat-, osmotic- and salt-stress/shock, mechanical stimuli, plant growth regulators and the action of bacteria through either pathogen infection or symbiotic interactions (Knight *et al.*, 1991, 1996; Poovaiah and Reddy, 1993; Bowler *et al.*, 1994; Bressan *et al.*, 1998; Sanders *et al.*, 1999; Knight, 2000). Because these increases in Ca^{2+} are by necessity short-lived, it implies that restoration to prestimulus Ca^{2+}_{cyt} concentrations must also be rapid and tightly regulated.

The primary intracellular site for calcium storage and retrieval in the plant is the vacuole. Estimations of Ca^{2+}_{vac} concentrations indicate that levels may be in the mM range, a value more than 10,000 times higher than that of the cytoplasm (Felle, 1989a). Therefore, removal of Ca^{2+}_{cyt} following a signaling event requires active Ca^{2+} transport across the tonoplast against a large Ca^{2+} concentration gradient. In fact, the tonoplast of plant cells contains two transporters capable of energized Ca^{2+} uptake, a Ca^{2+}/H^+ antiporter and a Ca^{2+}-ATPase.

The activity of a Ca^{2+}/H^+ antiporter has been detected in tonoplast vesicles from a number of species, suggesting that it is ubiquitous in plants (Schumaker and Sze, 1985; Blumwald and Poole, 1986; Blackford *et al.*, 1990; Chanson, 1991; Hirschi *et al.*, 1996; Ueoka-Nakanishi *et al.*, 1999). Ca^{2+}/H^+ antiport activity is driven by the proton motive force (pmf), across the tonoplast, generated by the activity of the two primary vacuolar H^+-translocating pumps, V-ATPase and V-PPase. To function as a Ca^{2+} uptake/retrieval system under physiological conditions, rather than in the reverse for Ca^{2+} release, the stoichiometric ratio of H^+ to Ca^{2+} must be greater than 2, as has been demonstrated for Ca^{2+}/H^+ exchange in *B. vulgaris* (Blackford *et al.*, 1990). The K_m for Ca^{2+}

of the Ca^{2+}/H^+ antiport is reportedly in the low μM range (Schumaker and Sze, 1985; Blumwald and Poole, 1986), in contrast to that of the Ca^{2+}-ATPase, which is in the high nM range (Evans and Williams, 1998). This more than 10-fold difference in affinity for the same substrate suggests that the Ca^{2+}/H^+ antiport acts as a low affinity Ca^{2+} transport system on the tonoplast.

Evidence for distinct roles in Ca^{2+}_{cyt} regulation of the two tonoplast Ca^{2+} transport proteins comes from recent studies in yeast. Like plants, the vacuolar membrane of *S. cerevisiae* contains a Ca^{2+}-ATPase (encoded by the gene *Pmc1p*; Cunningham and Fink, 1994) and a Ca^{2+}/H^+ antiport (encoded by the gene *Vcx1Δ/Hum1Δ*; Cunningham and Fink, 1996; Pozos *et al*., 1996), both shown to mediate Ca^{2+} sequestration into the yeast vacuole. Investigating the response of the yeast mutants, *vcx1Δ* and *pmc1Δ*, lacking the Ca^{2+}/H^+ antiport and Ca^{2+}-ATPase, respectively, to a short-term external Ca^{2+} shock, Miseta *et al*. (1999) were able to demonstrate that the loss of Pmc1p had no significant effect on the cells ability to restore cytosolic Ca^{2+} concentrations to resting levels. In contrast, loss of Vcx1 severely handicapped the cells, which were impaired in their ability to rapidly lower Ca^{2+}_{cyt} upon exposure to high extracellular Ca^{2+} (Miseta *et al*., 1999). These findings suggested that Vcx1, alone, may act to rapidly restore resting state levels of Ca^{2+}_{cyt} following a Ca^{2+}-induced signaling event.

Further insight into the discrete roles played by the vacuolar Ca^{2+} transporters can be gained from the knowledge that their regulation (again demonstrated in yeast), in response to changes in Ca^{2+}_{cyt} differs fundamentally. Initial work demonstrated that yeast mutants, lacking Ca^{2+}-ATPase activity (*pmc1Δ*) regained tolerance to high external concentrations of Ca^{2+} following the inhibition of calcineurin, a Ca^{2+}/calmodulin-dependent protein phosphatase, by FK506, an immunosuppressive inhibitor of calcineurin (Cunningham and Fink, 1994). Similar results were then reported for yeast strains lacking a functional calcineurin (*cnb1Δ*) (Cunningham and Fink, 1996). In both cases, suppression of *pmc1Δ* by inhibition/mutation of calcineurin, was dependent on the activity of Vcx1p (Cunningham and Fink, 1996). This calcineurin downregulation of Ca^{2+}/H^+ antiport activity occurs mainly by post-translational modifications and may be important for the initiation of a Ca^{2+} signaling event (Cunningham and Fink, 1996). Calcineurin regulation of the plant Ca^{2+}/H^+ antiport has not yet been demonstrated; however, it is implicated in mediating salt tolerance/adaptation in plants (Liu and Zhu, 1998; Pardo *et al*., 1998), and mRNA for an *A. thaliana* calcineurin B-like protein has been shown to respond to drought, cold and wounding stress (Kudla *et al*., 1999).

Functional cloning of plant Ca^{2+}/H^+ antiports was made possible by transformation and suppression of the Ca^+ hypersensitive *S. cerevisiae* mutant strain *vcx1Δ* and *pmc1Δ* using an *A. thaliana* cDNA library (Hirschi *et al*., 1996). Two *A. thaliana* genes, designated CAX1 and CAX2 (calcium exchangers), were shown to suppress the yeast double mutant and demonstrated structural and functional similarities to previously cloned Ca^{2+}/H^+ exchangers from yeast and bacteria (Hirschi *et al.*, 1996). CAX1, in particular, showed high affinity Ca^{2+}/H^+ antiport activity similar to that previously described for the plant tonoplast antiport (Hirschi *et al.*, 1996), and expression was induced by treatment of plants with high extracellular Ca^{2+} (80 mM for 16 h) (Hirschi, 1999). A CAX1 homolog has since been identified and cloned in *Vigna radiata* (VCAX1) (Ueoka-Nakanishi *et al.*, 1999), and peptide specific antibodies against VCAX1 were shown to cross-react with a tonoplast polypeptide of 42 kDa. However, the exclusivity of a tonoplast location for VCAX1 was not demonstrated through the analysis of other membrane fractions or by the demonstration of tonoplast purity.

When expressed in tobacco, *A. thaliana* CAX1 caused symptoms of Ca^{2+} deficiency, which could be partially suppressed by addition of Ca^{2+} to the medium, although measurements of total cellular Ca^{2+} in *CAX1*-expressing plants indicated that these plants contained significantly higher concentrations of Ca^{2+} in both root and shoot tissue (Hirschi, 1999), presumably sequestered in the vacuole through the activity of CAX1 Ca^{2+}/H^+ exchange. Measurements of Ca^{2+}/H^+ exchange activity in *CAX1*-expressing tobacco plants were significantly higher as compared to control plants (Hirschi, 1999); however, surprisingly, the author states that values for both K_m and V_{max} were shown not to differ and, as such, it is difficult to understand how such large changes in activity would not be reflected in changes in the kinetic parameters of the exchanger.

9.3.3 Cd^{2+}/H^+ antiporter

The vacuole of plant cells is also the major detoxifying compartment for the accumulation of the heavy metal, Cd^{2+} (Vögeli-Lange and Wagner, 1990). Salt and Wagner (1993) demonstrated that Cd^{2+} uptake could be driven by the inside acidic pH gradient generated by the V-ATPase or by the artificial formation of a ΔpH, using K^+ and nigericin. Initial rates of Cd^{2+} accumulation showed saturation kinetics with respect to external Cd^{2+} concentrations, with a K_m of 5.5 μM and a V_{max} of 14 nmol mg protein^{-1} min^{-1}. The possibility that Cd^{2+} shares a common divalent cation co-transporter with Ca^{2+} has been suggested (Barkla and Pantoja, 1996). In an attempt to address this possibility and distinguish between Ca^{2+} versus Cd^{2+} antiport activity, Gonzalez *et al.* (1999) employed known inhibitors of Ca^{2+} transport in plants and measured both $^{45}Ca^{2+}$ and $^{109}Cd^{2+}$ ΔpH-dependent uptake into oat root tonoplast vesicles. Both the Ca^{2+} channel blockers, verapamil and ruthenium red (100 μM concentration), were shown to inhibit Ca^{2+} exchange to a greater extent than Cd^{2+} (Gonzalez *et al.*, 1999). However, in view of the nonspecificity of the inhibitors used (Terry *et al.*, 1992; Allen *et al.*, 1995; Pottosin *et al.*, 1999), these results are difficult to evaluate. One possible candidate for a Cd^{2+}/H^+ antiport could be the *A. thaliana* CAX2 exchanger, which demonstrates low-affinity Ca^{2+}/H^+ antiport activity when expressed in yeast, but has been suggested to be a high-affinity, high-capacity heavy metal cation/H^+ antiporter (Hirschi *et al.*, 1996).

9.3.4 *Mg^{2+} and Zn^{2+}/H^+ antiporter*

Two other essential cations with roles in enzyme function, structure and regulation, shown to be accumulated in vacuoles via cation/H^+ antiport mechanisms, are Mg^{2+} and Zn^{2+}. Activity of an Mg^{2+}/H^+ antiport was measured in the tonoplast of maize roots (Pfeiffer and Hager, 1993) and has also been described in lutoids (specialized lysosomal vacuoles) from latex of the rubber tree, *Hevea brasiliensis* (Amalou *et al.*, 1992). In experiments using solubilized and reconstituted lutoid membrane protein, ΔpH-dependent Mg^{2+} and Zn^{2+} exchange was detected, however Ca^{2+}/H^+ antiport activity, present in the native vacuoles, was absent (Amalou *et al.*, 1994). This observation raised the initial possibility that Mg^{2+} (and Zn^{2+}) were transported via a distinct exchanger to that of Ca^{2+}. Evidence for ΔpH-dependent Zn^{2+} exchange has also been detected in oat root tonoplast (Gonzalez *et al.*, 1999).

The cloning of a plant $Mg^{2+}(Zn)^{2+}/H^+$ antiporter was made possible by screening for an *A. thaliana* homolog of NCX1, a mammalian Na^+/Ca^{2+} exchanger (Shaul *et al.*, 1999). Although the gene, designated *AtMHX1*, showed 36% identity to NCX1 at the amino acid level and protein structural features were shared, functional studies detected no similarities. Functional analysis of AtMHX1 was undertaken by overexpressing the gene in transformed tobacco and employing the patch-clamp technique on giant membrane patches of isolated vacuoles. Large inward currents, dependent upon the presence of Mg^{2+} on the cytoplasmic side, were activated by acidification of the lumenal side of the membrane patch in vacuoles isolated from *AtMHX1*-transformed tobacco plants (Shaul *et al.*, 1999). Exchange of Ca^{2+} for Mg^{2+} resulted in no current activation, however Zn^{2+} and Fe^{2+} were able to substitute for Mg^{2+}, suggesting electrogenic $Mg^{2+}(Zn^{2+}; Fe^{2+})/H^+$ exchange. Inclusion of Ca^{2+} (50 μM to 2 mM) on the cytoplasmic side of the membrane patch stimulated the outward current indicative of a regulatory role for Ca^{2+} on antiport activity (Shaul *et al.*, 1999). Independent evidence supported a tonoplast location for AtMHX1; protoplasts from *AtMHX1*-transformed tobacco plants showed no Mg^{2+}-induced electrogenic currents, and membrane fractionation studies of *A. thaliana* root membranes, using AtMHX1 peptide antibodies, localized AtMHX1 to the tonoplast fraction (Shaul *et al.*, 1999). *In situ* hybridization of AtMHX1 mRNA indicated expression in regions bordering xylem tracheary elements, suggesting a possible role in Mg^{2+}/Zn^{2+} partitioning (Shaul *et al.*, 1999).

9.4 Conclusions

Substantial progress has been made on the characterization of tonoplast ion channels and ion co-transporters through application of a diverse range of techniques. However, it is clear that, while success at the molecular level has been achieved by the cloning of ion co-transporters, there has so far been no report on the cloning of tonoplast ion channels. One obvious area of focus is to use the information gained from the genome sequencing projects to identify possible channel homologs. On the other hand, we may have to return to a more traditional biochemical approach for tonoplast channel protein identification.

Acknowledgments

The authors acknowledge financial support from DGAPA-UNAM and CONACyT.

References

Alexandre, J., Lasalles, J. and Kado, R. (1990) Opening of Ca^{2+} channels in isolated red beet root vacuole membrane by inositol 1,4,5-trisphosphate. *Nature*, **343** 567-70.

Allen, G.J. and Sanders, D. (1996) Control of ionic currents in guard cell vacuoles by cytosolic and luminal calcium. *Plant J.*, **10** 1055-69.

Allen, G.J., Muir, S.R. and Sanders, D. (1995) Release of Ca^{2+} from individual plant vacuoles by both $InsP_3$ and cyclic ADP-ribose. *Science*, **268** 735-37.

Amalou, Z., Gibrat, R., Brugidou, C., Trouslot, P. and d'Auzac, J. (1992) Evidence for an amiloride-inhibited $Mg^{2+}/2H^+$ antiporter in lutoid (vacuolar) vesicles from latex of *Hevea brasiliensis*. *Plant Physiol.*, **100** 255-60.

Amalou, Z., Gibrat, R., Trouslot, P. and d'Auzac, J. (1994) Solubilization and reconstitution of the $Mg^{2+}/2H^+$ antiporter of the lutoid tonoplast from *Hevea brasiliensis*. *Plant Physiol.*, **106** 79-85.

Amodeo, G., Escobar, A. and Zeiger, E. (1994) A cationic channel in the guard cell tonoplast of *Allium cepa*. *Plant Physiol.*, **105** 999-1006.

Apse, M.P., Aharon, G.S., Snedden, W.A. and Blumwald, E. (1999) Salt tolerance conferred by overexpression of a vacuolar Na^+/H^+ antiport in *Arabidopsis*. *Science*, **285** 1256-58.

Ballesteros, E., Blumwald, E., Donaire, J.P. and Belver, A. (1997) Na^+/H^+ antiport activity in tonoplast vesicles isolated from sunflower roots induced by NaCl stress. *Physiol. Plant.*, **99** 328-34.

Barkla, B.J. and Blumwald, E. (1991) Identification of a 170 kDa protein associated with the vacuolar Na^+/H^+ antiport of *Beta vulgaris*. *Proc. Natl. Acad. Sci. USA*, **88** 11177-81.

Barkla, B.J. and Pantoja, O. (1996) Physiology of ion transport across the tonoplast of higher plant cells. *Annu. Rev. Plant Physiol. Plant Mol. Biol.*, **47** 159-84.

Barkla, B.J., Zingarelli, L., Blumwald, E. and Smith, J.A.C. (1995) Tonoplast Na^+/H^+ antiport activity and its energization by the vacuolar H^+-ATPase in the halophytic plant *Mesembryanthemum crystallinum* L. *Plant Physiol.*, **109** 549-56.

Barkla, B.J., Vera-Estrella, R., Maldonado-Gamma, M. and Pantoja, O. (1999) Abscisic acid induction of vacuolar H^+-ATPase activity in *Mesembryanthemum crystallinum* is developmentally regulated. *Plant Physiol.*, **120** 811-19.

Blackford, S., Rea, P.A. and Sanders, D. (1990) Voltage sensitivity of H^+/Ca^{2+} antiport in higher plants tonoplast suggests a role in vacuolar calcium accumulation. *J. Biol. Chem.*, **265** 9617-20.

Blumwald, E. and Poole, R. (1985) Na^+/H^+ antiport in isolated tonoplast vesicles from storage tissue of *Beta vulgaris*. *Plant Physiol.*, **78** 163-67.

Blumwald, E. and Poole, R. (1986) Kinetics of Ca^{2+}/H^+ antiport in isolated tonoplast vesicles from storage tissue of *Beta vulgaris* L. *Plant Physiol.*, **80** 727-31.

Blumwald, E., Cragoe Jr. E.J. and Poole, R. (1987) Inhibition of Na^+/H^+ antiport activity in sugar beet tonoplast by analogs of amiloride. *Plant Physiol.*, **85** 30-33.

Bowler, C., Neuhaus, G., Yamagata, H. and Chua, N.-H. (1994) Cyclic GMP and calcium mediate phytochrome phototransduction. *Cell*, **77** 73-81.

Bray, E.A. (1997) Plant responses to water deficit. *Trends Plant Sci.*, **2** 48-54.

Bressan, R.A., Hasegawa, P.M. and Pardo, J.M. (1998) Plants use calcium to resolve salt stress. *Res. News*, **3** 411-12.

Cerana, R., Giromini, L. and Colombo, R. (1995) Malate-regulated channels permeable to anions in vacuoles of *Arabidopsis thaliana*. *Aust. J. Plant Physiol.*, **22** 115-21.

Chanson, A. (1991) A Ca^{2+}/H^+ antiport system driven by the tonoplast pyrophosphate-dependent proton transport of maize root membranes. *Plant Physiol.*, **84** 1431-36.

Cheffings, C.M., Pantoja, O., Ashcroft, F.M. and Smith, J.A.C. (1997) Malate transport and vacuolar ion channels in CAM plants. *J. Exp. Bot.*, **48** 623-31.

Counillon, L. and Pouysségur, J. (2000) The expanding family of eucaryotic Na^+/H^+ exchangers. *J. Biol. Chem.*, **275** 1-4.

Coyaud, L., Kurkdjian, A., Kado, R. and Hedrich, R. (1987) Ion channels and ATP-driven pumps involved in ion transport across the tonoplast of sugarbeet vacuoles. *Biochim. Biophys. Acta*, **902** 263-68.

Cunningham, K.W. and Fink, G.R. (1994) Calcineurin-dependent growth control in *Saccharomyces cerevisiae* mutants lacking PMC1, a homolog of plasma membrane Ca^{2+}-ATPases. *J. Cell Biol.*, **124** 351-63.

Cunningham, K.W. and Fink, G.R. (1996) Calcineurin inhibits VCX1-dependent H^+/Ca^{2+} exchange and induces Ca^{2+}-ATPase in *Saccharomyces cerevisiae*. *Mol. Cell Biol.*, **16** 2226-37.

Cushman, J.C., Michalowski, C.B. and Bohnert, H.J. (1990) Developmental control of crassulacean acid metabolism inducibility by salt stress in the common ice plant. *Plant Physiol.*, **94** 1137-42.

Davies, J.M. and Sanders, D. (1995) ATP, pH and Mg^{2+} modulate a cation current in *Beta vulgaris* vacuoles: a possible shunt conductance for the vacuolar H^+-ATPase. *J. Membr. Biol.*, **145** 75-86.

Evans, D.E. and Williams, L.E. (1998) P-type calcium ATPases in higher plants: biochemical, molecular and functional properties. *Biochim. Biophys. Acta*, **1376** 1-25.

Fan, T.W.-M., Higashi, R., Norlyn, J. and Epstein, E. (1989) *In vivo* ^{23}Na and ^{31}P NMR measurements of a tonoplast Na^+/H^+ exchange process and its characteristics in two barley cultivars. *Proc. Natl. Acad. Sci. USA*, **86** 9856-60.

Felle, H. (1989a) Ca^{2+}-selective microelectrodes and their application to plant cells and tissues. *Plant Physiol.*, **91** 1239-42.

Felle, H. (1989b) pH as a second messenger in plants, in *Second Messengers in Plant Growth and Development* (eds. D.J. Morré and W.F. Boss), Allan R. Liss, Inc., pp. 145-66.

Flowers, T.J., Troke, P.F. and Yeo, A.R. (1977) The mechanism of salt tolerance in halophytes. *Annu. Rev. Plant Physiol.*, **28** 89-121.

Frommer, W.B., Ludewig, U. and Rentsch, D. (1999) Taking transgenic plants with a pinch of salt. *Science*, **285** 1222-23.

Fukada, A., Yazaki, Y., Ishikawa, T., Koike, S. and Tanaka, Y. (1998) Na^+/H^+ antiporter in tonoplast vesicles from rice roots. *Plant Cell Physiol.*, **39** 196-201.

Fukada, A., Nakamura, A. and Tanaka, Y. (1999) Molecular cloning and expression of the Na^+/H^+ exchanger gene in *Oryza sativa*. *Biochim. Biophy. Acta*, **1446** 149-55.

Galione, A. (1994) Cyclic ADP-ribose, the ADP-ribosyl cyclase pathway and calcium signaling. *Mol. Cell. Endocrinol.*, **98** 125-31.

Galione, A., Lee, H.C. and Busa, W.B. (1991) Ca^{2+}-induced Ca^{2+} release in sea urchin egg homogenates: modulation by cyclic ADP-ribose. *Science*, **253** 1143-46.

Galione, A., McDougall, A., Busa, W.B., Willmott, N.G., Gillot, I. and Whitaker, M. (1993) Redundant mechanisms of calcium-induced calcium release underlying calcium waves during fertilization of sea urchin eggs. *Science*, **261** 348-52.

Gambale, F., Bregante, M., Stragapede, F. and Cantu, A.M. (1996) Ionic channels of the sugar beet tonoplast are regulated by a multi-ion single-file permeation mechanism. *J. Membr. Biol.*, **154** 69-79.

Garbarino, J. and Dupont, F.M. (1988) NaCl induces a Na^+/H^+ antiport in tonoplast vesicles from barley roots. *Plant Physiol.*, **86** 231-36.

Garbarino, J. and Dupont, F.M. (1989) Rapid induction of Na^+/H^+ exchange activity in barley root tonoplast. *Plant Physiol.*, **89** 1-4.

Gaxiola, R.A., Rao, R., Sherman, A., Grisafi, P., Alper, S.L. and Fink, G.R. (1999) The *Arabidopsis thaliana* proton transporters, AtNhx1 and Avp1, can function in cation detoxification in yeast. *Proc. Natl. Acad. Sci. USA*, **96** 1480-85.

Glenn, E.P., Brown, J.J. and Blumwald, E. (1999) Salt tolerance and crop potential of halophytes. *Crit. Rev. Plant Sci.*, **18** 227-55.

Glick, D.L., Hellmich, M.R., Beushausen, S., Tempst, P., Bayley, H. and Strumwasser, F. (1991) Primary structure of a molluscan egg-specific NADase, a second-messenger enzyme. *Cell Reg.*, **2** 211-18.

Gonzalez, A., Koren'kov, V. and Wagner, G.J. (1999) A comparison of Zn, Mn, Cd and Ca transport mechanisms in oat root tonoplast vesicles. *Physiol. Plant.*, **106** 203-209.

Guern, J., Mathieu, Y., Kurkdjian, A., Manigault, P., Manigault, J., Gillet, B., Beloeil, J.-C. and Lallemand, J.-Y. (1989) Regulation of vacuolar pH of plant cells. II. A ^{31}P NMR study of the modifications of vacuolar pH in isolated vacuoles induced by proton pumping and cation/H^+ exchanges. *Plant Physiol.*, **89** 27-36.

Harris, C. and Fliegel, L. (1999) Amiloride and the Na^+/H^+ exchange protein: mechanism and significance of inhibition of the Na^+/H^+ exchanger. *Int. J. Mol. Med.*, **3** 315-21.

Hedrich, R. and Neher, E. (1987) Cytoplasmic calcium regulates voltage-dependent ion channels in plant vacuoles. *Nature*, **329** 833-36.

Hedrich, R., Flügge, U.I. and Fernandez, J.M. (1986) Patch-clamp studies of ion transport in isolated plant vacuoles. *FEBS Lett.*, **204** 228-32.

Hirschi, K.D. (1999) Expression of *Arabidopsis* CAX1 in tobacco: altered calcium homeostasis and increased stress sensitivity. *Plant Cell*, **11** 2113-22.

Hirschi, K.D., Zhen, R.-G., Cunningham, K.W., Rea, P.A. and Fink, G.R. (1996) CAX1, an H^+/Ca^{2+} antiporter from *Arabidopsis*. *Proc. Natl. Acad. Sci. USA*, **93** 8782-86.

Irving, H.R., Gehring, C.A. and Parish, R.W. (1992) Changes in cytosolic pH and calcium of guard cells precede stomatal movements. *Proc. Natl. Acad. Sci. USA*, **89** 1790-94.

Iwasaki, I., Arata, H., Kijima, H. and Nishimura, M. (1992) Two types of channels involved in the malate ion transport across the tonoplast of a crassulacean acid metabolism plant. *Plant Physiol.*, **98** 1494-97.

Knight, H. (2000) Calcium signaling during abiotic stress in plants. *Int. Rev. Cytol.*, **195** 269-324.

Knight, H., Trewavas, A.J. and Knight, M.R. (1996) Cold calcium signaling in *Arabidopsis* involves two cellular pools and a change in calcium signature after acclimation. *Plant Cell*, **8** 489-503.

Knight, M.R., Campbell, A.K., Smith, S.M. and Trewavas, A.J. (1991) Transgenic plant aequorin reports the effects of touch and cold shock and elicitors on cytoplasmic calcium. *Nature*, **353** 524-26.

Kudla, J., Xu, Q., Harter, K., Gruissem, W. and Luan, S. (1999) Genes for calcineurin-B-like proteins in *Arabidopsis* are differentially regulated by stress signals. *Proc. Natl. Acad. Sci. USA*, **96** 4718-23.

Leckie, C.P., McAinsh, M.R., Allen, G.J., Sanders, D. and Hetherington, A. (1998) Abscisic acid-induced stomatal closure mediated by cyclic ADP-ribose. *Proc. Natl. Acad. Sci. USA*, **95** 15837-42.

Linz, K.W. and Köhler, K. (1994) Vacuolar ion currents in the primitive green alga, *Eresmophaera viridis*: the electrical properties are suggestive of both the Characeae and higher plants. *Protoplasma*, **179** 34-45.

Liu, J. and Zhu, J.-K. (1998) A calcium sensor homolog required for plant salt tolerance. *Science*, **280** 1943-45.

Marigo, G., Bouyssou, H. and Laborie, D. (1988) Evidence for malate transport into vacuoles isolated from *Catharanthus roseus* cells. *Bot. Acta*, **101** 187-91.

Martinoia, E. and Rentsch, D. (1994) Malate compartmentation-responses to a complex metabolism. *Annu. Rev. Plant Physiol. Plant Mol. Biol.*, **45** 447-67.

Martinoia, E., Flügge, U.I., Kaiser, G., Heber, U. and Heldt, H.W. (1985) Energy-dependent uptake of malate into vacuoles isolated from barley mesophyll protoplasts. *Biochim. Biophys. Acta*, **806** 311-19.

Martinoia, E., Vogt, E., Rentsch, D. and Amrhein, N. (1991) Functional reconstitution of the malate carrier of barley mesophyll vacuoles in liposomes. *Biochim. Biophys. Acta*, **1062** 271-78.

Matoh, T., Ishikawa, T. and Takahashi, E. (1989) Collapse of ATP-induced pH gradient by sodium ions in microsomal membrane vesicles prepared from *Atriplex gmelini* leaves: possibility of Na^+/H^+ antiport. *Plant Physiol.*, **89** 180-83.

Mennen, H., Jacoby, B. and Marschner, H. (1990) Is sodium proton antiport ubiquitous in plant cells? *J. Plant Physiol.*, **137** 180-83.

Miseta, A., Kellermayer, R., Aiello, D.P., Fu, L. and Bedwell, D.M. (1999) The vacuolar Ca^{2+}/H^+ exchanger Vcx1p/Hum1p tightly controls cytosolic Ca^{2+} levels in *S. cerevisiae*. *FEBS Lett.*, **451** 132-36.

Nass, R. and Rao, R. (1998) Novel localization of a Na^+/H^+ exchanger in a late endosomal compartment of yeast: implications for vacuole biogenesis. *J. Biol. Chem.*, **273** 21054-60.

Nass, R., Cunningham, K.W. and Rao, R. (1997) Intracellular sequestration of sodium by a novel Na^+/H^+ exchanger in yeast is enhanced by mutations in the plasma membrane H^+-ATPase: insights into mechanisms of sodium tolerance. *J. Biol. Chem.*, **272** 26145-52.

Niemietz, C. and Willenbrink, J. (1985) The function of tonoplast ATPase in intact vacuoles of red beet is governed by direct and indirect ion effects. *Planta*, **166** 545-49.

Numata, M., Petrecca, K., Lake, N. and Orlowski, J. (1998) Identification of a mitochondrial Na^+/H^+ exchanger. *J. Biol. Chem.*, **273** 6951-59.

Pantoja, O., Dainty, J. and Blumwald, E. (1992a) Cytoplasmic chloride regulates cation channels in the vacuolar membrane of plant cells. *J. Membr. Biol.*, **125** 219-29.

Pantoja, O., Gelli, A. and Blumwald, E. (1992b) Characterization of vacuolar malate and K^+ channels under physiological conditions. *Plant Physiol.*, **100** 1137-41.

Pantoja, O., Gelli, A. and Blumwald, E. (1992c) Voltage-dependent calcium channels in plant vacuoles. *Science*, **255** 1567-70.

Pardo, J.M., Reddy, M.P., Yang, S., Maggio, A., Huh, G.-H., Matsumoto, T., Coca, M.A., Paino-D'Urzo, M., Koiwa, H., Yun, D.-J., Watad, A.A., Bressan, R.A. and Hasegawa, P.M. (1998) Stress signaling through Ca^{2+}/calmodulin-dependent protein phosphatase calcineurin mediates salt adaptation in plants. *Proc. Natl. Acad. Sci. USA*, **95** 9681-86.

Pei, Z.-M., Ward, J.M., Harper, J.F. and Schroeder, J.I. (1998) A novel chloride channel in *Vicia faba* guard cell vacuoles activated by the serine/threonine kinase, CDPK. *EMBO J.*, **15** 6564-74.

Pei, Z.-M., Ward, J.M. and Schroeder, J.I. (1999) Magnesium sensitizes slow vacuolar channels to physiological cytosolic calcium and inhibits fast vacuolar channels in fava bean guard cell vacuoles. *Plant Physiol.*, **121** 977-86.

Pfeiffer, W. and Hager, A. (1993) A Ca^{2+}-ATPase and a Mg^{2+}/H^+ antiporter are present on tonoplast membranes from roots of *Zea mays* L. *Planta*, **191** 377-85.

Poovaiah, B.W. and Reddy, A.S.N. (1993) Calcium and signal transduction in plants. *Crit. Rev. Plant Sci.*, **12** 185-211.

Pottosin, I.I., Tikhonova, L.I., Hedrich, R. and Schonknecht, G. (1997) Slowly activating vacuolar channels can not mediate Ca^{2+}-induced Ca^{2+} release. *Plant J.*, **12** 1387-98.

Pottosin, I.I., Dobrovinskaya, O.R. and Muñíz, J. (1999) Cooperative block of the plant endomembrane ion channel by ruthenium red. *Biophys. J.*, **77** 1973-79.

Pozos, T., Sekler, I. and Cyert, M. (1996) The product of *HUM1*, a novel yeast gene, is required for vacuolar Ca^{2+}/H^+ exchange and is related to mammalian Na^+/Ca^{2+} exchangers. *Mol. Cell Biol.*, **16** 3730-41.

Ratajczak, R., Kemna, I. and Lüttge, U. (1994) Characteristics, partial purification and reconstitution of the vacuolar malate transporter of the CAM plant *Kalanchoë daigremontiana* Hamet et Perrier de la Bathie. *Planta*, **195** 226-36.

Rentsch, D. and Martinoia, E. (1991) Citrate transport into barley mesophyll vacuoles: comparison with malate-uptake activity. *Planta*, **184** 532-37.

Rusinko, N. and Lee, H.C. (1989) Widespread occurrence in animal tissues of an enzyme catalyzing the conversion of NAD^+ into a cyclic metabolite with intracellular Ca^{2+}-mobilizing activity. *J. Biol. Chem.*, **264** 11725-31.

Salt, D.E. and Wagner, G.J. (1993) Cadmium transport across tonoplast vesicles from oat roots: evidence for a Cd^{2+}/H^+ antiport activity. *J. Biol. Chem.*, **268** 12297-302.

Sanders, D., Brownlee, C. and Harper, J.F. (1999) Communicating with calcium. *Plant Cell*, **11** 691-706.

Schulz-Lessdorf, B. and Hedrich, R. (1995) Protons and calcium modulate SV-type channels in the vacuolar-lysosomal compartment: channel interaction with calmodulin inhibitors. *Planta*, **197** 655-71.

Schumaker, K.S. and Sze, H. (1985) A Ca^{2+}/H^+ system driven by the proton electrochemical gradient of a tonoplast H^+-ATPase from oat roots. *Plant Physiol.*, **79** 1111-17.

Shaul, O., Hilgemann, D.W., de-Almeida-Engler, J., Van Montagu, M., Inzé, D. and Galili, G. (1999) Cloning and characterization of a novel Mg^{2+}/H^+ exchanger. *EMBO J.*, **18** 3973-80.

Staal, M., Maathuis, F.J.M., Elzenga, T.M., Overbeek, H.M. and Prins, H.B.A. (1991) Na^+/H^+ antiport activity in tonoplast vesicles from roots of the salt-tolerant *Plantago maritima* and the salt-sensitive *Plantago media*. *Physiol. Plant.*, **82** 179-84.

Terry, B.R., Findlay, G.P. and Tyerman, S.D. (1992) Direct effects of Ca^{2+}-channel blockers on plasma membrane cation channels of *Amaranthus tricolor* protoplasts. *J. Exp. Bot.*, **43** 1457-73.

Tikhonova, L.I., Pottosin, I.I., Dietz, K.-J. and Schonknecht, G. (1997) Fast-activating cation channel in barley mesophyll vacuoles: inhibition by calcium. *Plant J.*, **11** 1059-70.

Ueoka-Nakanishi, H., Nakanishi, Y., Tanaka, Y. and Maeshima, M. (1999) Properties and molecular cloning of Ca^{2+}/H^+ antiporter in the vacuolar membrane of mung bean. *Eur. J. Biochem.*, **262** 417-25.

Vögeli-Lange, R. and Wagner, G.J. (1990) Subcellular localization of cadmium and cadmium-binding peptides in tobacco leaves: implication of a transport function for cadmium-binding peptides. *Plant Physiol.*, **92** 1086-93.

Wakabayashi, S., Shigekawa, M. and Pouyssegur, J. (1997) Molecular physiology of vertebrate Na^+/H^+ exchangers. *Physiol. Rev.*, **77** 51-74.

Ward, J.M. and Schroeder, J.I. (1994) Calcium-activated K^+ channels and calcium-induced calcium release by slow vacuolar ion channels in guard cell vacuoles implicated in the control of stomatal closure. *Plant Cell*, **6** 669-83.

Ward, J.M., Pei, Z.-M. and Schroeder, J.I. (1995) Roles of ion channels in initiation of signal transduction in higher plants. *Plant Cell*, **7** 833-44.

White, P.J. and Smith, J.A.C. (1989) Proton and anion transport at the tonoplast in crassulacean acid metabolism plants: specificity of the malate influx system of *Kalanchoë daigremontiana*. *Planta*, **179** 265-74.

Wu, Y., Kuzma, J., Marechal, E., Graeff, R., Lee, H., Foster, R. and Chua, N.-H. (1997) Abscisic acid signaling through cyclic ADP-ribose in plants. *Science*, **278** 2126-30.

10 Vacuolar transport of secondary metabolites and xenobiotics

E. Martinoia, M. Klein, M. Geisler,
R. Sánchez-Fernández and P.A. Rea

10.1 Introduction

This review is concerned with the mechanisms responsible for the vacuolar sequestration of secondary metabolites and xenobiotics, with special regard to the ways in which these compounds are modified before uptake, the types of transporters involved in their passage across the vacuolar membrane and the processes that promote their retention after uptake into the vacuole.

The relevance of vacuolar sequestration for a general understanding of plant function is evident at several levels. Firstly, the chemical diversity of plants is largely attributable to interspecies differences in secondary metabolite composition achieved by the modification of a relatively limited number of core structures. A reflection of this is the fact that some 50,000 different secondary metabolites have been identified in plants, and in an organism such as *Arabidopsis thaliana* two of the key classes of enzyme involved in secondary metabolism, cytochrome P450-dependent monooxygenases and glucosyltransferases, are coded by more than 300 and 80 genes, respectively. Secondly, plant secondary metabolites are implicated in a multitude of processes (Harborne, 1993; Dixon and Paiva, 1995). Examples include: the elaboration of lignins as rigid polymers for mechanical support; the production of herbivore antifeedants, antifungal agents and UV light screens; the secretion of flavonoids by roots for the attraction of symbiotic bacteria and modification of the spectrum of microorganisms associated with the rhizosphere; and the synthesis of pigments and fragrances to direct and promote plant-herbivore interactions (Turlings and Benrey, 1998). Thirdly, most secondary metabolites are toxic to the plant itself and necessitate effective mechanisms for detoxification and/or compartmentation. Secondary metabolites, such as phenolics, are active in their native state and inactivate enzymes and permeabilize membranes. Others, such as the *Sorghum* cyanogenic glucoside, dhurrin, are innocuous in their native state but are rapidly converted to toxic derivatives as a result of tissue damage (Kojima *et al.*, 1979). Fourthly, many secondary metabolites must be accumulated to relatively high levels to fulfil their functions: they must be stored in a compartment of large volume and/or high capacity. The UV light-protective and insect-attractive actions of flavonoids and anthocyanins necessitate epidermal or subepidermal concentrations in the millimolar range. The cyanogenic glucoside, dhurrin, accounts for as much as 30% of the dry weight of sorghum (Saunders and Conn, 1978; Kojima *et al.*, 1979), which, because of its restriction to the epidermis, represents cellular concentrations in excess of 300 mM. Compartmentation of the toxic compound itself or of the enzymes that convert its nontoxic precursors to toxic derivatives either in the vacuole or apoplast is imperative if plants cells are not to poison themselves (Matile, 1987; Wink, 1997). Finally, the cellular machinery required for the modification of secondary metabolites also participates in xenobiotic detoxification. The main enzyme classes involved in secondary metabolism—cytochrome P450-dependent monooxygenases and glutathione, glucose, glucuronate, malonate and sulfate transferases—also play a central role in xenobiotic detoxification.

The diversity, biochemical versatility, inherent toxicity and abundance of secondary metabolites in plants makes the mechanisms by which they are accumulated and sequestered from other cellular components of paramount importance. The fact that the cellular machinery responsible for the detoxification of xenobiotics is analogous to, if not equivalent to, that responsible for the modification and accumulation of secondary metabolites means that insights gained into the mechanistic basis of one process will probably have implications for our understanding of the other.

10.2 Common pathways for detoxification

The detoxification of exogenous and endogenous cytotoxins is a four-phase process (Ishikawa, 1992; Sanderman, 1992; Kreuz *et al.*, 1996; Ishikawa *et al.*, 1997) (Figure 10.1). Phase I (activation) is the introduction or exposure of functional groups of the appropriate reactivity for phase II enzymes. Cytochrome P450-dependent monooxygenases and mixed function oxidases, responsible for the hydroxylation of aromatic rings, alkyl groups or heteroatom release, are examples of phase I enzymes. Phase II (conjugation) is covalent attachment of the activated derivative to glutathione, glucose, glucuronic acid or sulfate by glutathione *S*-transferases, glucosyltransferases, glucuronyltransferases or sulfotransferases, respectively. Phase III (elimination) is transport of the conjugates out of the cytosol into intracellular compartments and/or the extracellular space. Phase IV (transformation) comprises the processes responsible for the further substitution, degradation and/or partial salvage of the transported conjugates. Whether the target compound is a secondary metabolite or xenobiotic, the underlying principle is that the conjugation reaction increases the hydrophilicity and/or net charge of the compound, while introducing a specific chemical tag for recognition and transport by vacuolar and/or other conjugate

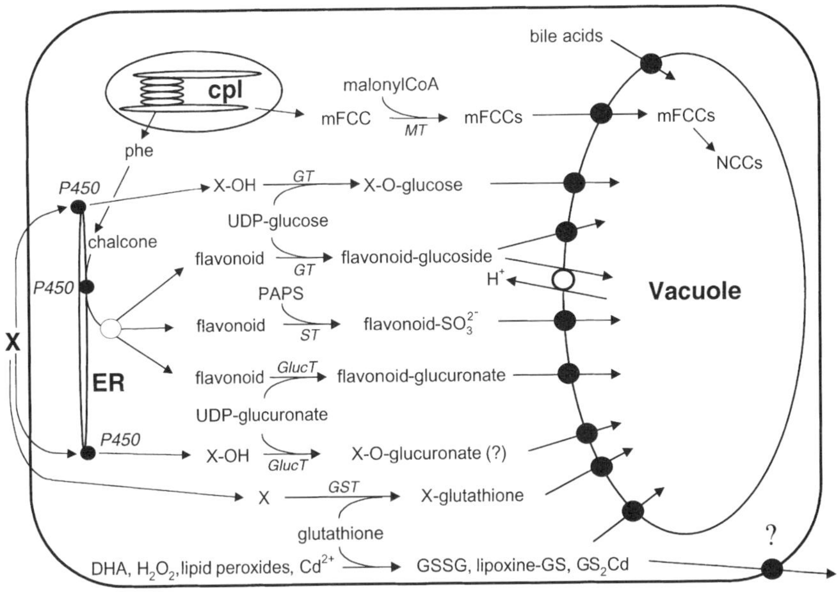

Figure 10.1 Examples of secondary metabolites and xenobiotics destined for transport into the vacuole. The terminal stages of secondary compound metabolism and xenobiotic (X) detoxification are depicted. Abbreviations: cpl, chloroplast; ER, endoplasmic reticulum; *P450*, cytochrome P450-dependent monooxygenases; *MT*, malonyltransferases; *GT*, glucosyltransferases; *ST*, sulfotransferases; *GlucT*, glucuronyl transferases; *GST*: glutathione *S*-transferases; mFCC, modified fluorescent chlorophyll catabolites; NCC, nonfluorescent chlorophyll catabolites; DHA: dehydroascorbate (A); UDP, uridine 5′-diphosphate; GSSG, oxidised glutathione; PAPS, adenosine 3′-phosphoadenosine-5′-phosphosulfate.

transporters (Ishikawa, 1992; Sanderman, 1992; Kreuz *et al.*, 1996; Ishikawa *et al.*, 1997).

Having drawn these parallels, two points should be clarified. The first is that, whilst it is clear that the same classes of enzymes are involved, the participation of a particular phase I, II or III enzyme in both secondary metabolism and xenobiotic detoxification has yet to be demonstrated (Zimmerlin and Durst, 1992). The seemingly broad and overlapping substrate specificities of the enzymes concerned often prohibit ascription of an enzyme to a particular reaction *in vivo*, even if the enzyme is capable of catalyzing the reaction in question *in vitro*. The second is that because so little is known of the enzymology of phase IV, other than that both secondary metabolites (Sharma and Strack, 1985) and xenobiotics (Wolf *et al.*, 1996) often undergo modification after delivery into the vacuole, any parallels that can be drawn for this phase are necessarily tentative.

10.3 Secondary energized transporters and ABC transporters

10.3.1 Strategies for vacuolar accumulation of secondary metabolites

Four principal mechanisms for the vacuolar accumulation of secondary metabolites have been described: H^+-antiport and electrogenic uniport; acid or ion trapping; conformational trapping; and direct energization by MgATP (Figure 10.1). Of these, the first three mechanisms depend on a preexistent H^+-electrochemical potential difference across the vacuolar membrane, established by the vacuolar-type H^+-transporting adenosine triphosphatase (V-ATPase) and vacuolar-type H^+-transporting pyrophosphatase (V-PPase) (Rea and Sanders, 1987; Rea and Poole, 1993; Zhen *et al.*, 1997; Sze *et al.*, 1999). Both pumps catalyze electrogenic H^+-translocation from the cytosol into the vacuole to generate an inside-acid H^+ chemical potential difference (ΔpH) and inside-positive electrical potential difference ($\Delta\psi$). ΔpH typically has a value of 1–2 units but in some systems, such as crassulacean acid metabolism (CAM) plants after dark fixation (Lüttge, 1987) or grape berries (Terrier *et al.*, 1998), for which vacuolar pH values of 3 and less than 2.5 units have been estimated, ΔpH values in excess of 5 units can be achieved. $\Delta\psi$ generally has a value of between 20 and 50 mV (cytosol negative) (Rea and Sanders, 1987).

Depending on the secondary metabolite concerned, the H^+-electrochemical potential difference may drive uptake in several ways. Cationic metabolites, for example strongly basic alkaloids, may undergo exchange for luminal H^+ in response to the inside-acid ΔpH (H^+-antiport), while anionic metabolites, for example carboxylates, may undergo electrophoretic uptake in response to the inside-positive $\Delta\psi$ (electrogenic uniport). Alternatively, or in addition, the metabolite may passively equilibrate across the vacuolar membrane but be luminally restricted as a result of a change in its dissociation and/or conformational state at the low pH values that prevail within the vacuole. Lipophilic, weakly basic alkaloids may be subject to the former, 'ion trap', mechanism (Matile *et al.*, 1970; Guern *et al.*, 1987). Compounds capable of assuming different conformational states depending on pH may be subject to accumulation by the latter, 'conformational' or 'isomer' trap, mechanism (Matern *et al.*, 1983; Rataboul *et al.*, 1985).

H^+-coupled transport and acid trapping are reasonably well established as mechanisms for the vacuolar sequestration of some secondary metabolites and xenobiotics but the full significance of conformational trapping remains to be determined. Based on the finding that some secondary metabolites, such as apigenin 7-*O*-(6-*O*-malonylglucoside) (Matern *et al.*, 1983) and coumaric acid glucoside (Rataboul *et al.*, 1985), isomerize when acidified, the concept is that only the higher pH (cytosolic) isomer is capable of acting as a substrate for transport. In the case of apigenin 7-*O*-(6-*O*-malonylglucoside), the malonylated glucose residue shifts its position with respect to the planar C15 core structure at low pHs to yield an isomer that cannot exit the vacuole on the same transporter as it entered (Matern *et al.*, 1983). In the case of coumaric acid glucoside, low pH values promote conversion of the transport-active *trans*-isomer to the transport-inactive *cis*-isomer (Rataboul *et al.*, 1985).

The one known mechanism for the accumulation of secondary metabolites that is not obligatorily dependent on the acidity of the vacuole is the most recently discovered: MgATP-energized transport by ATP-binding cassette (ABC) transporters (Rea *et al.*, 1998; Rea, 1999). First defined in the context of xenobiotic detoxification (Martinoia *et al.*, 1993; Li *et al.*, 1995a), it is now clear that many plant secondary metabolites are also amenable to vacuolar accumulation by ABC transporters.

The salient properties of ABC transporter-catalyzed vacuolar uptake are: i) Direct energization by MgATP—uptake is driven by MgATP but not by free ATP or non-hydrolyzable ATP analogs, such as 5′-(β, γ-imino)triphosphate (AMP-PNP). Other nucleoside triphosphates, for example guanosine 5′-triphosphate (GTP) or uridine 5′-triphosphate (UTP), can partially substitute for ATP but inorganic pyrophosphate, the primary energy source for the V-PPase, cannot. ii) Insensitivity to the transmembrane H^+-electrochemical potential difference—abolition of the inside-acid ΔpH, that would otherwise be established by the V-ATPase and/or V-PPase, through the addition of protonophores, ionophores or H^+-phosphohydrolase-specific inhibitors, such as the potent V-ATPase inhibitor, bafilomycin A_1 (Bowman *et al.*, 1988), does not inhibit uptake. iii) Exquisite sensitivity to vanadate–micromolar concentrations of the phosphoryl transition-state inhibitor, vanadate, a compound employed to inhibit transporters that form an acylphosphate during their catalytic cycles, strongly inhibit uptake.

In outlining the differences between these mechanisms, it should be appreciated that they are not necessarily mutually exclusive. For instance, uniport may be a prelude to ion trapping and MgATP-dependent uptake may be a prelude to conformational trapping, as appears to be the case for coumaric acid glucoside (Alibert *et al.*, 1982). By the same token, these mechanisms do not preclude other processes that may contribute to the maintenance of high vacuolar secondary metabolite levels. Subsidiary processes, such as the complexation of anthocyanins by metal cations or the cooperative stacking of anthocyanins and flavonoids ('co-pigmentation') (Dangles and Brouillard, 1992), may also contribute to the vacuolar retention of secondary metabolites.

10.3.2 Vacuolar uptake of secondary metabolites and conjugated xenobiotics

10.3.2.1 Alkaloids

Two mechanisms have been proposed to explain vacuolar alkaloid accumulation. In the first, the acid trap hypothesis, accumulation is attributed to the passive diffusion of alkaloid across the vacuolar membrane in its lipophilic, neutral form, followed by intravacuolar protonation and trapping of the alkaloid as its lipid-insoluble, cationic, conjugate acid (reviewed in Matile, 1987). In the second model, the alkaloid is envisaged to undergo carrier-mediated H^+-antiport across the vacuolar membrane in response to the H^+-electrochemical potential difference generated by the vacuolar H^+ pumps (Deus-Neumann and Zenk, 1984, 1986; Homeyer and Roberts, 1984; Mende and Wink, 1987; Roberts *et al.*, 1991).

On the one hand, both the stereospecificity and energy requirements of several alkaloid transport processes are consistent with H^+-antiport. (S)-reticulin but not (R)-reticulin is taken up at high affinity ($K_m = 0.2$–$2.5\ \mu M$) by vacuoles of *Fumaria capreolata* (Deus-Neumann and Zenk, 1986); lupanin uptake by vacuoles from *Lupinus polyphyllus* can be resolved into high affinity, saturable and low affinity, linear components (Mende and Wink, 1987); and, while its *N*-oxide and senecione itself are taken up by cell suspension cultures of *Senecio vulgaris*, vacuoles from preloaded protoplasts retain only the plant-specific compound, senecione *N*-oxide (Ehmke *et al.*, 1988). In the case of (S)-reticulin uptake in *Fumaria*, lupanin uptake in *Lupinus*, and morphine uptake into *Papaver somniferum* latex vacuoles, the stimulatory action of MgATP is abolished by protonophores and NH_4^+ (Homeyer and Roberts, 1984; Deus-Neumann and Zenk, 1986; Mende and Wink, 1987). Observations on isolated intact vacuoles to the effect that uptake of alkaloids, such as morphine, is not stimulated by MgATP (Roberts *et al.*, 1991) may be attributable to the small surface area:volume ratio of these organelles and their high capacity for maintaining transmembrane pH differences, even under non-energized conditions. On the other hand, while these findings point to the existence of stereospecific alkaloid transporters, they do not disprove the participation of ion trap mechanisms for all alkaloids because of the diverse stereochemical, acid-base and lipid-solubility characteristics of this class of compounds and the probability that transport across the vacuolar membrane and retention within the vacuole are complementary processes. Moreover, despite the implied involvement of H^+-alkaloid antiport in stereospecific uptake, none of the putative transport elements have been identified at the protein level either chemically or molecularly, and in some cases the precise role of the H^+-electrochemical gradient has not been defined. For example, inclusion of KCl in the uptake medium has been found to increase the rates of alkaloid uptake by 7–22-fold (Deus-Neumann and Zenk, 1984; Mende and Wink, 1987), but it is not known if this effect results from modification of $\Delta\psi$, stimulation of the V-ATPase by Cl^- (Rea and Sanders, 1987), or is indicative of a role for K^+ as an active component in ion trapping.

One challenge posed by the intracellular transport of alkaloids in plants is the chemical diversity of the compounds involved and the observation that in some systems, for example latex vacuoles from *Papaver* and *Chelidonium majus*, steady state alkaloid concentrations as high as 200–500 mM have been reported (Hauser and Wink, 1990; Roberts *et al.*, 1991). Mammalian cells appear to have overcome the problem of alkaloid diversity by the utilization of ABC transporters, multidrug resistance (MDR) proteins, with extraordinarily broad substrate specificities (Endicott and Ling, 1989). It has, therefore, been suggested that because of their possession of MDR-like genes, plants may deploy similar mechanisms for dealing with the same problem. Attractive as this hypothesis is, all attempts to measure MgATP-energized, protonophore-insensitive alkaloid uptake into vacuoles or vacuolar membrane vesicles have yielded negative results (Hörtensteiner and Martinoia, unpublished; Li and Rea, unpublished), suggesting that if the MDR-like transporters of plants have similar transport capabilities to those of mammals, they are not on the vacuolar membrane.

10.3.2.2 Glucosylated secondary metabolites and xenobiotics

Most of the products of secondary metabolism in plants are glycosylated (Harborne, 1993), as are many herbicides after their initial modification by cytochrome P450-dependent monooxygenases (Kreuz *et al.*, 1996) (Figure 10.1). An impressive array of glycosylated secondary metabolites, including coumaryl glucosides, flavonoids, anthocyanins, cardenolides, saponins, cyanogenic glucosides, glucosinolates and betalins (for extensive list, see Wink, 1997) are known to be stored within the vacuole, while their respective uridine 5′-diphosphate (UDP)-glucose transferases have a cytosolic localization. It is therefore generally assumed that these bulky amphipathic compounds undergo carrier-mediated uptake into the vacuole. Only in a few cases, however, have the transport mechanisms involved been investigated. Matern *et al.* (1986) and Hopp and Seitz (1987) have shown that acylation is a prerequisite for vacuolar uptake of the *Daucus carota* anthocyanin and apigenin 7-*O*-(6-*O*-malonylglucoside) from parsley, but in both cases the mechanism has not been defined. A similar acylation-dependent isomerization has also been proposed for the vacuolar uptake of coumaric acid glucoside (Rataboul *et al.*, 1985). Although these authors suggest that accumulation is contingent on a pH-dependent change in the conformation of the acylated glucoside because abolition of the transmembrane ΔpH markedly inhibits uptake, the alternative or supplementary possibility of antiport has not been excluded.

Comparisons of the energy requirements for uptake of isovitexin, a flavone *C*-glucoside derived from apigenin (Figure 10.2) present in minor amounts in barley, and hydroxyprimisulfuron-glucoside (HPS-glucoside) (Figure 10.2), a sulfonylurea herbicide *O*-glucoside, by barley vacuoles implicate two transport mechanisms for glucosides: direct energization by MgATP and H^+-antiport (Klein *et al.*, 1996). Uptake of the endogenous glucoside, isovitexin, is postulated to proceed by H^+-antiport. Transport of isovitexin is stereospecific, in that orientin, a luteolin derivative glucosylated at the C8 position of the A ring

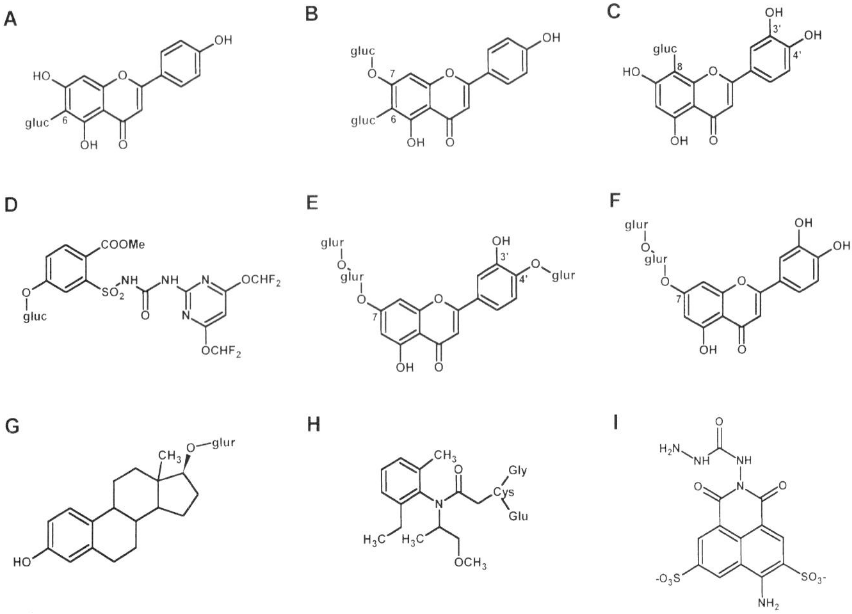

Figure 10.2 Chemical structures of some of the secondary metabolites and conjugated xenobiotics described in this review. The structures shown are those of typical glucosides (A–D), glucuronides (E–G), glutathione derivatives (H) and sulfonated derivatives (I) used in vacuolar uptake experiments. A) isovitexin (apigenin 6-*C*-glucoside); B) saponarin (apigenin 6-*C*-glucoside-(7-*O*-glucoside)); C) orientin (luteolin 8-*C*-glucoside); D) hydroxyprimisulfuron glucoside; E) luteolin 7-*O*-diglucuronyl-(4′-*O*-glucuronide); F) luteolin 7-*O*-diglucuronide; G) β-estradiol 17-(β-D-glucuronide); H) metolachlor-glutathione; I) lucifer yellow CH.

and differentially hydroxylated on the B ring (Figure 10.2), is not taken up. Bafilomycin A_1 and protonophores abolish uptake, as would be expected for a ΔpH-driven process, yet isovitexin is not amenable to acid-promoted conformational changes because it lacks acyl side-chains. The results from preliminary measurements of uptake of saponarin (Figure 10.2), the principal flavonoid of barley, which is derived from isovitexin by *O*-glucosylation at position 7 of the A ring, indicate that the monoglucoside and the diglucoside are recognized by the same transporter (Frangne, Weissenböck, Martinoia and Klein, unpublished).

An intriguing feature of the barley glucoside uptake system is that while barley vacuoles contain both apigenin and luteolin derivatives, levels of the latter are appreciable only after the exposure of primary leaves to UV light (Reuber *et al.*, 1996). Lack of luteolin transport by vacuoles isolated from plants grown under a normal lighting regime may, therefore, reflect a requirement for synthesis of endogenous substrate for activation and/or induction of the luteolin transport pathway. Alternatively, luteolin derivatives and their respective vacuolar carrier(s) may be restricted to the epidermis.

In direct contrast to the results obtained with isovitexin and related compounds, uptake of HPS-glucoside is subject to direct energization by MgATP (Klein *et al.*, 1996). HPS-glucoside uptake is more than 60% insensitive to bafilomycin A_1 and protonophores, but is inhibited by vanadate in a manner consistent with the direct participation of an ABC transporter. Further experiments with a broader range of glucosides are required before firm conclusions can be drawn, but there is a possibility that simple glycosylated phenolics and endogenous flavonoid-glucosides enter the vacuole by H^+-antiport, while complex glucosides, such as glucosylated xenobiotics, are either not transported into the vacuole or enter via ABC transporters.

10.3.2.3 Glutathione conjugates

Recognition that plants contain transporters that are directly energized by MgATP (Martinoia *et al.*, 1993; Li *et al.*, 1995a) came from investigations of the uptake of two model glutathione (GS-) conjugates, *N*-ethylmaleimide-GS and *S*-(2,4-dinitrophenyl)-GS (DNP-GS), and of the glutathionated chloroacetanilide herbicide, metolachlor (metolachlor-GS) (Figure 10.2) by isolated vacuoles and vacuolar membrane vesicles. So far, two major classes of candidate endogenous substrates for the vacuolar GS-conjugate pump have been proposed: anthocyanins and isoflavonoid phytoalexins (Rea *et al.*, 1998).

It has been known for some time that the bronze coloration of maize *Bronze-2* (*bz2*) mutants is due to accumulation of anthocyanin, cyanidin-3-glucoside (C3G), derivatives in the cytosol. In wild-type plants, anthocyanins are accumulated in the vacuole as purple or red derivatives but in *bz2* mutants they are restricted to the cytosol, where they undergo oxidation and cross-linking to generate brown products. However, the biochemical basis of this lesion was not known. Incisive, therefore, are experiments showing that *Bz2* encodes a type III glutathione *S*-transferase capable of conjugating C3G with glutathione (Marrs *et al.*, 1995). 1-Chloro-dinitrobenzene-conjugating glutathione *S*-transferase activity is diminished in *bz2* versus *Bz2* plants, and transient expression of *Bz2* results in increased extractable glutathione *S*-transferase activity concomitant with an enhancement of vacuolar anthocyanin accumulation (Marrs *et al.*, 1995). Hence, a rational basis for the *bz2* phenotype follows. Being defective in the glutathionation of anthocyanins, *bz2* mutants are unable to pump these pigments from the cytosol into the vacuole via the vacuolar GS-conjugate pump. In agreement with this notion, exposure of wild-type protoplasts to the potent GS-conjugate pump inhibitor, vanadate, phenocopies the *bz2* mutation (Marrs *et al.*, 1995) and, of the various GS-conjugates tested as *in vitro* substrates for the vacuolar GS-conjugate pump, C3G-GS is one of the most efficacious (Li, Alfenito, Rea and Walbot, unpublished). Whereas unconjugated C3G undergoes negligible uptake into vacuolar membrane vesicles isolated from roots of maize or hypocotyls of mung bean, C3G-GS is subject to high rates of MgATP-dependent, uncoupler-insensitive uptake. The K_m values for uptake are in the micromolar range (40–45 μM), the V_{max} values (45–80 nmol/mg/10 min) are 7–40-fold greater than those measured with the model transport substrate, DNP-GS, and the concentrations of DNP-GS and vanadate required for 50% inhibition of C3G-GS uptake are commensurate with those defined previously for the transport of other GS-conjugates by the vacuolar GS-conjugate pump (Li *et al.*, 1995a,b).

Although the ability of the sequence-divergent petunia *An9*-encoded type I glutathione *S*-transferase to complement the maize *Bz2* mutation and vice versa (Alfenito *et al.*, 1998) indicates a wide distribution for GS-conjugate pump-mediated vacuolar anthocyanin uptake in plants, two perplexing problems remain to be clarified. The first is the lack of direct evidence that the *Bz2* and *An9* gene products are actually competent in the glutathionation of anthocyanins *in vitro*. The second is the seeming absence of glutathionated, minimally sulfurylated, anthocyanin derivatives from plant extracts.

One group of antimicrobial compounds for which there is an association between vacuolar accumulation and glutathionation are the isoflavonoid phytoalexins of legumes. Elicitation of infection is associated with both *de novo* synthesis of isoflavonoid phytoalexins from distant precursors and/or release of isoflavone precursors from vacuolar stores (Dixon and Paiva, 1995). An increase in glutathione *S*-transferase transcripts and activities coincides with the production of isoflavonoid phytoalexins in elicitor-treated or infected cells (Levine *et al.*, 1994). Medicarpin, the major isoflavonoid phytoalexin of alfalfa and several other legumes, is amenable to glutathione *S*-transferase-catalyzed glutathionation *in vitro*, and the product of this reaction, medicarpin-GS, is the most efficacious (glutathionated) substrate for the plant vacuolar GS-X pump known (Li *et al.*, 1997a). The K_m for vacuolar GS-conjugate pump-mediated medicarpin-GS uptake (<20 μM) is more than 2–25-fold lower than that estimated for DNP-GS and C3G-GS in the same system, and its V_{max} (=78 nmol/mg/10 min) is similar to that for C3G-GS (Li *et al.*, 1997a). One function of the glutathione *S*-transferase induced following the hypersensitive response to avirulent fungal pathogens may, therefore, be to promote the vacuolar sequestration of antimicrobial compounds in the healthy cells surrounding the hypersensitive lesion, thereby limiting the spread of infection and tissue damage (Li *et al.*, 1997a).

It is instructive to note that the results of recent studies demonstrate that the addition of glutathione to unsaturated phenylpropanoids and related compounds is probably attributable to a second activity associated with at least some glutathione *S*-tranferases, namely that of a glutathione peroxidase (Dean and Devarenne, 1997). Some GS-conjugates, for example those of cinnamic acid, may not be the product of an electrophilic Michael addition (reviewed in Marrs, 1996) but instead the product of a radical addition. If generally applicable, such a reaction mechanism might explain both how maize mixed glutathione *S*-transferases generate medicarpin-GS derivatives (Li *et al.*, 1997a), as well as the observation that, upon fungally elicited synthesis of isoflavonoids and pterocarpans by *Cicer arietinum* cell cultures, externally applied cinnamic acid is metabolized to GS-conjugates (Barz and Mackenbrock, 1994).

10.3.2.4 Glucuronate conjugates

The biosynthesis and transport of glucuronate conjugates in animals, where they participate in the catabolism of heme, steroids and xenobiotics, has been studied extensively (Loe *et al.*, 1996), but in plants it is thought that conjugates of this type have a limited species distribution by comparison with their glutathionyl and glucosyl counterparts. Currently, the most extensively investigated system with regard to glucuronide transport is the rye primary leaf, in which glucuronated flavonoids are abundant (Anhalt and Weissenböck, 1992). Mesophyll cells of rye primary leaf contain two dominant glucuronated derivatives of the flavone luteolin: luteolin 7-*O*-diglucuronyl-4′-*O*-glucuronide (R1) and luteolin 7-*O*-diglucuronide (R2) (Figure 10.2). By contrast, the flavonoids found in the epidermal layer of rye leaves are predominantly glucosylated.

Studies with the model glucuronate conjugate, β-estradiol 17-β-glucuronide (β-EG) (Figure 10.2), have established that this compound is subject to MgATP-energized, vanadate-inhibitable uptake by rye and barley vacuoles and by vesicles containing heterologously expressed plant multidrug resistance-associated protein (MRP)-type ABC transporters (Klein *et al.*, 1998; Liu *et al.*, 2000). Direct energization has also been observed for the vacuolar uptake of the two rye luteolin glucuronides, R1 and R2 (Klein *et al.*, 2000), by vacuoles isolated from rye, as well as from plants, such as barley and broccoli, that are unable to synthesize glucuronidated secondary compounds. In all three membrane preparations, glucuronide uptake is strongly modulated by GS-conjugates and moderately modulated by $\Delta\psi$. The model GS-conjugate, DNP-GS, consistently promotes uptake, while other glutathione derivatives, such as oxidized glutathione (GSSG) and decyl-GS, act as activators or inhibitors depending on the membrane source and glucuronide under investigation. The effects of artificially imposed $\Delta\psi$ values confirm that valinomycin inhibits glucuronide uptake by dissipating the inside-positive $\Delta\psi$ consequent on V-ATPase activity, rather than through competition with glucuronides for uptake (Klein *et al.*, 2000).

10.3.2.5 Sulfated and sulfonated compounds

Results similar to those obtained with glucuronides have been obtained with the disulfonated 4-aminophthalimide, lucifer yellow CH (LYCH) (Figure 10.2) (Klein *et al.*, 1997). LYCH uptake is subject to direct energization by MgATP but inhibited by valinomycin. As would be predicted, if LYCH and glucuronides are transported by the same moiety, not only sulfated compounds but also glucuronides compete with LYCH for uptake. It is not yet known if sulfated secondary metabolites (Varin *et al.*, 1992), such as those of flavonoids or glucosinolates are recognized by the same transporter, but the results of preliminary experiments indicate that synthetic anthraquinones, used for the chemical synthesis of a broad range of dyes, are taken up into vacuoles in a manner similar to that of LYCH (Buser, Schwitzguebel and Martinoia, unpublished). Native plants with, or plants engineered for, enhanced uptake might therefore find application in the removal of these and related sulfonated compounds from polluted waste waters.

10.3.2.6 Bile acids

One of the first steps in the secretion of bile acids, a process critical for the hydrolysis and absorption of lipids by animals, is their ABC transporter-catalyzed transport across the canalicular membrane of the liver into the bile canaliculus (Kitamura *et al.*, 1990; Stieger *et al.*, 1992; Oude-Elferink *et al.*, 1995). In view of the fact that plants are not known to contain bile acids, it is perhaps surprising that vacuoles from barley are competent in the MgATP-energized, protonophore-insensitive uptake of bile acids, such as taurocholate (Hörtensteiner *et al.*, 1993). As judged by its insensitivity to inhibition by GS-conjugates and differential sensitivity to oligomycin, the vacuolar bile transporter appears to be distinct from the GS-conjugate pump, as is the case for its nearest equivalent in the canalicular membrane of mammalian liver (Oude-Elferink *et al.*, 1995). Candidate endogenous substrates for transporters of this type in plants might be digitonins and/or brassinosteroids.

10.3.2.7 Chlorophyll catabolites

Conversion of mesophyll chloroplasts into gerontoplasts during leaf senescence and yellowing is accompanied by the disassembly of thylakoid pigment-protein complexes and breakdown of chlorophyll into phytol and water-soluble porphyrin derivatives (Matile *et al.*, 1999). The water-soluble porphyrins are exported from gerontoplasts and accumulated in the vacuoles of the senescing cells. Although chlorophyll catabolites are not strictly secondary metabolites, they are nonetheless subject to the same types of chemical modifications, malonylation and glucosylation, as flavonoids and anthocyanins. Efficient export of photoactive tetrapyrroles from the cytosol is imperative if extensive photo-oxidative damage is to be avoided.

Three of the principal vacuolar tetrapyrroles of rape (*Brassica napus*), *Bn*-NCC-1, *Bn*-NCC-2 and *Bn*-NCC-3, have been purified and structurally characterized, and one, *Bn*-NCC-1, has been shown to undergo MgATP-energized, vanadate-inhibitable uptake into barley vacuoles (Hinder *et al.*, 1996). However, given that the K_m for *Bn*-NCC-1 transport is relatively high (110 μM) and *Hv*-NCC-2, a flurorescent precursor of the NCCs, is a potent competitor of *Bn*-NCC-1 uptake (K_i = 60 μM), it is not known if the *in vivo* substrate for the chlorophyll catabolite transporter is the non-fluorescent end-product or its fluorescent precursor. Malonylation does not seem to be an essential requirement for NCC transport because *Bn*-NCC-2, a non-malonylated, glucosylated derivative, is transported at rates comparable to those of *Bn*-NCC-1, the malonylated derivative.

10.3.2.8 Modulation of transport activities by endogenous substrates

Stress factors, including UV irradiation, phosphate starvation, herbivore or pathogen attack, and oxidative damage can have marked effects on the production of secondary metabolites. Moreover, in many cases, these factors do not modulate overall metabolic pathways but, instead, elicit the formation of compounds that are not present or are present in only minor amounts under other conditions. For instance, UV irradiation of barley and rape plants induces a shift in the pattern of flavonoid synthesis from the production of 3′-monohydroxylated flavonoids, which are the major compounds of this class in the unstressed state, to the synthesis of 3′,4′-dihydroxylated derivatives (Reuber *et al.*, 1996; Olsson *et al.*, 1998). While phenomena of this type offer the prospect of assessing, or at least gaining an initial indication of, the effects of a particular compound or subset of compounds on vacuolar transport activity *in vivo*, there are very few reports in which this property has been exploited.

The few studies that have been performed indicate that transport is or is not affected depending on the compounds concerned. Thus, on the one hand, the capacity of barley vacuoles for flavonoid transport is decreased in mutants impaired in flavonoid biosynthesis because of a deficiency in chalcone synthase (Frangne, Eggmann, Weissenböck, Martinoia and Klein, in preparation), suggesting that transport substrates can feed-forward activate transport *in vivo*. On the other hand, the facility of rye vacuoles for the transport of flavonoid glucosides is relatively invariant, despite wide fluctuations in the steady state levels of these compounds in the leaf (Klein *et al.*, 2000).

The situation with xenobiotics is slightly better defined, in that pretreatment of plants with herbicide safeners increases the activity of the vacuolar GS-conjugate pump and increases the transport of HPS-glucoside *in vitro* (Gaillard *et al.*, 1994), while pretreatment with cytotoxins, such as 1-chloro-2,4-dinitrochlorobenzene, that are detoxified exclusively via the glutathione-dependent pathway, increases the capacity of vacuolar membrane vesicles isolated from pretreated mung bean seedling for GS-conjugate transport (Li *et al.*, 1995b). Although the mechanisms underlying these effects are not known, they provide compelling evidence in support of the contention that vacuolar transporters are integral components of plant detoxification systems (Gaillard *et al.*, 1994; Li *et al.*, 1995b).

10.3.2.9 Vesicle-mediated transport

The delivery of secondary metabolites into the vacuole may not be mediated exclusively by membrane transporters. Pathways akin to those proposed for products of the phenylpropanoid pathway, in which transport of secondary compounds is via endoplasmic reticulum (ER)-derived vesicles that fuse with the vacuolar membrane, may apply in some cases. Three lines of evidence support such a scheme for phenylpropanoid biosynthesis: i) externally applied radiolabeled phenylalanine (PAL) or *trans*-cinnamic acid do not freely equilibrate with the cellular pools of these compounds, implying that the endogenously produced intermediates are utilized preferentially (Czichi and Kindl, 1975, 1977; Margna and Margna, 1978); ii) cell fractionation experiments demonstrate co-localization of PAL and cinnamic acid 4-hydroxylase (C4H), a cytochrome P450-dependent monooxygenase, to membranes of the ER (Czich and Kindl, 1975; Hrazdina and Wagner, 1985); and iii) ectopic expression of bean PAL in tobacco cell cultures perturbs a metabolic pool distinct from that acted on by the endogenous membrane-associated complex of PAL isozymes and C4H (Rasmussen and Dixon, 1999).

Direct transfer of the products of a multienzyme complex into the ER lumen and subsequent vesicular transport is an attractive concept, and could also explain the rapid formation of vesicles replete with anthocyanin phytoalexins after inoculation of sorghum leaves with a fungal pathogen (Snyder and Nicholson, 1990). Similar mechanisms have been postulated for alkaloid synthesis and accumulation (Amann *et al*., 1986).

10.4 Characteristics of vacuolar ABC transporters

The ABC superfamily is the largest protein family known (Decottignies and Goffeau, 1997; Henikoff *et al*., 1997) and most, but not all, are transport-active membrane proteins ('ABC transporters'). In excess of 130 representatives have been identified in species ranging from the archaea to humans (Higgins, 1992). Among the substances known to be transported by ABC transporters from non-plant sources are alkaloids, amino acids, heavy metal chelates, inorganic ions, lipids, peptides and sugars (reviewed in Higgins, 1992). Some ABC proteins, such as bacterial amino acid and sugar permeases, are highly specific for a single compound; others, such as the mammalian MDR proteins, are capable of transporting a broad range of seemingly unrelated lipophilic compounds.

10.4.1 ABC transporters from plants

Results from two different types of investigation were instrumental in prompting studies of ABC transporters in plants. The first observation was the molecular cloning of a MDR-like gene (*AtPGP1*) from *Arabidopsis* (Dudler and Hertig, 1992) and the subsequent independent isolation of other MDR homologs from the same and other plant species (potato, Wang *et al*., 1996; barley, Davies *et al*., 1997). Since all of these genes encode ABC transporters bearing strong sequence similarities to one another and to the animal *MDR* gene products, it seemed likely that ATP-dependent, primary active transport functions, analogous to those catalyzed by the MDRs from other organisms, were also operative in plants. The second observation was that intact vacuoles and vacuolar membrane vesicles from plants mediate the MgATP-dependent, H^+ gradient-independent accumulation of GS-conjugates (Martinoia *et al.*, 1993; Li *et al.*, 1995a).

Of these two classes of finding, investigations of vacuolar GS-conjugate transport have yielded the most definitive data. It is now firmly established that the vacuolar GS-conjugate pump is an MRP-type ABC transporter. Although several plant *MDR*s, and their close relatives the *PDR*s, have been cloned in their entirety (Dudler and Hertig, 1992; Smart and Fleming, 1996; Wang *et al.*, 1996; Davies *et al.*, 1997), and the various plant genome databases abound with *MDR*-like sequences, definition of the transport capabilities of the proteins encoded by these genes has proved to be less tractable than those for products of genes encoding GS-conjugate pumps. Thus, while AtPGP1 is the most thoroughly characterized MDR from a plant source, and elegant investigations of *AtPGP1*-manipulated plants have shown this ABC transporter to be a plasma membrane protein involved in light-dependent hypocotyl elongation (Dudler and Sidler, 1998; Sidler *et al.*, 1998), it is not known how its effects on growth are exerted. By necessity, the treatment that follows focuses exclusively on plant MRPs, specifically those from *Arabidopsis*, the AtMRPs.

10.4.2 Structure of ABC transporters

All ABC transporters contain one or two copies each of two core structural elements: a hydrophobic, membrane spanning domain (MSD) containing multiple (usually four or six) transmembrane spans; and a cytosolically oriented ATP-binding domain (nucleotide binding fold, NBF) (Higgins, 1992). The NBFs, the most diagnostic feature of ABC proteins, are 30–40% identical among family members over a span of about 200 amino acid residues, and each encompasses one copy each of three idiotypic sequence motifs. These are a Walker A box ($GX_4GK[ST]$) and a Walker B box ($[RK]X_3GX_3L[hydrophobic]_3$), separated by approximately 120 amino acids (Walker *et al.*, 1982), and an ABC signature (or C motif), situated between the two Walker boxes, consisting of the amino acids $[LIVMFY]S[SG]GX_3[RKA][LIVMYA]X[LIVFM][AG]$ (Bairoch, 1992), or a variant thereof. Walker motifs are found in other nucleotide-binding proteins, including P-, F- and V-ATPases, G proteins, adenylate kinase, myosin, phosphofructokinase and ATP/ADP exchangers (Walker *et al.*, 1982), but ABC signature motifs are unique to the NBFs of ABC proteins (Higgins, 1992).

The MSDs are considered to form the pathway for solute movement across the phospholipid bilayer (or between bilayer leaflets) and, in those proteins investigated in sufficient detail, appear to determine or at least contribute to the substrate selectivity of the transporter. However, by comparison with the marked similarities among the NBFs, the MSDs of even closely-related ABC transporters exhibit little or no discernible sequence similarity, other than a preponderance of hydrophobic residues, with a distribution consistent with structures of approximately equivalent topology.

A feature of ABC transporters, evident from a survey of the superfamily, is their modular construction. The four core domains (or modules)—two NBFs and two MSDs—may be expressed as separate polypeptides or as multidomain proteins (Higgins *et al.*, 1986). In some ABC proteins, primarily those from bacterial sources, the four domains reside on different polypeptides ('quarter molecules') composed of individual NBFs and MSDs. In other ABC proteins, the domains are fused in various combinations. In the ribose permease system of *Escherichia coli*, the two NBFs are fused; in the mammalian TAP1-TAP2 peptide transporter, the *Schizosaccharomyes pombe* HMT1 gene product and the *Drosophila melanogaster* pigment transporters *white*, *brown* and *scarlet*, one MSD is fused with an NBF to form a 'half molecule'. In the majority of eukaryotic ABC transporters, the core domains are contiguous on a single multidomain polypeptide in a 'forward' MSD1-NBF1-MSD2-NBF2 or 'reverse' NBF1-MSD1-NBF2-MSD2 orientation.

10.4.3 The MRP subfamily

10.4.3.1 Structure of MRPs

Computer-assisted hydropathy and sequence alignments between the AtMRPs, the mammalian MRPs and one of their yeast equivalents, YCF1, reveal a fundamental equivalence of domain organization among these transporters and structures that distinguish them from other members of the ABC superfamily. The domains identified include two NBFs (NBF1 and NBF2), and two MSDs (MSD2 and MSD3) in a forward, MSD2-NBF1-MSD3-NBF2, orientation, and two subfamily-specific structures: a putative 'regulatory' (R) or 'connector' domain contiguous with NBF1, rich in charged amino acid residues, common to the MRP and cystic fibrosis transmembrane regulator (CFTR) subfamilies but truncated in the former class (117–161 versus >250 amino acids); and an approximately 200 amino acid residue hydrophobic N-terminal extension (NTE), encompassing MSD1, that is absent from the CFTR subfamily.

A current topological model of the MRPs, exemplified by AtMRP3, showing the relative disposition of these domains is depicted in Figure 10.3. Although derived largely from analyses of HmMRP1 and AtMRP2 (Hipfner *et al.*, 1997; Rea *et al.*, 1998), this model is probably as applicable to other AtMRPs as it is to the mammalian MRPs, given the equivalence of the hydropathy profiles of all the AtMRPs (AtMRP6 excepted; Figure 10.4) and their close correspondence with those of other members of the MRP subfamily. The structure shown consists of a cytosolically oriented R domain, two cytosolically oriented NBFs, a five-span NTE containing MSD1 and an extracytoplasmic N-terminus, one

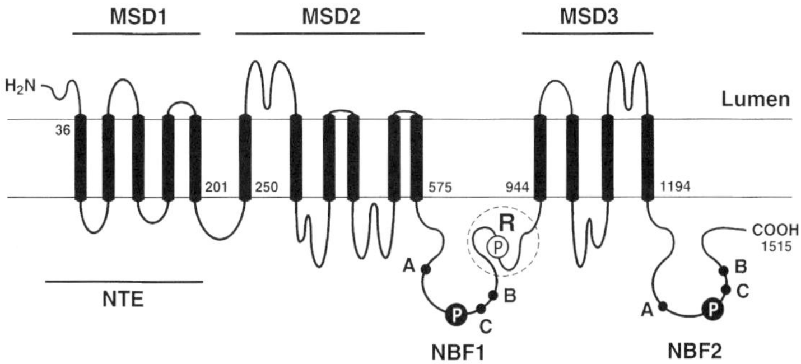

Figure 10.3 Tentative topological model of AtMRP3. The model, which is based on that deduced for HmMRP1 (Hipfner *et al.*, 1997) and AtMRP2 (Rea *et al.*, 1998), is probably applicable to most AtMRPs. The numbers delimit the membrane-spanning domains, MSD1, MSD2 and MSD3. The positions of putative transmembrane spans 1–15 were computed using programs, TMpred (http://www.ch.embnet.org/software/TMPRED_form.html), DAS (http://www.biokemi.su.se/~server/DAS/) and TMHMM (http://www.cbs.dtu.dk/services/TMHMM-1.0/), and the most likely consensus domains are shown as black bars. Putative PKC recognition sites identified using the ExPASy ScanProsite tool (http://www.expasy.ch/tools/scnpsite.html) that are shared between plant, yeast and human MRPs are denoted by a white P against a black background, those that are mainly conserved in AtMRP1 to AtMRP9 are denoted by a black P against a white background. The position of the Walker A (A) and Walker B (B) motifs and the ABC transporter signature (C) are indicated. Abbreviations: R, 'regulatory' or 'connector' domain; MSD, membrane spanning domain; NBF, nucleotide binding fold; NTE, N-terminal extension. Note that the precise topology for MSD3 is uncertain, in that it is not known if it contains four or six transmembrane spans.

MSD (MSD2) containing six transmembrane spans, a third MSD (MSD3) containing four or six spans and a C-terminal extension (C domain). The precise topology of MSD3 is uncertain because the results of comparative hydropathy analyses and sequence alignments are equally consistent with six or four transmembrane spans in this region (Hipfner *et al.*, 1997). This is not a problem unique to the MRPs, however. A re-evaluation of the results of previous studies of mammalian MDRs, archetypal 6 + 6 proteins, show them also to be consistent with four or six transmembrane spans in the region corresponding to MSD3 of the MRPs.

10.4.3.2 Arabidopsis *MRPs*

So far, nine AtMRPs have been identified in *Arabidopsis* (Table 10.1). *AtMRP1* and *AtMRP2* have been isolated as full-length genomic and cDNA clones, functionally expressed in yeast and physically mapped (to chromosomes I and II, respectively) (Lu *et al.*, 1997, 1998); *AtMRP4* has been isolated as full-length genomic and cDNA clones and functionally expressed in yeast (Sánchez-Fernández *et al.*, 1998; Sánchez-Fernández and Rea, unpublished); *AtMRP3* (Tommasini *et al.*, 1998) and *AtMRP5* (Gaedeke, Klein, Muller-Rober and

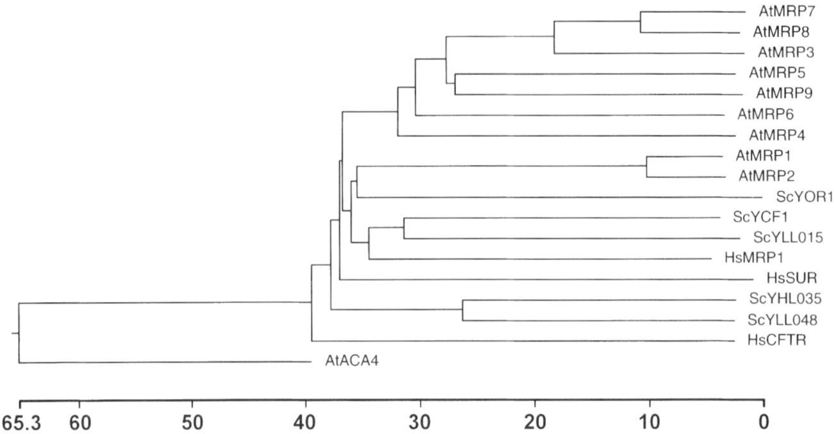

Figure 10.4 Phylogenetic analysis of known multidrug resistance-associated proteins (MRPs) from *Arabidopsis thaliana* and representative MRPs from nonplant sources. The analyses were based on deduced amino acid sequence alignments performed using the ClustalW 1.7 or PAUP methods (Wisconsin Sequence Analysis Package). Qualitatively similar results were obtained by both procedures, but only the single tree retrieved using the MEGALIGN program of Clustal, with a gap penalty of 20 and gap length penalty of 10, is shown here. *Arabidopsis* P-type Ca^{2+}-ATPase AtACA4 (accession number AF200739) was used as the outgroup. The sequences shown are those from *A. thaliana* (At), *Homo sapiens* (Hs) and *S. cerevisiae* (Sc). They are: AtMRP1 (accession number AAF19743), AtMRP2 (accession number AAC16268), AtMRP3 (accession number AAC49791), AtMRP4 (accession number AAC63634), AtMRP5 (accession number AAC16754), AtMRP9 (accession number CAB75931), HsMRP1 (accession number P33527), HsSUR (accession number Q09428), HsCFTR (accession number NP_000483), ScYOR1 (accession number NP_011797), ScYCF1 (accession number P39109), ScYLL015 (accession number Z73120), ScYHL035 (accession number U11583), and ScYLL048 (accession number Z73153). The amino acid sequences of AtMRP6 (AB023045), AtMRP7 and AtMRP8 (AP00375) were assembled by comparing their genomic sequences with those of related *Arabidopsis* MRPs and manually deducing the splice sites. The assembled sequence of AtMRP6 is N-terminally truncated because the translation start site could not be identified unambiguously from the genomic sequence data available.

Martinoia, unpublished) have been isolated as full-length cDNAs and functionally expressed in yeast. *AtMRPs 6–9* have been identified as putative MRP coding sequences in the *Arabidopsis* genome database. AtMRPs 1 and 2 cluster with human MRP1, yeast YOR1, yeast YLL015 and yeast YCF1, as judged by phylogenetic analyses of the full-length deduced amino acid seqences of AtMRPs 1–9 together with those of MRPs from nonplant sources, whereas all of the remaining AtMRPs (3–9) form their own cluster (Figure 10.4). As is evident from a detailed analysis of the C-terminal domains, a subgroup comprising AtMRPs 3, 8 and 7 is also evident within the AtMRP 3–9 cluster (Figure 10.4).

10.4.3.3 Transport activities of Arabidopsis *MRPs*

AtMRP1, *AtMRP2*, *AtMRP3*, *AtMRP4* and *AtMRP5* encode functional GS-conjugate pumps. Transformation of *S. cerevisiae ycf1Δ* strains from which

Table 10.1 Inventory of known MRPs from *Arabidopsis thaliana*

	Chr	BAC/P1 clone	Genomic sequence		cDNA	Protein
			AtDB	Other		
AtMRP1	I	F26G16	AC009917	AF008125	AF008124	AAF19743
AtMRP2	II	T29F13	AC003096	AF020289	AF020288 AF014960	AAC16268
AtMRP3	III	MJG9	AP000375	–	U92650	AAC49791
AtMRP4	II	F17A22	AC005309	AJ002584	AF243509	AAC63634
AtMRP5	I	F20D22	AC002411	–	Y11250	AAC16754 CAA72129
AtMRP6	III	MXL8	AB023045	–	–	–
AtMRP7	III	MJG9	AP000375	–	–	–
AtMRP8	III	MJG9	AP000375	–	–	–
AtMRP9	III	F18P9	AL138658	–	–	CAB75931

The nine known *AtMRP* genes are listed, together with their chromosome locations (Chr), the BAC/P1 clones on which they are situated, their genomic accession numbers from the *Arabidopsis thaliana* genome database (AtDB), their protein accession numbers and the genomic and cDNA accession numbers of clones isolated by individual laboratories. Note that the genes for AtMRP3, AtMRP7 and AtMRP8 map to the same contig (MJG9) and are arranged in tandem in the order AtMRP3-AtMRP8-AtMRP7.

more than 95% of the coding sequence of the *ScYCF1* gene has been deleted (Szczypka *et al.*, 1994) and high-affinity MgATP-energized vacuolar GS-conjugate transport is grossly impaired (Li *et al.*, 1996; Tommasini *et al.*, 1996; Li *et al.*, 1997b) with individual expression vectors containing the entire ORFs of *AtMRP1*, *AtMRP2*, *AtMRP3*, *AtMRP4* or *AtMRP5* restores GS-conjugate transport (Lu *et al.*, 1997, 1998; Tommasini *et al.*, 1998; Sánchez-Fernández and Rea, unpublished; Gaedeke, Klein, Muller-Rober and Martinoia, unpublished).

A direct comparison of the transport activities for a range of glutathionated compounds for AtMRP1 and AtMRP2, the only two AtMRPs for which quantitative transport data are available, however, shows marked differences in transport capacity and substrate preference for at least some AtMRPs. The overall transport capacity of AtMRP2 exceeds that of AtMRP1 by 3–6-fold, and the substrate specificity of AtMRP1 better approximates that of the endogenous vacuolar GS-conjugate pump than does AtMRP2. Whereas the rates of uptake of DNP-GS, oxidized glutathione (GSSG), metolachlor-GS and the glutathionated anthocyanin, cyanidin-3-glucoside-GS (C3G-GS) by AtMRP1 fall into the same rank order as for mung bean vacuolar membrane vesicles (Li *et al.*, 1995a; Lu *et al.*, 1997) (C3G-GS $>$ metolachlor-GS $>$ DNP-GS $\geq$ GSSG; uptake ratio for AtMRP1 $= 2.7:1.5:1.0:0.7$; uptake ratio for mung bean $= 5.5:3.9:1.0:0.8$), that of AtMRP2 is quite distinct (metolachlor-GS $>$ C3G-GS $\gg$ GSSG $>$ DNP-GS; uptake ratio $= 7.8:6.4:6.1:2.2:1.0$) (Lu *et al.*, 1997, 1998). Moreover, AtMRP2, unlike AtMRP1, has the facility for high-affinity, high-capacity transport of the nonglutathionated chlorophyll catabolite *Bn*-NCC-1 (Lu *et al.*, 1998). With the exception of GSSG, whose K_m for AtMRP1-mediated uptake (ca. 220 μM) is three times greater than that for AtMRP2-mediated uptake (ca. 66 μM), the K_m values for all of the glutathionated compounds examined are very similar for AtMRP1 and AtMRP2 (60–75 μM). The transport activities of AtMRP3 approximate those of AtMRP2. AtMRP3 is competent in the transport of DNP-GS and chlorophyll catabolites and transport of the former is competitively inhibited by metolachlor-GS and GSSG (Tommasini *et al.*, 1998). The detailed functional characteristics of AtMRP4 and AtMRP5 have not yet been defined.

10.4.3.4 Enhanced AtMRP multifunctionality

In addition to implicating vacuolar GS-conjugate pumps in transport of chlorophyll catabolites during leaf senescence, the results of investigations of *Bn*-NCC-1 transport yield crucial new information into the functional characteristics of the AtMRPs. Firstly, they clearly demonstrate that at least some AtMRPs are not only able to transport GS-conjugates but also other bulky amphipathic anions, such as the linearized tetrapyrrole, *Bn*-NCC-1. Secondly, of all the compounds screened against AtMRP2 so far, *Bn*-NCC-1 is transported with the highest affinity ($K_m = 15\,\mu M$) (Lu *et al.*, 1998). Indeed, *Bn*-NCC-1 is the most efficacious known substrate for a plant GS-conjugate pump ($V_{max}/K_m = 4.2\,nmol/mg/10\,min/\mu M$ versus $1.8\,nmol/mg/10\,min/\mu M$ for the transport of metolachlor-GS by AtMRP2; Lu *et al.*, 1998). Thirdly, both AtMRP2 and AtMRP3 appear to be competent in the parallel transport of *Bn*-NCC-1 and GS-conjugates (Lu *et al.*, 1998; Tommasini *et al.*, 1998). AtMRP2-dependent uptake of *Bn*-NCC-1 is nearly threefold less sensitive to inhibition by DNP-GS than would be predicted if this GS-conjugate were a simple competitive inhibitor. Reciprocally, the concentration-dependence of AtMRP2-mediated DNP-GS uptake is not affected appreciably by inclusion of DNP-GS in the uptake medium (Lu *et al.*, 1998).

It was initially surprising to find that AtMRP2 can transport both *Bn*-NCC-1 and GS-conjugates, because the functionality, and by implication the pump, responsible for MgATP-energized *Bn*-NCC-1 uptake by barley vacuoles had previously been considered to be different from that responsible for GS-conjugate uptake, since GS-conjugates did not compete with *Bn*-NCC-1 for uptake (Hinder *et al.*, 1996). However, it is now clear, at least in the case of some AtMRPs, that lack of competition between two candidate transport substrates does not automatically preclude transport by the same transporter. Rather, the data available indicate that AtMRP2 and AtMRP3, and presumably their equivalents in barley vacuoles, each contain two semi-autonomous functional modules. One module transports GS-conjugates, while the other transports compounds such as *Bn*-NCC-1, and transport through each module is largely independent of transport through the other. An indication that substrate recognition mechanisms of this type may operate comes from the finding that the substrate specificity of chimeras constructed from HsMRP1 and its murine ortholog assume the glucuronide-selectivity of the MRP from which the C-terminal third of the molecule is derived (Stride *et al.*, 1999).

The general applicability of these findings remains to be determined but Swanson *et al.* (1998) observed that the transport of halobimanes and the protease substrate 7-amino-4-chloromethylcoumarin benzyloxycarbonyl-L-phenylalanyl-L-arginine amide hydrochloride into barley aleurone vacuoles requires glutathionation, whereas uptake of the pH-sensitive probe 2,7′-bis(2-carboxyethyl)-5(6)-carboxyfluorescein does not, yet transport of all three compounds proceeds via an MgATP-dependent, vanadate-inhibitable pathway. These authors appropriately interpret their results in terms of the operation of two ABC transporters, one for GS-conjugates and another for nonglutathionated anions. However, in the light of what is now known of the properties of the AtMRPs, their findings are equally compatible with the operation of a multifunctional AtMRP2- or AtMRP3-like transporter capable of handling both types of compound.

Recent investigations of heterologously expressed AtMRP2 extend these principles to show that a single MRP is not only competent in the transport of GS-conjugates and tetrapyrroles but also glucuronate conjugates, and that some substrates reciprocally activate each other's transport (Liu *et al.*, 2000). Based on the markedly increased uptake of DNP-GS by β-EG and vice versa, on double-label experiments demonstrating that the two substrates are subject to simultaneous inward transport, and on the results of preloading experiments which suggest that the effects seen derive from *cis*, not *trans*, interactions, it is inferred that some GS-conjugates and some glucuronides are transported via distinct but coupled binding sites. Parallel experiments on mammalian and yeast MRPs establish that these properties are specific to AtMRP2 (Liu *et al.*, 2000). The near equivalence of the multiple substrate interactions detected with heterologously expressed AtMRP2 to those detected in native barley vacuoles (Klein *et al.*, 2000) is consistent with the notion that a single moiety could be responsible for the complex behavior of the latter, without the necessity of invoking interactions with other transporters or protein regulatory elements.

10.4.3.5 MRP homologs from other plant species

It is now clear that MRP-like vacuolar transport activities are ubiquitous in plants (Rea *et al.*, 1998; Rea, 1999). However, except for an isolated report of an M_r 170,000 polypeptide, enriched in the vacuolar membrane fractions from herbicide- and herbicide safener-treated plants (Theodoulou *et al.*, 1998), very little is known of the molecular basis of MRP-mediated transport in plants other than *Arabidopsis*. Despite this paucity of molecular physiological information for other plants, it is nevertheless evident that they contain AtMRP homologs, and that these homologs can be assigned a specific identity on the basis of their equivalence to a particular AtMRP or to members of a particular AtMRP subgroup.

The analyses summarized in Tables 10.1–10.3, which are, respectively, an inventory of the nine known AtMRPs, their alignments with each other, and a sample of the results of expressed sequence tag (EST) database searches for AtMRP homologs in other plant species, clearly illustrate this point. The sequences from individual AtMRPs (Table 10.1) were compared with each other (Table 10.2) and then used as query sequences not only to identify AtMRP homologs but also to gain an indication of the particular type of AtMRP identified in the target organism (Table 10.3). To simplify the search and preclude double hits for a given gene, the C-terminal portions of the AtMRPs, corresponding to the second NBF and the C-terminal domain, were used for the searches. C-terminal sequences were chosen because 3′ coding sequences are more proportionately represented than 5′ sequences on a gene-for-gene expression basis in EST databases. To offset the inevitable bias of database searches for ABC proteins in favor of NBFs, only those hits for which the alignments were contiguous with the C-terminal domain were considered further.

The results of these analyses are very informative. Firstly, within the AtMRP family itself, sequence comparisons of the C-terminal portions of the AtMRPs used as query sequences for searching the EST database confirm and further extend the notion of subgroups inferred previously from global domain comparisons of AtMRPs 1–5 (Rea, 1999). Comparisons between AtMRPs 1–9 reaffirm that AtMRP1 and AtMRP2 constitute a small subgroup distinct from all other known AtMRPs, but in addition disclose another subgroup, comprising AtMRP3, AtMRP7 and AtMRP8, whose members are as sequence-related (83–91% identical; 90–95% similar) as those in the AtMRP1/AtMRP2 subgroup (Table 10.2). Furthermore, it was determined from inspection of the *Arabidopsis* sequence database that, whereas the genes encoding AtMRP1 and AtMRP2 map to different chromosomes (I and II, respectively) (Lu *et al.*, 1997, 1998), those encoding AtMRP3, AtMRP7 and AtMRP8 are organized as a tandem cluster on chromosome III (Table 10.1). Apparently, the genes encoding AtMRP1 and AtMRP2 and those encoding AtMRP3, AtMRP7 and AtMRP8 arose from interchromosomal and intrachromosomal gene duplications, respectively, that postdated the initial radiation of this subfamily of ABC transporters.

Secondly, extension of this approach to other plant species by employing the AtMRP sequences used for the intraspecies comparisons to interspecies EST database searches enables the identification of ESTs from tomato, soybean and maize whose translation products are strict homologs, in that they align more closely with a specific AtMRP or members of a specific AtMRP subgroup than the AtMRPs align with each other (Table 10.3). Thus: tomato ESTs AW092564 and AW219577, share greater than 82% identity (91% similarity) with members of the AtMRP1/AtMRP2 subgroup; tomato EST AW216929 shares greater than 82% identity (91% similarity) with members of the AtMRP3/AtMRP7/AtMRP8 subgroup; and, soybean EST AW278374, maize EST AI668454, tomato EST AW034253 and soybean EST AW396018 share greater than 80%

Table 10.2 Alignments between the C-terminal portions of AtMRPs 1–9

AtMRP1		AtMRP2		AtMRP3		AtMRP4		AtMRP5		AtMRP6		AtMRP7		AtMRP8		AtMRP9		
% I	% S	% I	% S	% I	% S	% I	% S	% I	% S	% I	% S	% I	% S	% I	% S	% I	% S	
100	100	88	93	52	71	54	73	56	74	50	66	54	70	52	69	52	69	AtMRP1
		100	100	54	73	54	74	58	75	52	67	54	70	52	68	53	69	AtMRP2
				100	100	62	81	75	86	59	78	85	92	83	90	65	81	AtMRP3
						100	100	61	78	65	78	63	83	61	80	61	80	AtMRP4
								100	100	63	76	68	81	69	82	69	82	AtMRP5
										100	100	43	60	49	66	47	65	AtMRP6
												100	100	91	95	60	72	AtMRP7
														100	100	66	78	AtMRP8
																100	100	AtMRP9

The percentage identities (%I) and percentage similarities (%S) for the sequences encompassed by the second nucleotide binding fold (NBF) and the C-terminal domain are presented. For AtMRPs 1–5, C-terminal sequences starting at residues 1245, 1245, 1280, 1284 and 1280 were employed for the comparisons; for AtMRPs 6–9, the last 250 amino acid residues of the putative translation product were employed.

Table 10.3 Alignments between the C-terminal portions of AtMRPs 1–9 and those of the putative translation products of expressed sequence tags (ESTs) from plant species other than *Arabidopsis*

	AtMRP1		AtMRP2		AtMRP3		AtMRP4		AtMRP5		AtMRP6		AtMRP7		AtMRP8		AtMRP9	
	% I	% S	% I	% S	% I	% S	% I	% S	% I	% S	% I	% S	% I	% S	% I	% S	% I	% S
AW092564	84	**92**	85	**91**	54	71	56	77	58	76	40	55	54	71	52	69	54	74
AW219577	82	**93**	82	**91**	45	63	50	66	51	69	50	64	48	61	48	62	49	66
AW216929	51	68	51	68	85	**93**	64	81	70	83	57	74	83	**90**	82	**90**	69	82
AW278374	51	74	52	74	60	81	83	**92**	63	83	57	81	60	81	58	79	60	80
AI668454	55	71	57	66	74	86	60	77	88	**97**	31	50	68	82	70	84	70	82
AW034253	44	66	46	67	58	80	61	77	57	75	80	**88**	58	78	55	76	55	70
AW396018	51	68	54	70	63	78	57	74	68	78	51	68	66	81	61	77	82	**94**

Abbreviations: %I, percentage identity; %S, percentage similarity.
The same C-terminal AtMRP sequences as those listed in Table 10.2 were used as query sequences for BLAST searches of the EST databases. The results shown are for ESTs from tomato (AW092564, AW219577, AW216929, AW034253), soybean (AW278374, AW396018) and maize (AI668454).

identity (88% similarity) with AtMRP4, AtMRP5, AtMRP6 and AtMRP9, respectively (Table 10.3). In all cases, as was found for the intraspecific AtMRP comparisons (Table 10.2), the closeness of the alignments for the next best fit are consistently markedly lower than for the best fit (Table 10.3).

Three conclusions, therefore, follow: i) The numerical designations for the AtMRPs and their homologs from other plant species have meaning. With the exception of the AtMRP1/AtMRP2 and AtMRP3/AtMRP7/AtMRP8 subgroups, whose members are so similar to each other as to prohibit assignment of a specific number to a homolog from another plant species, each of the AtMRPs have identifiable cognates in other species. A given MRP from a plant source other than *Arabidopsis* more closely resembles a particular AtMRP or members of an AtMRP subgroup (AtMRP1/AtMRP2 subgroup, AtMRP3/AtMRP7/AtMRP8 subgroup, AtMRP4, AtMRP5, AtMRP6, AtMRP9) than do the AtMRPs from different subgroups resemble each other. Consequently, comparisons based on the C-terminal sequences used in these analyses not only enable the identification of MRPs in other species but might also provide an indication of the type of MRP identified. ii) Since all of the AtMRPs examined so far—AtMRP1 (Lu *et al.*, 1997), AtMRP2 (Lu *et al.*, 1997, 1998), AtMRP3 (Tommasini *et al.*, 1998), AtMRP4 (Sánchez-Fernández *et al.*, 1998; Sánchez-Fernández and Rea, unpublished) and AtMRP5 (Gaedeke, Klein, Müller-Röber and Martinoia, unpublished)—are GS-conjugate pumps and catalyze the MgATP-dependent, vanadate-inhibitable transport of GS-conjugates and other amphipathic anions, it is tempting to speculate that their equivalents from other plants are also GS-conjugate pumps. iii) Sequence information, alone, may eventually prove sufficient to gain an indication of the type-specific capabilities of a new MRP. When more is known of the substrate selectivities, membrane localizations, tissue distributions and modes of regulation of individual AtMRPs, it is conceivable that by direct sequence comparisons of a new MRP from a given plant species, for example a species of agronomic value, with the AtMRP dataset, it will be practicable to predict the functions and potential utility of the new MRP.

10.4.3.6 Localization of MRPs

It is now firmly established that at least a subset of plant GS-conjugate pumps deliver their transport substrates into the vacuolar compartment of protoplasts and intact cells. Coleman *et al.* (1997) have demonstrated *in situ* cytosolic glutathionation of monochlorobimane to its fluorescent derivative, bimane-GS, and accumulation of the latter in the vacuoles of protoplasts and cell suspension cultures. Wolf *et al.* (1996) have found that exposure of barley leaves to the herbicide, alachlor, results in a greater than 50-fold vacuolar accumulation of alachlor-GS. Swanson *et al.* (1998) have shown MRP-like transport of both glutathionated and nonglutathionated compounds into the vacuoles of barley aleurone cells. Moreover, using antibody raised against a synthetic peptide corresponding to the C-terminal portion of an MRP homolog (tentatively that of AtMRP4) from wheat, Theodoulou *et al.* (1998) report an immunoreactive M_r 170,000 band that is enriched in vacuolar membrane fractions from herbicide safener-treated but not untreated plants. Despite this compelling evidence for a vacuolar localization for at least a subset of these transporters, it is nonetheless important to recognize that some herbicide metabolites derived from the glutathione- and glucose-dependent detoxification pathways are deposited as 'bound residues' in the extracellular matrix (Sandermann, 1992) and most plant membrane density fractionation procedures have an inherent bias in favor of the isolation of sealed (transport-competent) vacuolar membrane vesicles by comparison with transport-competent plasma membrane vesicles (Rea and Turner, 1990). Therefore, the results showing localization of some GS-conjugate pumps to the vacuolar membrane do not automatically preclude the localization of other members of the subfamily to other membranes. The need for caution is reinforced by the recent cloning of *AtMRP1* by screens of an *Arabidopsis* expression library with polyclonal antisera raised against total plasma membranes (Galaud *et al.*, 1999), which may or may not be attributable to contamination of the plasma membrane fraction with vacuolar membranes, and by the functional contiguity between the GS-conjugate pump activities on the endomembranes and exomembranes of some mammalian cell types (reviewed in Ishikawa *et al.*, 1997) and possibly yeast (discussed in Li *et al.*, 1996). Recent results have shown that the difficulties associated with low expression and/or high homology between different ABC transporters can be overcome using epitope-tag or green fluorescent protein (GFP)-fusion-based techniques (Sidler *et al.*, 1998). It will therefore be interesting to determine, using these approaches whether different MRPs reside on different endomembranes, or, as shown for aquaporins (Jauh *et al.*, 1999), are differentially partitioned between vacuole types (Paris *et al.*, 1996).

10.4.3.7 Regulation of MRPs

There are increasing intimations from mammalian systems that at least some MRPs and MRP-related proteins might be subject to kinase-mediated phosphorylation. The most thoroughly investigated example is CFTR, whose R domain contains multiple consensus phosphorylation recognition sites for cAMP-dependent protein kinases, Ca^{2+}-dependent protein kinases (PKCs) and type II cGMP-dependent protein kinases. In many but not all cases, these sites have been shown to contribute to channel gating (Akabas *et al.*, 2000). In view of their structural similarities with CFTR, analogous processes have been suggested for the MRPs (Ma *et al.*, 1995; Rea, 1999), although so far there is no direct evidence for this, or the alternative possibility that phosphorylation influences membrane trafficking, as has been found for some yeast ABC transporters (Decottignies *et al.*, 1999).

With these qualifications in mind, it is nonetheless noteworthy that inspection of the sequences of the known AtMRPs discloses several conserved PKC-like recognition sites ([ST]-X-[RK] motifs) in NBFs 1 and 2, two of which are adjacent to the ABC signature motif (Figure 10.3). One of these putative PKC recognition sites may be of special interest because it is located in the R domain and is specific to plant MRPs. Moreover, an analogous motif in an equivalent position in the human P-glycoprotein has been shown to undergo PKC-mediated phosphorylation (Chambers *et al.*, 1993).

By comparison, transcriptional modulation of some of the AtMRPs in response to stress factors is reasonably well established. The steady state levels of transcripts derived from *AtMRP2*, *AtMRP3* and *AtMRP4* (probed with ESTs) was shown to be elevated by the addition of CDNB, metolachlor, primisulfuron or IRL 1803 to *Arabidopsis* root cultures (Tommasini *et al.*, 1997). *AtMRP3* and *AtMRP4* transcript levels were elevated by more than five fold by addition of the catalase inhibitor, 3-amino-1,2,4-triazole, and the superoxide dismutase inhibitor, menadione, respectively, to *Arabidopsis* cell cultures (Sánchez-Fernández *et al.*, 1998).

10.5 Conclusions and outlook

An extraordinary range of secondary metabolites and xenobiotics are transported into the vacuole. It is reasonably well established that the energy-dependent uptake of glutathione and glucuronate conjugates is catalyzed by ABC transporter-like pumps, whereas the energy-dependent uptake of glucosylated or malonylated compounds is catalyzed by ABC transporters and/or H^+-antiporters. In the case of substrates that are not transported by ABC transporters, the onus is now to determine the molecular identities and precise energy requirements of the elements involved, so as to generate tools for their mechanistic and physiological analysis. In the case of substrates that are transported by ABC transporters, the onus is to determine which ABC transporters are responsible for which processes. Specifically, the presence of at least nine MRPs in *Arabidopsis* raises the question of functional diversification. Although it has been shown that AtMRPs 1–5 are all competent in GS-conjugate transport, implying that the remaining four or more have the same capability, this does not exclude the possibility of functional divergence, as appears to be the case for AtMRPs 1, 2 and 3. Moreover, it does not exclude the possibility of localization of different MRPs to different membranes or cell types, as is the case for human and rodent MRPs (König *et al.*, 1999).

Note

Since the preparation of this manuscript five new MRP homologs have been entered into the *Arabidopsis thaliana* genome database, giving a total *AtMRP* gene count of 14 as of August 2000. Of these 14 genes, 7 are known to be transcriptionally active as evidenced by the identification of their cognate ESTs (Sánchez-Fernández, Davies, Coleman and Rea, unpublished).

Acknowledgments

The authors would like to thank T. Eggmann for his help in the computational analysis of the MRPs. Work from the Martinoia laboratory was funded by a grant from the Swiss National Foundation, the Humboldt Foundation (M.K., M.G) and the Novartis Foundation (M.G). Work from the Rea laboratory was funded by grants from the United States Department of Agriculture (NRICGP grant 93-37304-8932) and United States Department of Energy (DE-FG02-91ER20055).

References

Akabas, M.H. (2000) Cystic fibrosis transmembrane conductance regulator: structure and function of an epithelial chloride channel. *J. Biol. Chem.*, **275** 3729-32.

Alfenito, M.R., Souer, E., Goodman, C.D., Buell, R., Mol, J., Koes, R. and Walbot, V. (1998) Functional complementation of anthocyanin sequestration in the vacuole by widely divergent glutathione *S*-transferases. *Plant Cell*, **10** 1135-49.

Alibert, G., Boudet, M. and Rataboul, P. (1982) Transport of *o*-coumaric acid glucosides in isolated vacuoles of sweet clover, in *Plasmalemma and Tonoplast: Their Function in the Plant Cell* (eds. D. Marmé, E. Marré and R. Hertel), Elsevier Biomedical Press, Amsterdam, pp. 193-200.

Amann, M., Wanner, G. and Zenk, M.H. (1988) Intracellular compartmentation of two enzymes of berberine biosynthesis in plant cell cultures. *Planta*, **167** 310-20.

Anhalt, S. and Weissenböck, G. (1992) Subcellular localization of luteolin glucuronides and related enzymes in rye mesophyll. *Planta*, **187** 83-88.

Bairoch, A. (1992) PROSITE: a dictionary of sites and patterns in proteins. *Nucleic Acids Res.*, **20** 2013-18.

Barz, W. and Mackenbrock, U. (1994) Constitutive and elicitation induced metabolism of isoflavones and pterocarpanes in chickpea (*Cicer arietinum*) cell suspension cultures. *Plant Cell Tissue Organ Cult.*, **38** 199-211.

Bowman, E.J., Siebers, A. and Altendorf, K.H. (1988) Bafilomycins: a class of inhibitors of membrane ATPases from microorganisms, animal cells and plant cells. *Proc. Natl. Acad. Sci. USA*, **85** 7972-76.

Chambers, T.C., Pohl, J., Raynor, R.L. and Kuo, J.F. (1993) Identification of specific sites in human P-glycoprotein phosphorylated by protein kinase C. *J. Biol. Chem.*, **268** 4592-95.

Coleman, J.O.D., Randall, R. and Blake-Kalff, M.M.A. (1997) Detoxification of xenobiotics in plant cells by glutathione conjugation and vacuolar compartmentalization: a fluorescent assay using monochlorobimane. *Plant Cell Environ.*, **20** 449-60.

Czichi, U. and Kindl, H. (1975) Formation of *p*-coumaric acid and *o*-coumaric acid from phenylalanine by microsomal membrane fractions from potato: evidence of membrane-bound membrane complexes. *Planta*, **125** 115-25.

Czichi, U. and Kindl, H. (1977) Phenylalanine amonia lyase and cinnamic acid hydroxylases as assembled consecutive enzymes on microsomal membranes of cucumber cotyledons: cooperation and subcellular distribution. *Planta*, **134** 133-43.

Dangles, O. and Brouillard, R. (1992) Polyphenol interactions: the co-pigmentation case—thermodynamic data from temperature variation and relaxation kinetics—medium effect. *Can. J. Chem.*, **70** 2174-89.

Davies, T.G.E., Theodoulou, F.L., Hallahan, D.L. and Forde, B.G. (1997) Cloning and characterisation of a novel P-glycoprotein homolog from barley. *Gene*, **199** 195-202.

Dean, J. V. and Devarenne, T. P. (1997) Peroxidase-mediated conjugation of glutathione to unsaturated phenylpropanoids: evidence against glutathione *S*-transferase involvement. *Physiol. Plant.*, **99** 271-78.

Decottignies, A. and Goffeau, A. (1997) Complete inventory of the yeast ABC proteins. *Nat. Genet.*, **15** 137-45.

Decottignies, A., Owsianik, G. and Ghislain, M. (1999) Casein kinase I-dependent phosphorylation and stability of the yeast multidrug transporter, Pdr5P. *J. Biol. Chem.*, **274** 37139-46.

Deus-Neumann, B. and Zenk, M.H. (1984) A highly selective alkaloid uptake system in vacuoles of higher plants. *Planta*, **162** 250-60.

Deus-Neumann, B. and Zenk, M.H. (1986) Accumulation of alkaloids in plant vacuoles does not involve an ion-trap mechanism. *Planta*, **167** 44-53.

Dixon, R.A. and Paiva, N.L. (1995) Stress-induced phenylpropanoid metabolism. *Plant Cell*, **7** 1085-97.

Dudler, R. and Hertig, C. (1992) Structure of an MDR-like gene from *Arabidopsis thaliana*: evolutionary implications. *J. Biol. Chem.*, **267** 5882-88.

Dudler, R. and Sidler, M. (1998) *Arabidopsis* MDR genes: molecular cloning and protein chemical aspects. *Methods Enzymol.*, **292** 162-73.

Ehmke, A., von Borstel, K. and Hartmann, T. (1988) Alkaloid *N*-oxides as transport and vacuolar storage compounds of pyrrolizidine alkaloids in *Senecio vulgaris* L. *Planta*, **176** 83-90.

Endicott, J.A. and Ling, V. (1989) The biochemistry of P-glycoprotein mediated multidrug resistance. *Annu. Rev. Biochem.*, **58** 137-71.

Gaillard, C., Dufaud, A., Tommasini, R., Kreuz, K., Amrhein, N. and Martinoia, E. (1994) A herbicide antidote (safener) induces the activity of both the herbicide detoxifying enzyme and of a vacuolar transporter for the detoxified herbicide. *FEBS Lett.*, **352** 219-21.

Galaud, J.P., Carriere, M., Pauly, N., Canut, H., Chalon, P., Caput, D. and Pont-Lezica, R.F. (1999) Construction of two ordered cDNA libraries enriched in genes encoding plasmalemma and tonoplast proteins from a high-efficiency expression library. *Plant J.*, **17** 111-18.

Guern, J., Renaudin, J.-P. and Barbier-Brygoo, H. (1987) Accumulation of organic solutes in plant vacuoles: the interpretation of data is not so easy, in *Plant Vacuoles* (ed. B. Marin), NATO ASI Series, pp. 305-14.

Harborne, J. (1993) *Introduction to Ecological Biochemistry*, 4th edition, Academic Press, London, UK.

Hauser, M.-T. and Wink, M. (1990) Uptake of alkaloids by latex vesicles and isolated mesophyll vacuoles of *Chelidonium majus* (Papaveraceae). *Z. Naturforsch.*, **45c** 949-57.

Henikoff, S., Greene, E.A., Pietrokovski, S., Bork, P., Attwood, T.E. and Hood, L. (1997) Gene families: the taxonomy of protein paralogs and chimeras. *Science*, **278** 609-14.

Higgins, C.F. (1992) ABC-transporters: from microorganisms to man. *Annu. Rev. Cell. Biol.*, **8** 67-113.

Higgins, C., Hiles, I., Salmond, G., Gill, D., Downie, J., Evans, I., Holland, I., Gray, L., Buckel, S., Bell, A. and Hermondson, M. (1986) A family of related ATP-binding subunits coupled to many distinct biological processes in bacteria. *Nature*, **323** 448-50.

Hinder, B., Schellenberg, M., Rodoni, S., Ginsburg, S., Vogt, E., Martinoia, E., Matile, P. and Hörtensteiner, S. (1996) How plants dispose of chlorophyll catabolites. *J. Biol. Chem.*, **271** 27233-36.

Hipfner, D.R., Almquist, K.C., Leslie, E.M., Gerlach, J.H., Grant, C.E., Deeley, R.G. and Cole, S.P.C. (1997) Membrane topology of the multidrug resistance protein (MRP): a study of glycosylation-site mutants reveals an extracytoplasmic NH_2-terminus. *J. Biol. Chem.*, **272** 23623-30.

Homeyer, B.C. and Roberts, M.F. (1984) Alkaloid sequestration by *Papaver somniferum* latex. *Z. Naturforsch.*, **39c** 876-81.

Hopp, W. and Seitz, H.U. (1987) The uptake of acylated anthocyanin into isolated vacuoles from a cell suspension culture of *Daucus carota*. *Planta*, **170** 74-85.

Hörtensteiner, S., Vogt, E., Hagenbuch, B., Meier, P.I., Amrhein, N. and Martinoia, E. (1993) Direct energization of bile acid transport into plant vacuoles. *J. Biol. Chem.*, **268** 18446-49.

Hrazdina, G. and Wagner, G. (1985) Metabolic pathways as enzyme complexes: evidence for the synthesis of phenylpropanoids and flavonoids on membrane-associated enzyme complexes. *Arch. Biochem. Biophys.*, **237** 88-100.

Ishikawa, T. (1992) The ATP-dependent glutathione *S*-conjugate export pump. *Trends Biochem. Sci.*, **17** 433-38.

Ishikawa, T., Li, Z.-S., Lu, Y.-P. and Rea, P.A. (1997) The GS-X pump in plants, yeast and animal cells: structure, function and gene expression. *Biosci. Rep.*, **17** 189-207.

Jauh, G.Y., Phillips, T.E. and Rogers, J.C. (1999) Tonoplast intrinsic protein isoforms as markers for vacuolar functions. *Plant Cell*, **11** 1867-82.

Kitamura, T., Jansen, P., Hadenbrook, C., Kamimoto, Y., Gatmaitan, Z. and Arias, I.M. (1990) Defective ATP-dependent bile canalicular transport of organic anions in mutant (TR-) rats with conjugated hyperbilirubinemia. *Proc. Natl. Acad. Sci. USA*, **87** 3557-61.

Klein, M., Weissenböck, G., Dufaud, A., Gaillard, C., Kreuz, K. and Martinoia, E. (1996) Different energization mechanisms drive the vacuolar uptake of a flavonoid glucoside and a herbicide glucoside. *J. Biol. Chem.*, **271** 29666-71.

Klein, M., Martinoia, E. and Weissenböck, G. (1997) Transport of Lucifer Yellow CH into plant vacuoles: evidence for direct energization of a sulphonated substance and implications for the design of new molecular probes. *FEBS Lett.*, **420** 86-92.

Klein, M., Martinoia, E. and Weissenböck, G. (1998) Directly energized uptake of β-estradiol 17-(β-D-glucuronide) in plant vacuoles is strongly stimulated by glutathione conjugates. *J. Biol. Chem.*, **273** 262-70.

Klein, M., Martinoia, E., Hoffmann-Thoma, G. and Weissenböck, G. (2000) A membrane-potential dependent, ABC-like transporter mediates the vacuolar uptake of rye flavone glucuronides. *Plant J.*, **21** 289-304.

Kojima, M., Poulton, J.E., Thayer, S.S. and Conn, E.E. (1979) Tissue distributions of dhurrin and of enzymes involved in its metabolism in leaves of *Sorghum bicolor. Plant Physiol.*, **63** 1022-28.

König, J., Nies, A.T., Cui, Y., Leier, I. and Keppler, D. (1999) Conjugate export pumps of the multidrug resistance protein (MRP) family: localization, substrate specificity, and MRP2-mediated drug resistance. *Biochim. Biophys. Acta*, **1461** 377-94.

Kreuz, K., Tommasini, R. and Martinoia, E. (1996) Old enzymes for a new job. herbicide detoxification in plants. *Plant Physiol.*, **111** 349-53.

Levine, A., Tenhaken, R., Dixon, R. and Lamb, C. (1994) H_2O_2 from the oxidative burst orchestrates the plant hypersensitive disease resistance response. *Cell*, **79** 583-93.

Li, Z.-S., Zhao, Y. and Rea, P.A. (1995a) Magnesium adenosine 5′-trisphosphate-energized transport of glutathione *S*-conjugates by plant vacuolar membrane vesicles. *Plant Physiol.*, **107** 1257-68.

Li, Z.-S., Zhen, R.-G. and Rea, P.A. (1995b) 1-chloro-dinitrobenzene-elicited increase in vacuolar glutathione *S*-conjugate transport activity. *Plant Physiol.*, **109** 177-85.

Li, Z.-S., Szczypka, M., Lu, Y.-P., Thiele, D.J. and Rea, P.A. (1996) The yeast cadmium factor protein (YCF1) is a vacuolar glutathione *S*-conjugate pump. *J. Biol. Chem.*, **271** 6509-17.

Li, Z.-S., Alfenito, M., Rea, P.A., Walbot, V. and Dixon, R.A. (1997a) Vacuolar uptake of the phytoalexin, medicarpin, by the glutathione conjugate pump. *Phytochemistry*, **45** 689-93.

Li, Z.-S., Lu, Y.P., Zhen, R.-G., Szczypka, M., Thiele, D.J. and Rea, P.A. (1997b) A new pathway for vacuolar cadmium sequestration in *Saccharomyces cerevisae*: YCF1-catalyzed transport of *bis*(glutathionato)cadmium. *Proc. Natl. Acad. Sci. USA*, **94** 42-47.

Liu, G., Li, Z.-S., Bartholomew, D.M. and Rea, P.A. (2000) Enhanced multispecificity of *Arabidopsis* MRP-type ABC transporter, AtMRP2. *Plant Physiol.*, in press.

Loe, D.W., Almquist, K.C., Cole, S.P.C. and Deeley, R. (1996) ATP-dependent 17 β-estradiol 17-(β-D-glucuronide) transport by multidrug resistance protein (MRP): inhibition by cholestatic steroids. *J. Biol. Chem.*, **271** 9683-89.

Lu, Y.-P., Li, Z.-S. and Rea, P.A. (1997) *AtMRP1* gene of *Arabidopsis* encodes a glutathione-*S*-conjugate pump: isolation and functional definition of a plant ABC-transporter gene. *Proc. Natl. Acad. Sci. USA*, **94** 8243-48.

Lu, Y.-P., Li, Z.-S., Drozdowicz, Y., Hörtensteiner, S., Martinoia, E. and Rea, P.A. (1998) AtMRP2, an *Arabidopsis* ATP binding cassette transporter able to transport glutathione *S*-conjugates and chlorophyll catabolites: functional comparisons with AtMRP1. *Plant Cell*, **10** 267-82.

Lüttge, U. (1987) Carbon dioxide and water demand: crassulacean acid metabolism (CAM), a versatile ecological adaption exemplifying the need for integration in ecophysiological work. *New Phytol.*, **106** 593-629.

Ma, L., Nandigama, K. and Center, M.S. (1995) Phosphorylation of the multidrug resistance associated protein gene encoded protein, P190. *Biochemistry*, **34** 3338-43.

Margna, U. and Margna, E. (1978) Incorporation of exogenous L-phenylalanine into C-glucosylflavones in buckwheat cotyledons. *Biochem. Physiol. Pflanzen.*, **173** 2-10.

Marrs, K.A. (1996) The functions and regulation of glutathione *S*-transferases in plants. *Annu. Rev. Plant Physiol. Plant Mol. Biol.*, **47** 127-57.

Marrs, K., Alfenito, M.R., Lloyd, A.M. and Walbot, V. (1995) A glutathione *S*-transferase involved in vacuolar transfer encoded by the maize gene, *Bronze-2*. *Nature*, **375** 397-400.

Martinoia, E., Grill, E., Tommasini, R., Kreuz, K. and Amrhein, N. (1993) An ATP-dependent glutathione *S*-conjugate 'export' pump in the vacuolar membrane of plants. *Nature*, **364** 247-49.

Matern, U., Heller, W. and Himmelspach, K. (1983) Conformational changes of apigenin 7-*O*-(6-*O*-malonylglycoside), a vacuolar pigment from parsley, with solvent composition and proton concentration. *Eur. J. Biochem.*, **133** 439-48.

Matern, U., Reichenbach, C. and Heller, W. (1986) Efficient uptake of flavonoids into parsley (*Petroselinum hortense*) vacuoles requires acylated glycosides. *Planta*, **167** 183-89.

Matile, P. (1987) The sap of plant cells. *New Phytol.*, **105** 1-26.

Matile, P., Jans, B. and Rickenbacher, R. (1970) Vacuoles of *Chelidonium latex*: lysosomal property and accumulation of alkaloids. *Biochem. Physiol.*, **161** 447-58.

Matile, P., Hörtensteiner, S. and Thomas, H. (1999) Chlorophyll degradation. *Annu. Rev. Plant Physiol. Plant Mol. Biol.*, **50** 27-45.

Mende, P. and Wink, M. (1987) Uptake of the quinolizidine alkaloid, lupanine, by protoplasts and isolated vacuoles of suspension-cultured *Lupinus polyphyllus* cells: diffusion or carrier-mediated transport? *J. Plant Physiol.*, **129** 229-42.

Olsson, L.C., Veit, M., Weissenböck, G. and Bornman, J. (1998) Differential flavonoid response to enhanced UV-B radiation in *Brassica napus*: isolation, identification and quantification of flavonol glycosides. *Phytochem.*, **49** 1021-28.

Oude-Elferink, R.P.J., Meijer, D.K.F., Kuipers, F., Jansen, P.L.M., Groen, A.K. and Groothuis, G.M.M. (1995) Hepatobiliary secretion of organic compounds: molecular mechanisms of membrane transport. *Biochim. Biophys. Acta*, **1241** 215-68.

Paris, N., Stanley, C.M., Jones, R.L. and Rogers, J.C. (1996) Plant cells contain two functionally distinct vacuolar compartments. *Cell*, **85** 563-72.

Ramussen, S. and Dixon, R.A. (1999) Transgene-mediated and elicitor-induced perturbation of metabolic channeling at the entry point into the phenylpropanoid pathway. *Plant Cell*, **11** 1537-51.

Rataboul, P., Alibert, G., Boller, T. and Boudet, A.M. (1985) Intracellular transport and vacuolar accumulation of *o*-coumaric acid glucoside in *Melilotus alba* protoplasts. *Biochim. Biophys. Acta*, **816** 25-36.

Rea, P.A. (1999) MRP subfamily ABC transporters from plants and yeast. *J. Exp. Bot.*, **50** 895-913.

Rea, P.A. and Sanders, D. (1987) Tonoplast energization: two H^+ pumps, one membrane. *Physiol. Plant.*, **71** 131-41.

Rea, P.A. and Turner, J.C. (1990) Tonoplast adenosine triphosphatase and inorganic pyrophosphatase. *Methods Plant Biochem.*, **3** 385-405.

Rea, P.A. and Poole, R.J. (1993). Vacuolar H^+-translocating pyrophosphatase. *Annu. Rev. Plant Physiol. Plant Mol. Biol.*, **44** 157-80.

Rea, P.A., Li, Z.S., Lu, Y.P., Drozdowicz, Y.M. and Martinoia, E. (1998) From vacuolar GS-X pumps to multispecific ABC transporters. *Annu. Rev. Plant Physiol. Plant Mol. Biol.*, **49** 727-60.

Reuber, S., Bornman, J.F. and Weissenböck, G. (1996) A flavonoid mutant of barley (*Hordeum vulgare* L.) exhibits increased sensitivity to UV-B radiation in the primary leaf. *Plant Cell Environ.*, **19** 593-601.

Roberts, M.F., Homeyer, B.C. and Pham, T.D. (1991) Further studies of sequestration of alkaloids in *Papaver somniferum* L. latex vacuoles. *Z. Naturforsch.*, **46c** 377-88.

Sánchez-Fernández, R., Ardiles-Diáz, W., Van Montagu, M., Inzé, D. and May, M.J. (1998) Cloning and expression analyses of *AtMRP4*, a novel MRP-like gene from *Arabidopsis thaliana*. *Mol. Gen. Genet.*, **258** 655-62.

Sandermann, H. (1992) Plant metabolism of xenobiotics. *Trends Biochem. Sci.*, **17** 82-84.

Saunders, J.A. and Conn, E.E. (1978) Presence of the cyanogenic glucoside, dhurrin, in isolated vacuoles from *Sorghum*. *Plant Physiol.*, **61** 154-57.

Sharma, V. and Strack, D. (1985) Vacuolar localization of 1-sinapoylglucose: L-malate sinapoyltransferase in protoplasts from cotyledons of *Raphanus sativus*. *Planta*, **163** 563-68.

Sidler, M., Hassa, P., Hasan, S., Ringli, C. and Dudler, R. (1998) Involvement of an ABC transporter in a developmental pathway regulating hypocotyl cell elongation in the light. *Plant Cell*, **10** 1623-36.

Smart, C.C. and Fleming, A.J. (1996) Hormonal and environmental regulation of a plant PDR5-like ABC transporter. *J. Biol. Chem.*, **271** 19351-57.

Snyder, B.A. and Nicholson, R.L. (1990) Synthesis of phytoalexins in sorghum as a site-specific response to fungal ingress. *Science*, **248** 1637-39.

Stieger, B., O'Neill, B. and Meier, P.J. (1992) ATP-dependent bile salt transport in canalicular rat liver plasma membrane vesicles. *Biochem. J.*, **284** 67-74.

Stride, B.D., Cole, S.P.C. and Deeley, R.G. (1999) Localization of a substrate specificity domain in the multidrug resistance protein. *J. Biol. Chem.*, **274** 22877-83.

Swanson, S.J., Bethke, P.C. and Jones, R.L. (1998) Barley aleurone cells contain two types of vacuoles: characterization of lytic organelles by use of fluorescent probes. *Plant Cell*, **10** 685-98.

Szczypka, M.S., Wemmie, J.A., Moye-Rowley, W.S. and Thiele, D.S. (1994) A yeast metal resistance protein similar to human cystis fibrosis transmembrane conductance regulator (CFTR) and multidrug resistance-associated protein. *J. Biol. Chem.*, **269** 22853-57.

Sze, H., Li, H. and Palmgren, M. (1999) Energization of plant cell membranes by H^+-pumping ATPases: regulation and biosynthesis. *Plant Cell*, **11** 677-89.

Terrier, N., Deguilloux, C., Sauvage, F.-X., Martinoia, E. and Romieu, C. (1998) V-ATPase, inorganic pyrophosphatase and anion transport on the tonoplast of grape berries (*Vitis vinifera* L.). *Plant Physiol. Biochem.*, **36** 367-77.

Theodoulou, F., Clark, I.M., Pellett, K.E. and Hallahan, D.L. (1998) A herbicide safener-induced MRP homolog from wheat. *11th International Workshop on Plant Membrane Biology*, Cambridge, UK, 9–14 August 1998.

Tommasini, R., Evers, R., Vogt, E., Mornet, C., Zaman, G.H., Schinkel, A.H., Borst, P. and Martinoia, E. (1996) The human multidrug resistance-associated protein functionally complements the yeast cadmium factor 1. *Proc. Natl. Acad. Sci. USA*, **93** 6743-48.

Tommasini, R., Vogt, E., Fromenteau, M., Hörtensteiner, S., Matile, P., Amrhein, N. and Martinoia, E. (1998) An ABC-transporter of *Arabidopsis thaliana* has both glutathione-conjugate and chlorophyll catabolite transport activity. *Plant J.*, **13** 773-80.

Turlings, T.C. and Benrey, B. (1998) Effects of plant metabolites on the behaviour and development of parasitic wasps. *Ecoscience*, **5** 321-33.

Varin, L., DeLuca, V., Ibrahim, R.K. and Brisson, N. (1992) Molecular characterization of two plant flavonol sulfotransferases. *Proc. Natl. Acad. Sci. USA*, **89** 1286-90.

Walker, J.E., Saraste, M., Runswick, M.J. and Gay, N.J. (1982) Distantly related sequences in α- and β-subunits of ATP synthase, myosin, kinases and other ATP-requiring enzymes and a common nucleotide binding fold. *EMBO J.*, **1** 945-51.

Wang, W., Takezawa, D. and Poovaiah, B.W.A. (1996) Potato cDNA encoding a homolog of mammalian multidrug resistant P-glycoprotein. *Plant Mol. Biol.*, **31** 683-87.

Werner, C. and Matile, P. (1985) Accumulation of coumaryl glucosides in vacuoles of barley mesophyll protoplasts. *J. Plant Physiol.*, **118** 237-49.

Wink, M. (1997) Compartmentation of secondary metabolites and xenobiotics in plant vacuoles, in *The Plant Vacuole: Advances in Botanical Research*, Vol. 25 (eds. R.A. Leigh, D. Sanders and J.A. Callow), Academic Press, London, New York, pp. 141-70.

Wolf, A.E., Dietz, K.-J. and Schröder, P. (1996) Degradation of glutathione *S*-conjugates by a carboxypeptidase in the plant vacuole. *FEBS Lett.*, **384** 31-34.

Zhen, R.-G., Kim, E.J. and Rea, P.A. (1997) The molecular and biochemical basis of pyrophosphatase-energized proton translocation at the vacuolar membrane, in *The Plant Vacuole: Advances in Botanical Research*, Vol. 25 (eds. R.A. Leigh, D. Sanders and J.A. Callow), Academic Press, London, New York, pp. 297-337.

Zimmerlin, A. and Durst, F. (1992) Acyl hydroxylation of the herbicide, diclofop, by a wheat cytochrome P-450 monooxygenase: substrate specificity and physiological activity. *Plant Physiol.*, **100** 874-81.

11 GFP as a marker for vacuoles in plants

Jean-Marc Neuhaus

11.1 Introduction

Vacuolar targeting in plants has been studied mostly with natural vacuolar proteins that were modified and tested for their new location. These proteins were usually detected with antibodies, and properly sorted versus missorted fractions were quantified after separation by gel electrophoresis. After the pioneering work on ricin (Lord, 1985) and phytohemagglutinin (Tague and Chrispeels, 1987; von Schaewen and Chrispeels, 1993), other studies utilized barley lectin (Wilkins *et al.*, 1990; Bednarek and Raikhel, 1991), barley aleurain (Holwerda *et al.*, 1990, 1992) and sweet potato sporamin (Matsuoka *et al.*, 1990, Matsuoka and Nakamura, 1991). Only for tobacco chitinase was intrinsic enzyme activity used for quantitation, but this was complicated by the non-linearity of the dose-response curve and the presence of endogenous activity in the test plants (Neuhaus *et al.*, 1991). Fusion proteins constructed from reporter enzymes were of limited use. Invertase was useful in yeast, but in plants it affected sugar metabolism and became an interesting tool for metabolic studies (Sonnewald *et al.*, 1991). Chloramphenicol acetyl transferase found only limited use (Jung *et al.*, 1993). The popular reporter for gene expression, E. coli β-glucuronidase (GUS), was found to be to be inadequate for the secretory pathway, because this cytosolic protein harbors a cryptic N-glycosylation site (Firek *et al.*, 1994) and free cysteines, which tend to form inappropriate disulfide bonds in the oxidative environment of the endoplasmic reticulum (ER) (Piñeiro *et al.*, 1994). A rat β-glucuronidase gave promising results in protoplasts but was not expressed adequately in transgenic tobacco (Di Sansebastiano, 1999).

A new reporter protein with remarkable properties recently became available, the green fluorescent protein (GFP). Originally cloned from the jellyfish, *Aequorea victoria*, GFP is a reporter protein which can be detected in a non-invasive way in plants, animals, fungi and prokaryotes (Chalfie *et al.*, 1994). It is 238 amino acids long and folds into an 11-stranded β-barrel surrounding a central α-helix. The fluorophore forms by spontaneous cyclization and oxidation of the three amino acids, Ser65, Tyr66 and Gly67 (Niwa *et al.*, 1996). Only oxygen is required, and there is no need for an added co-factor. The wild-type GFP absorbs both UV light (absorbance band A, 395 nm) and blue light (absorbance band B, 475 nm) and emits green light (508 nm). GFP is small and compact and can be fused both N- and C-terminally to other proteins. N- and C-termini of GFP are close together on one end of the β-barrel, which allows the insertion of GFP even into an internal loop of another protein. Proteins may also be fused within a loop of GFP itself, as long as their ends are near each other (Abedi *et al.*, 1998). Spectral variants have been generated, which absorb either UV or visible light, and emit at variously shifted wavelengths compared to the original wild-type GFP ('blue', 'cyan', 'yellow'). For reviews on the use of GFP in plants, see Haseloff (1999) or Neuhaus and Boevink (2000).

GFP has only recently been used to study the plant secretory pathway. There were early problems linked with the expression of a jellyfish gene in plants, namely codon usage and the presence of a cryptic intron (Haseloff *et al.*, 1997). These two problems have been resolved and several variants with optimized codon usage are available (Pang *et al.*, 1996; Rouwendal *et al.*, 1997; Davis and Vierstra, 1998). The next problem was specific to the secretory pathway: it appears that GFP is more temperature sensitive in the ER than in the cytosol. Isolation of a thermostable GFP was the key to the successful use of GFP in the whole secretory pathway (Siemering *et al.*, 1996; Haseloff *et al.*, 1997).

Analysis of several soluble vacuolar proteins has led to the identification of three different types of vacuolar sorting determinants (VSDs): sequence-specific VSDs (ssVSD) containing an essential Ile or Leu and often, but not obligately, located within N-terminal propeptides (also known as NTPPs); C-terminal VSDs (ctVSDs, also known as CTPPs) with no specific motif but which can be inactivated by the addition of terminal Gly; and less well-characterized internal or protein structure-dependent VSDs (psVSDs), which may mediate aggregation or membrane association. The best characterized psVSD is probably the internal saposin-like domain of phytepsin, an aspartic proteinase from barley (Kervinen *et al.*,1999, for reviews, see Neuhaus and Rogers, 1998; and Matsuoka and Neuhaus, 1999). Examples for two of these VSDs have been fused to GFP as described below. Tonoplast proteins have also been fused to GFP but only preliminary results are available so far.

Only a few papers on the subject covered by this chapter had been published at the time of writing. The discussion thus presents mostly unpublished results from the author's laboratory, and from other researchers who agreed to provide unpublished data and pictures. Because this field of research is early in its development, there are some difficulties in comparing results from different systems. Each laboratory uses its own favorite experimental system but, as will be discussed, results may vary considerably between different species and different cell types. It can be expected that the next few years will bring considerable progress, as more results accumulate and become more readily comparable.

11.2 Microscopic evidence for the coexistence of two different vacuoles

Biochemists and molecular biologists have usually worked on the assumption that there is only a single type of vacuole in plant cells, at least in their experimental system. The experimental tools used previously did not allow one to distinguish whether all markers used to identify vacuoles were in the same compartment. In the presence of a single central vacuole, for example in mesophyll protoplasts, the likelihood that all markers would be together in that vacuole would seem obvious. It was, therefore, puzzling to find different types of vacuolar sorting determinants and even to detect differences in sensitivity to the drug, wortmannin (Matsuoka *et al.*, 1995). These results suggested that at least two of the VSDs corresponded to separate transport pathways. One microscopic study was devoted to the comparison of these two pathways. In transgenic tobacco plants expressing both barley lectin, a protein carrying a C-terminal VSD, and sweet potato sporamin, a protein carrying a sequence-specific VSD, both proteins were found in the same vacuoles (Schroeder *et al.*, 1993).

On the other hand, microscopists have observed in certain tissues of other plants vacuoles that obviously differed in content. For example, apple leaf cells harbor vacuoles of similar size, but only some contain an electron-dense material, presumably tannins (Michel *et al.*, 1992). A specialized tissue with different types of vacuole can be found in the touch-sensitive plant, *Mimosa pudica* L. The pulvini are the motor organs that control seismonasty, the rapid leaf movement induced by touch (for a review on these organs, see Fleurat-Lessard, 1988). Within pulvini, motor cells are characterized by the presence of two types of vacuole: the tannin vacuole, which appears first during tissue differentiation; and the colloidal vacuole, which develops later to become much larger in adult cells. The colloidal vacuole usually appears transparent under the light microscope and stains poorly for electron microscopy. Sometimes it contains fibrillar material. In contrast, the tannin vacuole stains with several reagents, which react with tannins, phenolic compounds and sugar condensates. It also accumulates neutral red, indicating an acidic pH. The tannin vacuole is located near the nucleus and is convex at the zone of close apposition of the two vacuoles. It seems to contain more osmotically active solutes as it swells under hypotonic conditions, while the colloidal vacuole concomitantly shrinks (Toriyama and Sato, 1968). The tannin vacuole contains electron-dense, i.e. osmiophilic, material that appears either to fill the tannin vacuole or to lie under the tonoplast, leaving a less stained vacuolar interior (Fleurat-Lessard and Millet, 1984). Upon stimulation, motor cells shrink within a second due to the massive efflux of potassium ions followed by water fluxes from the symplast to the apoplast. Shrinkage is accompanied by shape changes: the colloidal vacuole develops invaginations, while the tannin vacuole forms interconnected tubules. Immunoelectron microscopy revealed that an aquaporin detected by antibodies obtained against radish γ-tonoplast intrinsic protein (TIP), VM23, was at least 10 times more abundant per unit length in the tonoplast of the colloidal vacuole than in the tonoplast of the tannin vacuole (Fleurat-Lessard *et al.*, 1997). The same preferential labeling was observed with antibodies directed against the α

subunit of the vacuolar-type hydrogen-transporting ATPase (H^+-ATPase) from *Mesembryanthemum crystallinum*. These results were interpreted as reflecting the much more dramatic fluxes of solutes and water that must cross the tonoplast of the larger colloidal vacuole. It would of course be interesting to know the distribution of other TIPs in the same system.

In legume cotyledons, the protein storage vacuole develops as a new organelle distinct from the pre-existing lytic vacuole. It appears to develop originally from the ER, but the storage proteins are transported via the Golgi apparatus. Both vacuoles can be seen to coexist within a cell (Robinson and Hinz, 1997; see also Chapter 12 in this volume).

Immunolocalization of various TIPs and soluble markers has indicated that in root cells different small vacuolar compartments can coexist. In barley root cells, small vacuoles harbouring γ-TIP also contained aleurain, while small vacuoles harbouring α-TIP also contained the barley lectin (Paris *et al.*, 1996). In a more recent study, antibodies directed against three different TIPs revealed the possible coexistence of three different vacuolar compartments, or indeed up to seven, since all combinations of two or even three TIPs were observed (Jauh *et al.*, 1999; see also Chapter 12 in this volume).

In conclusion, there are cases when plant cells clearly contain different types of vacuole, though many cells seem to have only a single central vacuole or a few indistinguishable vacuoles. The possibility of a hybrid nature for a single central vacuole is suggested by the finding of vacuoles with two or three different TIPs, or the colocalization of soluble vacuolar proteins targeted by different vacuolar sorting determinants.

11.3 GFP targeted to a neutral vacuole

For many years, tobacco chitinase A has been used to study targeting to vacuoles. Comparison with secreted isoforms suggested that a C-terminal propeptide could be involved in this process. This seven amino-acids-long propeptide turned out to be both necessary and sufficient for vacuolar targeting in tobacco mesophyll protoplasts and in transgenic tobacco (Neuhaus *et al.*, 1991). Mutant analysis indicated that there was very little sequence specificity required for the function of this VSD, as the last six amino acids could be, for example, either all hydrophilic or all hydrophobic. Only a terminal Gly was detrimental to the function (Neuhaus *et al.*, 1994). This propeptide was sufficient to target several unrelated proteins to the vacuole in protoplasts: cucumber chitinase, hen egg white lysozyme and rat β-glucuronidase (Freydl, 1994; Di Sansebastiano, 1999). It was, therefore, a good first candidate for use in targeting of GFP to vacuoles.

The ER-targeted mgfp5-ER from Jim Haseloff (Haseloff *et al.*, 1997) was modified by replacing the C-terminal ER-retention signal with the C-terminal chitinase propeptide. This modified GFP is designated GFP-Chi. Expression in tobacco mesophyll protoplasts resulted in green fluorescent central vacuoles in a portion of the cells. This confirmed, once more, that these seven amino acids are sufficient to target otherwise secreted proteins to vacuoles. At early times of expression, after 6–8 h, GFP-Chi was observed to accumulate in the nuclear envelope and in the ER, before appearing in a few larger, peripheral compartments and finally in the central vacuole (Figure 11.1A and B) (Di Sansebastiano *et al.*, 1998).

However, in many protoplasts GFP-Chi further accumulated in the peripheral compartments and failed to appear in the central vacuole, even after prolonged incubation (Figure 11.1C and D). This pattern was found in a quarter of the protoplasts that contained many chloroplasts, presumably derived from palisade mesophyll, but in the majority of the other, chloroplast-poor protoplasts, which are a more heterogeneous population probably derived from several different cell types in the leaf. Staining of the protoplasts with neutral red, a dye trapped by protonation in acidic compartments and often used as a vacuole stain, revealed that GFP-Chi was excluded from the acidic vacuoles and accumulated in neutral vacuoles. It also indicated that in all protoplasts there was generally a central vacuole of one kind (acidic or neutral) and peripheral compartments of the other, depending on the cell type (Di Sansebastiano *et al.*, 1998). Because of their size

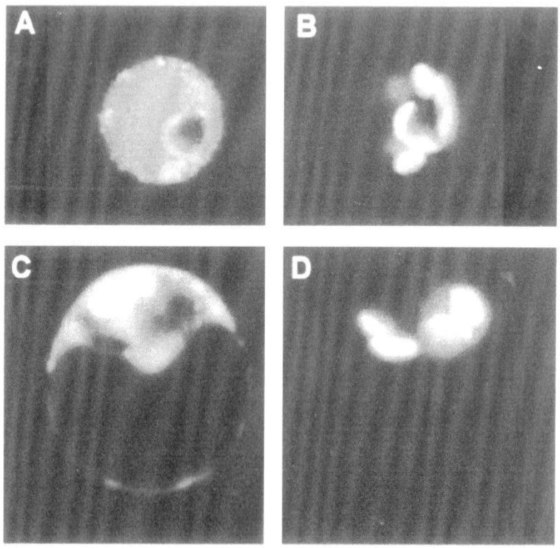

Figure 11.1 Green fluorescent protein (GFP) targeted to a neutral vacuole by the C-terminal vacuolar sorting determinant (ctVSD) of tobacco chitinase A. The figure shows tobacco mesophyll protoplasts that accumulate GFP-Chi in the large central vacuole (A and B) or in a small vacuole (C and D). (A) and (C) show the green fluorescence of GFP (C) and (D) show the red fluorescence of chloroplasts.

and longevity, the peripheral, neutral, GFP-Chi-accumulating compartments will be referred to here as 'peripheral vacuoles'. On the other hand, vacuoles accumulating GFP-Chi will be called 'neutral' vacuoles rather than 'storage' vacuoles, because it is not yet clear whether the same vacuoles would in all cases also accumulate storage proteins, such as those transported by smooth dense vesicles in legumes (Hohl *et al.*, 1996).

In BY-2 suspension cells, a tobacco cell line often used for cell biology experiments, GFP-Chi was never observed in the central vacuole, and only the ER became fluorescent (Gaire, 1998). In transgenic tobacco, only very few cells harbored small fluorescent compartments, mostly in young roots and in hypocotyl. In contrast, each guard cell contained a single round fluorescent vacuole, while the larger central vacuole was not fluorescent. When protoplasts were isolated from these plants, they quickly became highly fluorescent with the same patterns as previously observed in transient expression experiments. When transgenic calli were induced to form from leaves in tissue culture, they also fluoresced strongly, but the fluorescence was exclusively confined to many small compartments (preliminary observations).

11.4 GFP targeted to an acidic vacuole

It was an obvious next step to target a GFP reporter to the acidic vacuoles, but this was not in fact trivial. When fused to the N-terminal propeptide of barley aleurain, which contains a sequence-specific VSD (Holwerda *et al.*, 1992), the same variant of GFP (mgfp5, described above) failed to produce any fluorescent vacuoles in tobacco protoplasts. Only small fluorescent compartments could be observed at the periphery of the cell.

GFP has often been described as protease-resistant, based on experiments with animal cells. However, plants have different proteases that might more easily degrade this protein. In order to be able to detect, in lytic vacuoles, a smaller concentration of GFP that had not been degraded, the author and co-workers introduced into mGFP5 mutations that were found to enhance the fluorescence of GFP under blue light illumination at the expense of suppressing the UV absorption peak. These mutations, F64L and S65T, modify the fluorophore and change its protonation behavior (Cormack *et al.*, 1996). This new variant was called 'GFP6'. Haseloff independently produced the same variant and called it 'mGFP6' (Haseloff, 1999).

With this enhanced GFP, it became possible to visualize targeting of a barley aleurain propeptide-GFP6 fusion (called here Aleu-GFP) to the lytic vacuole. Transient expression in tobacco protoplasts led to the rapid appearance of small intensely fluorescing grains. After 24 h, about 15% of all green fluorescent protoplasts harboured a fluorescent central vacuole, irrespective of the cell type (Figure 11.2A and B). The fluorescence was much weaker than with GFP-Chi

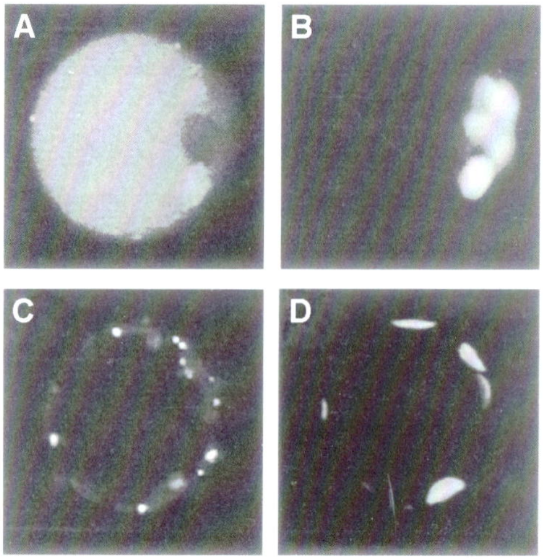

Figure 11.2 Green fluorescent protein (GFP) targeted to a lytic vacuole by the sequence-specific vacuolar sorting determinant (ssVSD) of barley aleurain. The figure shows tobacco mesophyll protoplasts that accumulate Aleu-GFP in the large central vacuole (A and B) or in small vacuoles (C and D). (A) and (C) show the green fluorescence of GFP (C) and (D) show the red fluorescence of chloroplasts.

and decayed rapidly. Whether the small green grains (Figure 11.2C and D) are intermediate compartments or small (peripheral) lytic vacuoles, such as those described in root cells (Paris *et al.*, 1996; Jauh *et al.*, 1999) cannot be determined at this stage. Upon staining with neutral red, however, it was possible to observe red vacuoles that were also labeled with Aleu-GFP. Interestingly, neutral red staining was much less efficient than in control protoplasts or protoplasts expressing GFP-Chi. This observation would be consistent with the concept that, when Aleu-GFP is targeted to acidic vacuoles, it in some way inhibits acidification (Di Sansebastiano *et al.*, 2000).

Vacuole regeneration in evacuolated protoplasts (miniprotoplasts) is a way to study the *de novo* formation of a vacuole. Regeneration of a central vacuole occurs within 2 days, and at this time increased levels of hydrolytic enzymes and of tonoplast H^+-pyrophosphatase can be measured. Neutral red accumulation is also restored (Hörtensteiner *et al.*, 1992). When GFP-Chi was expressed in miniprotoplasts, it accumulated in peripheral compartments similar in size to those seen in intact protoplasts. When the central vacuole became visible, it did not contain GFP-Chi, which continued to accumulate in peripheral vacuoles and remained there in the majority of the cells for several days. In about a third of the cells however, it suddenly appeared in the central vacuole, where a fluorescent gradient could sometimes be observed, suggesting a recent fusion between the two different vacuoles. When Aleu-GFP was expressed in miniprotoplasts, it rapidly appeared in small fluorescent grains as seen in protoplasts. When the central vacuole reappeared, it was intensely fluorescent. The fluorescence decayed over the following days. These results indicate that the regenerating central vacuole is always of the lytic and never of the neutral type. It also seems that, after the completed regeneration, there is a fusion of both types of vacuoles, the central lytic vacuole and the peripheral neutral vacuoles, in about a third of the cells (Di Sansebastiano *et al.*, 2000).

11.5 Identification of vacuoles in whole plants

Protoplasts are an artificial system with practical advantages, but extrapolation of results to whole plants is problematical. After all, what we want to know is the dynamics of the vacuolar system in differentiated tissues in whole plants. For this study transgenic plants are needed.

The author and co-workers have transformed tobacco with both types of vacuolar GFP reporter proteins, with unexpected results. They failed to obtain any fluorescent tobacco plants expressing Aleu-GFP. With GFP-Chi it appeared to be difficult to obtain any fluorescent plants until it was realized that seemingly negative plants had, in fact, fluorescent stomata. Close examination revealed the presence in each guard cell of a small round or oval green vacuole next to a non-fluorescent large vacuole. Screening other tissues, revealed a few cells in the hypocotyl, elongating roots or the bases of trichomes with one or a few small green vacuoles. As mentioned in Section 11.4, leaf protoplasts derived from such plants were strongly fluorescent with patterns similar to those obtained in transient expression.

Considering the short life of Aleu-GFP in the lytic vacuole of protoplasts, it is possible that the same protein produced at a lower rate in transgenic tissues is continuously degraded in lytic vacuoles of tobacco. On the other hand, reporter proteins targeted by the two types of VSD, that is, a ssVSD for sporamin and a ctVSD for barley lectin, were immunolocalized in the same vacuoles of transgenic tobacco (Schroeder *et al.*, 1993). Our hypothesis to explain these observations is that, in tobacco, essentially all cell types harbor hybrid vacuoles, where GFP is rapidly degraded by lytic enzymes. Only in special cell types, such as guard cells, or under special conditions, such as callus growth or protoplast isolation, do the two separate transport pathways terminate at separate vacuoles. Unfortunately, it appears that the popular tobacco cell suspension line, BY-2, also has a single hybrid vacuole type in which GFP is rapidly degraded.

Transgenic *Arabidopsis thaliana* plants expressing either vacuolar GFP marker have been generated. Here the results are quite different from those obtained in tobacco. Both GFPs are expressed visibly but with different patterns. The precise characterization is still in progress, but clear results were obtained with the following tissues. Aleu-GFP accumulates in the central vacuole of epidermal cells but not of mesophyll cells (Figure 11.3A and B). Aleu-GFP also accumulates in a large vacuole in trichomes (but with only a weak fluorescence),

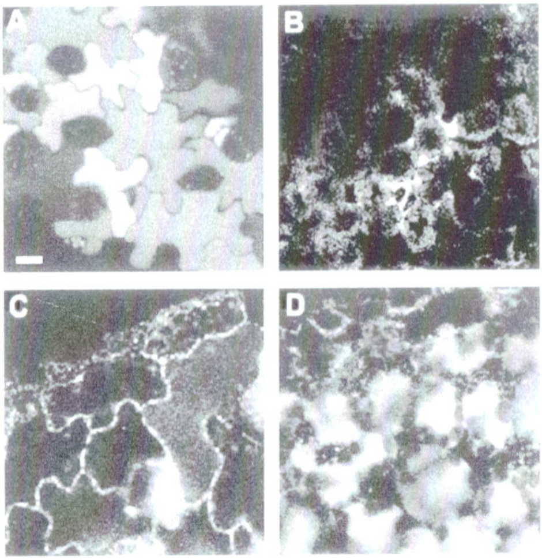

Figure 11.3 Leaves of transgenic *Arabidopsis thaliana* expressing a green fluorescent protein (GFP) targeted to the lytic vacuole (A and B) or to the neutral vacuole (C and D). (A) and (C) show a tangential section through the epidermis, while (B) and (D) show a tangential section through the mesophyll. Only the green fluorescence is shown. The differences in intensities between cells in (A) and (D) are due, in part, to the differences in depth of the section.

in root hairs, and in the root epidermis and endodermis, while other root tissues have only small green vacuoles. In mature leaves, GFP-Chi accumulates in the central vacuoles of mesophyll cells but not of the epidermal cells, where it is found instead at the periphery in small compartments (Figure 11.3C and D). In trichomes, root hairs and other root cells there are only small green vacuoles.

How can these results be compared with those obtained in tobacco? It appears that GFP is much more stable in *Arabidopsis* vacuoles, whether neutral or lytic, so *Arabidopsis* is a much better model plant to use for these studies. Additionally, there are several cell types that show separate vacuolar compartments of the two types, one forming the large central vacuole, while the other, if detected at all, is present as small peripheral compartments. There is also a negative consequence from this comparison. If a difference in the vacuole type in the same tissue of two different plants is observed using this assay system, we must assume that results obtained in one species do not necessarily apply to any other plant species.

11.6 Targeting of membrane-bound GFP to tonoplast and intermediate compartments of the secretory pathway

The sorting systems implicated in the transport of proteins to the tonoplast are much less well-characterized than those for soluble vacuolar proteins. It was proposed that tonoplast proteins and soluble vacuolar proteins are transported by different mechanisms, and that the tonoplast may be the default destination for membrane proteins in plant cells, as suggested also for yeast (Gomez and Chrispeels, 1993; Barrieu and Chrispeels, 1999). Other studies, however, have indicated the presence of at least two independent pathways for integral membrane protein traffic to vacuoles in plant cells (Jiang and Rogers, 1998). As there are many genes for TIPs (Weig *et al.*, 1997; Neuhaus and Roger, 1998), it is not always clear which isoforms are recognized by antibodies in immunolocalization studies, especially when the antibodies were obtained against a TIP from a different species. However, anti-peptide antibodies with isoform specificity have been generated (Jauh *et al.*, 1999). In this context, it was an obvious step to fuse GFP to tonoplast proteins to visualize their trafficking. Unpublished preliminary results obtained with a γ-TIP and a Ca^{2+}-channel were made available for this review.

TIP VM23 is an aquaporin of the γ-TIP subfamily that was cloned from radish (*Raphanus sativus*) and shown to have water-conductive properties (Higuchi *et al.*, 1998). GFP was fused to its N- or C-terminal extremities and the gene encoding this construct was introduced into tobacco BY-2 cells. With both hybrid proteins, transgenic BY-2 cells develop a beautiful pattern of green fluorescent membranes within the cells. All vacuoles seem to be labeled (Figure 11.4; and M. Maeshima, personal communication). The same TIP VM23 was used to produce a different fusion with GFP. In order to avoid interference between

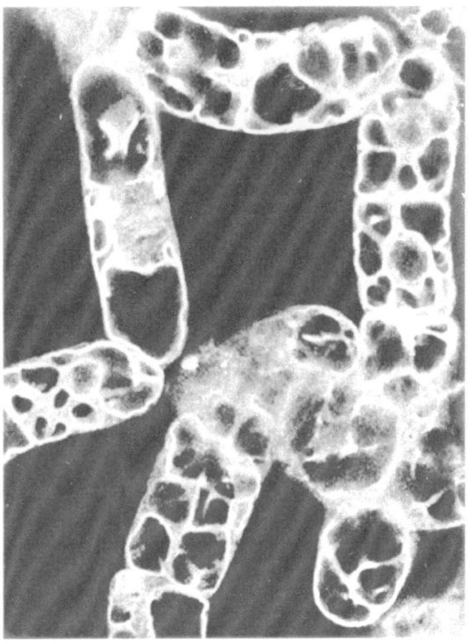

Figure 11.4 Tonoplast labeling in tobacco BY-2 cells by a green fluorescent protein (GFP) fused to the C-terminus of radish. γ-tonoplast intrinsic protein (γ-TIP VM23). All vacuoles are labeled (Internal scale bar = 10 μ). (The photograph was kindly provided by M. Maeshima).

the sorting machinery on the cytosolic face of secretory membranes and GFP, it was decided not to use the termini of the TIP, which are both cytosolic, but to insert the GFP into the third loop of the TIP, which is a variable region on the lumenal face of the tonoplast and which links the two three helix bundles that form the predicted structure of the aquaporins (Mathai, 1999). However, so far there is no proof that such a hybrid would still be functional. The hybrid was expressed in tobacco mesophyll protoplasts. After 1 day, the labeling was mostly reticular and perinuclear. After 2 days, the TIP-GFP hybrid labeled small round structures, which were sometimes large enough to allow clear separation of the labeled periphery from the unlabeled interior. In the same cell, large vacuole was unlabeled (Figure 11.5). Considering the short life of soluble GFP targeted to the lytic vacuole (see Section 11.4), it is possible that the label, in fact, slowly reached the large vacuole and was continuously degraded (N. Frangne, personal communication).

Another tonoplast protein was also fused with GFP, the *Arabidopsis* Ca^{2+}-ATPase isoform 4 (At-AC4), which had been cloned by homology with the cauliflower tonoplast Ca^{2+}-ATPase (Askerlund, 1997). In order to confirm the tonoplast localization of At-AC4, GFP was fused to its C-terminus, which is presumably cytosolic. While transgenic *Arabidopsis* plants were not obtained, transient expression was possible in *Arabidopsis* protoplasts, which were derived from cell suspension T87 (Axelos *et al.*, 1992). After 24 h expression, the

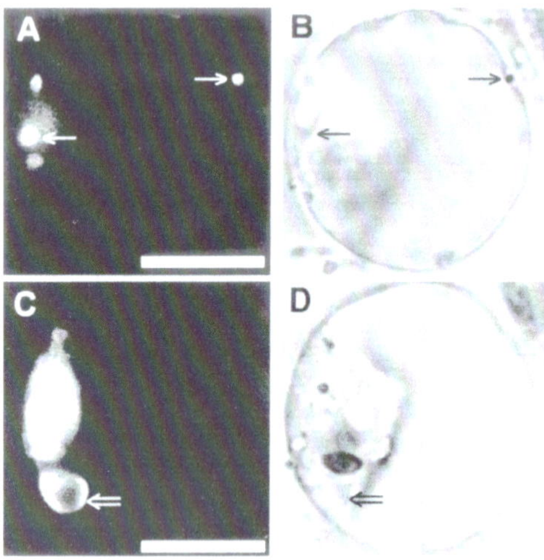

Figure 11.5 Tonoplast labeling in tobacco mesophyll protoplasts by green fluorescent protein (GFP) fused into the third loop of radish γ-tonoplast intrinsic protein (γ-TIP VM23). Two different sections (A+B, C+D) of the same protoplast are shown by fluorescence (A and C) and light imaging (B and D). Small round compartments (arrows) are labeled, while the central vacuole remains unlabeled. (The TIP clone was provided by M. Maeshima; the fusion and the expression in protoplasts were made by N. Frangne. Internal Scale bar = 10 μm.)

At-AC4-GFP protein labeled the periphery of small round compartments, but not the large vacuole (Figure 11.6; and M. Geisler, personal communication).

While these results are very promising, they are difficult to compare and interpret. They were obtained with different fusion protein constructs and in different cells. It has not yet been decided, whether the labeling of large vacuoles in BY-2 suspension culture cells and of small compartments in tobacco mesophyll protoplasts are conflicting observations, or whether the small round compartments labeled in tobacco and *Arabidopsis* are the same. A significant advance will be obtained as soon as several such GFP hybrids have been expressed in the same system and as soon as several transformation systems have been tested with the same hybrid.

It would be extremely useful to visualize not only the vacuoles but also the intermediate compartments through which the vacuolar proteins are transported. The ER was visualized by a GFP reporter protein carrying the ER-retention signal, HDEL (Boevink *et al.*, 1996; Haseloff *et al.*, 1997). The fine lace of mobile fluorescent ER is an amazing sight! The Golgi apparatus was visualized by the fusion of the N-terminal transmembrane segment of an animal sialyltransferase with GFP (Boevink *et al.*, 1998; Wee *et al.*, 1998), and with the coding sequence for soybean α1,2 mannosidase I (Nebenfuhr *et al.*, 1999). There is even a marker for both ER and Golgi, a GFP fusion with ERD2, the receptor for ER-retained proteins, which cycles between the two compartments, but is more concentrated in the Golgi. At optimum resolution, the crescent shape of the plant Golgi could be recognized (Boevink *et al.*, 1998). Films were produced that showed the movement of dictyosomes along tracks of ER, and staining of actin revealed that they are associated with microfilaments (http://www.brookes.ac.uk/schools/bms/research/molcell/hawes/gfp/gfp. html).

Further GFP fusions have been produced, especially with the putative vacuolar sorting receptor, VSR-Ps1 (Paris *et al.*, 1997; N. Paris, unpublished results). At this point, it is difficult to interpret the results, as small bright dots can be many

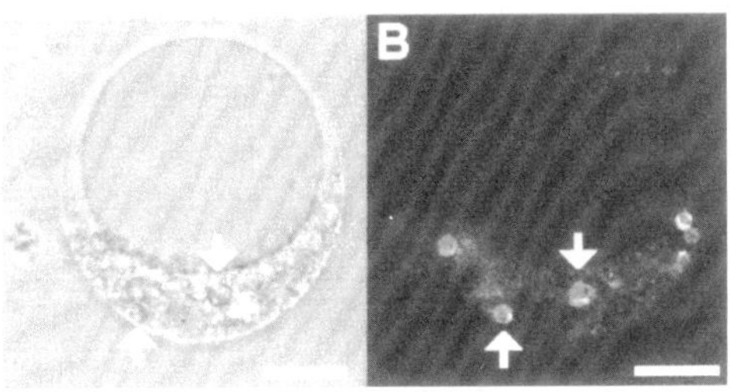

Figure 11.6 Tonoplast labeling in *Arabidopsis* protoplasts derived from cell suspension T87 by green fluorescent protein (GFP) fused to the C-terminus of the *Arabidopsis* Ca^{2+}-ATPase isoform 4. The same section of a protoplast is shown by light (A) and fluorescence imaging (B). Small round compartments (arrows) are labeled, while the central vacuole remains unlabeled. (The photograph was kindly provided by M. Geisler and M.G. Palmgren. Internal Scale bar = 10 μm.)

things, although they presumably correspond to Golgi and lytic prevacuolar compartments (Paris *et al.*, 1997; Jiang and Rogers, 1998). The aim of these approaches is to identify typical patterns for different compartments of plant cells and thereby to provide an easy comparison for the localization of new marker proteins, as can often be done in animal cells. This will, however, also require the use of different variants of GFP that can be co-expressed in the same cells and can be detected separately.

11.7 Conclusions

The use of GFP as a marker in plant cell biology is just beginning, and we must still establish solid bases for its use. For vacuoles, the results so far have already brought new insights. Firstly, it is clear that different types of vacuole can be found, not only in very young cells with many small vacuoles but also in more highly vacuolated cells. These highly vacuolated cells are present both in protoplasts and in whole plants, not only in special tissues like *Mimosa* pulvini or developing cotyledons, but also in the favorite model plant of molecular biologists, *Arabidopsis thaliana*. Secondly, the results differ strongly between different plants. This means that we have to be cautious when drawing general conclusions from experiments with one plant. Results obtained on tobacco, such as the colocalization of sporamin and barley lectin (Schroeder *et al.*, 1993) may not apply to *Arabidopsis*. The plasticity of the vacuolar system that these results suggest is opening new avenues of research. If TIPs reflect the identity of vacuoles, as has been suggested (Neuhaus and Rogers, 1998, Jauh *et al.*, 1999) we can expect a lot more complexity than was envisaged until recently. Furthermore, there is the possibility of combinations of two or three TIPs in a single vacuolar compartment, which may be linked to a different functional identity (Jauh *et al.*, 1999). Finally, expression pattern, and possibly function, of TIPs may differ from one plant species to another (Karlsson *et al.*, 2000). What, then, is the definition of a vacuole? Is it only a compartment larger than a certain limit, does the definition depend on the content of the organelle, or is any compartment harboring a TIP a vacuole, provided it is an end compartment of a transport pathway under given conditions? New approaches to the study of vacuole biogenesis, function, and interactions with other organelles, such as those described here, should help to clarify the situation in the next few years.

Acknowledgments

The author wishes to thank all colleagues who provided unpublished results, illustrations and discussions: Masayoshi Maeshima (Nagoya, Japan), Enrico Martinoia, Nathalie Frangne, Markus Geisler, Nadine Paris, Ricardo Flückiger (all Neuchâtel), Gian Pietro Di Sansebastiano (now at Wye College, England),

M.G. Palmgren (Copenhagen, Denmark). This work is supported by the Swiss National Science Foundation, grant 31-46926.96.

References

Abedi, M.R., Caponigro, G. and Kamb, A. (1998) Green fluorescent protein as a scaffold for intracellular presentation of peptides. *Nucleic Acids Res.*, **26** 623-30.

Askerlund, P. (1997) Calmodulin-stimulated Ca^{2+}-ATPases in the vacuolar and plasma membranes in cauliflower. *Plant Physiol.*, **114** 999-1007.

Axelos, M., Curie, C., Mazzolini, L., Bardet, C. and Lescure, B. (1992) A protocol for transient expression in *Arabidopsis thaliana* protoplasts isolated from cell suspension cultures. *Plant Physiol. Biochem.*, **30** 123-28.

Barrieu, F. and Chrispeels, M.J. (1999) Delivery of a secreted soluble protein to the vacuole via a membrane anchor. *Plant Physiol.*, **120** 961-68.

Bednarek, S. and Raikhel, N.V. (1991) The barley lectin carboxyl-terminal propeptide is a vacuolar protein sorting determinant in plants. *Plant Cell*, **3** 1195-206.

Boevink, P., Santa Cruz, S., Hawes, C., Harris, N. and Oparka, K.J. (1996) Virus-mediated delivery of the green fluorescent protein to the endoplasmic reticulum of plant cells. *Plant J.*, **10** 935-41.

Boevink, P., Oparka, K., Cruz, S.S., Martin, B., Betteridge, A. and Hawes, C. (1998) Stacks on tracks–the plant Golgi apparatus traffics on an actin/Er network. *Plant J.*, **15** 441-47.

Chalfie, M., Tu, Y., Euskirchen, G., Ward, W.W. and Prasher, D.C. (1994) Green fluorescent protein as a marker for gene expression. *Science*, **263** 802-805.

Cormack, B.P., Valdivia, R.H. and Falkow, S. (1996) FACS-optimized mutants of the green fluorescent protein (GFP). *Gene*, **173** 33-38.

Davis, S.J. and Vierstra, R.D. (1998) Soluble, highly fluorescent variants of green fluorescent protein (GFP) for use in higher plants. *Plant Mol. Biol.*, **36** 521-28.

Di Sansebastiano, G.-P. (1999). *Targeting of Soluble Proteins to Two Different Vacuoles in Plant Cells*, PhD thesis, University of Neuchâtel, 183 p.

Di Sansebastiano, G.-P., Paris, N., Marc-Martin, S. and Neuhaus, J.-M. (1998) Specific accumulation of GFP in a non-acidic vacuolar compartment via a C-terminal propeptide-mediated sorting pathway. *Plant J.*, **15** 449-57.

Di Sansebastiano, G.-P., Paris, N., Marc-Martin, S. and Neuhaus, J.-M. (2000) Regeneration of a lytic central vacuole and of neutral peripheral vacuoles can be visualised by GFP targeted to either type of vacuoles (submitted).

Firek, S., Whitelam, G.C. and Draper, J. (1994) Endoplasmic reticulum targeting of active modified β-glucuronidase (GUS) in transgenic tobacco plants. *Transgenic Res.*, **3** 326-31.

Fleurat-Lessard, P. (1988) Structural and ultrastructural features of cortical cells in motor organs of sensitive plants. *Biol. Rev.*, **63** 1-22.

Fleurat-Lessard, P. and Millet, B. (1984) Ultrastructural features of cortical parenchyma cells ('motor cells') in stamen filaments of *Berberis canadensis* Mill. and tertiary pulvini of *Mimosa pudica* L. *J. Exp. Bot.*, **35** 1332-41.

Fleurat-Lessard, P., Frangne, N., Maeshima, M., Ratajczak, R., Bonnemain, J. and Martinoia, E. (1997) Increased expression of vacuolar aquaporin and H^+-ATPase related to motor cell function in *Mimosa pudica* L. *Plant Physiol.*, **114** 827-34.

Freydl, E. (1994) *Analysis of the Vacuolar Targeting Peptide of Chitinase*, PhD thesis, University of Basel, 128 p.

Gaire, F. (1998) *Implication du système endomembranaire dans la réplication du virus du court-noué de la vigne (GFLV): Rôle de la protéine 2A dans la réplication du RNA2.*, PhD thesis, University Louis Pasteur, Strasbourg.

Gomez, L. and Chrispeels, M.J. (1993) Tonoplast and soluble vacuolar proteins are targeted by different mechanisms. *Plant Cell*, **5** 1113-24.

Haseloff, J. (1999) GFP variants for multispectral imaging of living cells. *Methods Cell Biol.*, **58** 139-51.

Haseloff, J., Siemering, K.R., Prasher, D.C. and Hodge, S. (1997) Removal of a cryptic intron and subcellular localization of green fluorescent protein are required to mark transgenic *Arabidopsis* plants brightly. *Proc. Nat. Acad. Sci. USA*, **94** 2122-27.

Higuchi, T., Suga, S., Tsuchiya, T., Hisada, H., Morishima, S., Okada, Y. and Maeshima, M. (1998) Molecular cloning, water channel activity and tissue specific expression of two isoforms of radish vacuolar aquaporin. *Plant Cell Physiol.*, **39** 905-13.

Hohl, I., Robinson, D.G., Chrispeels, M.C. and Hinz, G. (1996) Transport of storage proteins to the vacuole is mediated by vesicles without a clathrin coat. *J. Cell Sci.*, **109** 2539-50.

Holwerda, B.C., Galvin, N.J., Baranski, T.J. and Rogers, J. C. (1990) *In vitro* processing of aleurain, a barley vacuolar thiol protease. *Plant Cell*, **2** 1091-106.

Holwerda, B.C., Padgett, H.S. and Rogers, J.C. (1992) Proaleurain vacuolar targeting is mediated by short contiguous peptide interactions. *Plant Cell*, **4** 307-18.

Hörtensteiner, S., Martinoia, E. and Amrhein, N. (1992) Reappearance of hydrolytic activities and tonoplast proteins in the regenerated vacuole of evacuolated protoplasts. *Planta*, **187** 113-21.

Jauh, G.Y., Phillips, T.E. and Rogers, J.C. (1999) Tonoplast intrinsic protein isoforms as markers for vacuolar functions. *Plant Cell*, **11** 1867-82.

Jiang, L. and Rogers, J.C. (1998) Integral membrane protein sorting to vacuoles in plant cells: evidence for two pathways. *J. Cell Biol.*, **143** 1183-99.

Jung, R., Saalbach, G., Nielsen, N.C. and Müntz, K. (1993) Site-specific limited proteolysis of legumin chloramphenicol acetyl transferase fusions *in vitro* and in transgenic tobacco seeds. *J. Exp. Bot.*, **44** 343-49.

Karlsson, M., Johansson, I., Bush, M., McCann, M.C., Maurel, C., Larsson, C. and Kjellbom, P. (2000) An abundant TIP expressed in mature highly vacuolated cells. *Plant J.*, **21** 83-90.

Kervinen, J., Tobin, G.J., Costa, J., Waugh, D.S., Wlodawer, A. and Zdanov, A. (1999) Crystal structure of plant aspartic proteinase prophytepsin: inactivation and vacuolar targeting. *EMBO J.*, **18** 3947-55.

Lord, J.M. (1985) Synthesis and intracellular transport of lectin and storage protein precursors in endosperm from castor bean. *Eur. J. Biochem.*, **146** 403-409.

Mathai, J.C. and Agre, P. (1999) Hour-glass pore-forming domains restrict aquaporin-1 tetramer assembly. *Biochemistry*, **38** 923-28.

Matsuoka, K. and Nakamura, K. (1991) Propeptide of a precursor to a plant vacuolar protein required for vacuolar targetting. *Proc. Nat. Acad. Sci. USA*, **88** 834-38.

Matsuoka, K. and Neuhaus, J.-M. (1999) *Cis*-elements of protein transport to the vacuoles. *J. Exp. Bot.*, **50** 165-74.

Matsuoka, K., Matsumoto, S., Hattori, T., Machida, Y. and Nakamura, K. (1990) Vacuolar targeting and post-translational processing of the precursor to the sweet potato tuberous root storage protein in heterologous plant cells. *J. Biol. Chem.*, **265** 19750-57.

Matsuoka, K., Bassham, D.C., Raikhel, N. and Nakamura, K. (1995) Different sensitivity to wortmannin of two vacuolar sorting signals indicates the presence of distinct sorting machineries in tobacco cells. *J. Cell Biol.*, **130** 1307-18.

Michel, M., Gnägi, H. and Müller, M. (1992) Diamonds are a cryosectioner's best friend. *J. Microsc.*, **166** 43-56.

Nebenfuhr, A., Gallagher, L.A., Dunahay, T.G., Frohlick, J.A., Mazurkiewicz, A.M., Meehl, J.B. and Staehelin, L.A. (1999) Stop-and-go movements of plant Golgi stacks are mediated by the actomyosin system. *Plant Physiol.*, **121** 1127-42.

Neuhaus, J.M. and Rogers, J.C. (1998) Sorting of proteins to vacuoles in plant cells. *Plant Mol. Biol.*, **38** 127-44.

Neuhaus, J.-M. and Boevink, P. (2000). The green fluorescent protein (GFP) as reporter in plant cells, in *Plant Cell Biology* (eds. C. Hawes and B. Satiat-Jeunemaitre) IRL Press, Oxford (in press).

Neuhaus, J.-M., Sticher, L., Meins, F. and Boller, T. (1991) A short C-terminal sequence is necessary and sufficient for the targeting of chitinases to the plant vacuole. *Proc. Nat. Acad. Sci. USA*, **88** 10362-66.

Neuhaus, J.-M., Pietrzak, M. and Boller, T. (1994) Mutation analysis of the C-terminal vacuolar targeting peptide of tobacco chitinase: low specificity of the sorting system, and gradual transition between intracellular retention and secretion into the extracellular space. *Plant J.*, **5** 45-54.

Niwa, H., Inouye, S., Hirano, T., Matsuno, T., Kojima, S., Kubota, M., Ohashi, M. and Tsuji, F.I. (1996) Chemical nature of the light emitter of the Aequorea green fluorescent protein. *Proc. Nat. Acad. Sci. USA*, **93** 13617-22.

Pang, S.Z., DeBoer, D.L., Wan, Y., Ye, G.B., Layton, J.G., Neher, M.K., Armstrong, C.L., Fry, J.E., Hinchee, M.A.W. and Fromm, M.E. (1996) An improved green fluorescent protein gene as a vital marker in plants. *Plant Physiol.*, **112** 893-900.

Paris, N., Stanley, C.M., Jones, R.L. and Rogers, J.C. (1996) Plant cells contain two functionally distinct vacuolar compartments. *Cell*, **85** 563-72.

Paris, N., Rogers, S.W., Jiang, L., Kirsch, T., Beevers, L., Phillips, T.E. and Rogers, J.C. (1997) Molecular cloning and further characterization of a probable plant vacuolar sorting receptor. *Plant Physiol.*, **115** 29-39.

Piñeiro, M., García-Olmedo, F. and Diaz, I. (1994) Redox modulation of the expression of bacterial genes encoding cysteine-rich proteins in plant protoplasts. *Proc. Natl. Acad. Sci. USA*, **91** 3867-71.

Robinson, D.G. and Hinz, G. (1997) Vacuole biogenesis and protein transport to the plant vacuole: a comparison with the yeast vacuole and the mammalian lysosome. *Protoplasma*, **197** 1-25.

Rouwendal, G.J.A., Mendes, O., Wolbert, E.J.H. and Deboer, A.D. (1997) Enhanced expression in tobacco of the gene encoding green fluorescent protein by modification of its codon usage. *Plant Mol. Biol.*, **33** 989-99.

Schroeder, M.R., Borkhsenious, O.N., Matsuoka, K., Nakamura, K. and Raikhel, N.V. (1993) Co-localization of barley lectin and sporamin in vacuoles of transgenic tobacco plants. *Plant Physiol.*, **101** 451-58.

Siemering, K.R., Golbik, R., Sever, R. and Haseloff, J. (1996) Mutations that suppress the thermo-sensitivity of green fluorescent protein. *Curr. Biol.*, **6** 1653-63.

Sonnewald, U., Brauer, M., von Schaewen, A., Stitt, M. and Willmitzer, L. (1991) Transgenic tobacco plants expressing yeast-derived invertase in either the cytosol, vacuole or apoplast: a powerful tool for studying sucrose metabolism and sink/source interactions. *Plant J.*, **1** 95-106.

Tague, B.W. and Chrispeels, M.J. (1987) The plant vacuolar protein, phytohemagglutinin, is transported to the vacuole of transgenic yeast. *J. Cell Biol.*, **105** 1971-79.

Toriyama, H. and Satô, S. (1968) On the membrane of the tannin vacuole in the *Mimosa* motor cell. *Proc. Japan Acad.*, **44** 528-31.

von Schaewen, A. and Chrispeels, M.J. (1993) Identification of vacuolar sorting information in phytohemagglutinin, an unprocessed vacuolar protein. *J. Exp. Bot.*, **44** (Suppl.), 339-42.

Wee, E.G.-T., Sherrier, D.J., Prime, T.A. and Dupree, P. (1998) Targeting of active sialytransferase to the plant Golgi apparatus. *Plant Cell*, **10** 1759-68.

Weig, A., Deswarte, C. and Chrispeels, M.J. (1997) The major intrinsic protein family of *Arabidopsis* has 23 members that form three distinct groups with functional aquaporins in each group. *Plant Physiol.*, **114** 1347-57.

Wilkins, T.A., Bednarek, S.Y. and Raikhel, N.V. (1990) Role of propeptide glycan in post-translational processing and transport of barley lectin to vacuoles in transgenic tobacco. *Plant Cell*, **2** 310-13.

12 Post-Golgi, prevacuolar compartments

David G. Robinson, John C. Rogers
and Giselbert Hinz

12.1 Why do cells need prevacuolar compartments?

Acid hydrolases synthesized co-translationally at the endoplasmic reticulum (ER) exit the secretory pathway at the Golgi apparatus. This is a well-established fact for yeast and mammalian cells (Braulke, 1996; Conibear and Stevens, 1998). Recognition of these proteins, their segregation from secretory proteins and subsequential delivery to the lytic compartment of the cell are achieved through a combination of two molecular mechanisms: pH-regulated receptor-ligand interactions, and transport vesicle targeting. Although the receptors and the sorting motifs on the ligands in these two major groups 'of organisms differ considerably (see and compare Höning *et al.*, 1997, and Le Borgne and Hoflack, 1998, for mammals; with Conibear and Stevens, 1998, and Jørgensen *et al.*,1999, for yeast), their intracellular transport has three features in common. Firstly, the ligand (the acid hydrolase) leaves the Golgi apparatus attached to its integral membrane receptor, but the receptor is not found in the lytic compartment where the ligand is finally deposited. Thus, in mammalian cells the mannosyl 6-phosphate receptor (MPR) is characteristically not present in the lysosome (Kornfeld and Mellman, 1989), and Vps10p, the receptor for carboxypeptidase Y (CPY), is absent from the yeast vacuole (Cereghino *et al.*, 1995). Secondly, both MPR and Vps10p are intimately associated with the formation of Golgi-derived clathrin-coated vesicles (CCVs) (see Sahagian and Steer, 1985, and Le Borgne and Hoflack, 1998 for mammals; and Robinson and Hinz, 1997 for a summary of the situation in yeasts). Thirdly, the hydrolase receptor accompanies its ligand to an intermediate prelysosomal/prevacuolar compartment (PLC/PVC), where the ligand dissociates from its receptor, and the receptor recycles back to the Golgi (see Duncan and Kornfeld, 1988 for mammals; and Cooper and Stevens, 1996, for yeast).

As with receptor-mediated endocytosis, where internalized receptor-ligand complexes are delivered via CCV to an 'early' endosome, the separation of a ligand from its receptor is generally considered to result from exposure to an acid pH upon entering the respective uncoupling compartment (Dautry-Varsat *et al.*, 1983). Indeed, *in situ* measurements of the pH in the *trans*-Golgi network (TGN) of mammalian cells, where the receptor-ligand complex is formed, give values of around 6.2 ± 0.4, whereas the pH of the PLC is <5.5

(Anderson and Orci, 1988; Miesenböck *et al.*, 1998). The low pH of the PLC and other endosomal compartments is generated by the activity of a vacuolar-type adenosine triphosphatase (V-ATPase), which is characteristic of these compartments and of the vacuole (Nelson and Harvey, 1999). Raising the pH through the application of ionophores (e.g. monensin), lysomotopic agents (e.g. chloroquine) or V-ATPase inhibitors (e.g. bafilomycin), invariably results in a blockage of both endocytotic and biosynthetic transport pathways to the lysosome/vacuole (Mellman *et al.*, 1986; van Deurs *et al.*, 1996).

Since they are degradative compartments, it is obviously advantageous to separate the receptor from its ligand at some point before the ligand enters the mammalian lysosome/yeast vacuole. If, in addition, a mechanism exists for receptor recycling from the PLC/PVC to the TGN, this then constitutes a highly efficient system for the delivery of specific macromolecules to the lysosome/vacuole. Considering the similarity between plant and yeast vacuoles and the high degree of conservatism in vesicle-mediated protein trafficking throughout the eukaryotic kingdom (Robinson *et al.*, 1998d; Sanderfoot and Raikhel, 1999), it is not unreasonable to expect that plant cells will also possess a PVC. This chapter examines the evidence in support of this contention and dwells on the consequences of multiple vacuolar compartments, a feature unique to plant cells (Paris *et al.*, 1996; Jauh *et al.*, 1998, 1999; Jiang and Rogers, 1998).

12.2 The PLC/PVC in yeast and mammals: an overview

12.2.1 Morphology and molecular organization

In mammalian cells, the PLC is a 0.5–1 μm diameter organelle, which typically has internal 60–80 nm vesicles (e.g. Dunn *et al.*, 1986; van Deurs *et al.*, 1993). Sometimes, small plaques are present at the surface, from which a tuft of 5–10 nm long filaments are seen to emanate (Friend and Farquhar, 1967; Futter *et al.*, 1996). The internal vesicles are qualitatively different from the boundary membrane: the MPR and the epidermal growth factor receptor are concentrated on them (Griffiths *et al.*, 1988; Felder *et al.*, 1990), whereas the lysosomal membrane proteins, Lamp-1 and Lamp-2, are restricted to the surface membrane (Escola *et al.*, 1998). A recent report has identified an unusual lipid, lysobisphospatidic acid, which is only present on the internal vesicles (Kobayashi *et al.*, 1998). An antibody generated against this lipid prevents the recycling of MPR to the TGN.

Although, with *Saccharomyces cerevisiae*, good structural preservation at the electron-microscopic (EM) level is notoriously difficult to attain, it would nevertheless appear that the PVC in baker's yeast is also multivesiculate in nature (Hicke *et al.*, 1997; Odorizzi *et al.*, 1998; Mulholland *et al.*, 1999). In the class E group of vacuolar protein sorting (vps) mutants, the yeast PLC enlarges and/or proliferates and vacuole-destined proteins accumulate in it (Raymond *et al.*, 1992; Rieder *et al.*, 1996). Evidence has also been provided for a sorting of proteins between the internal vesicles and the limiting membrane of the yeast PVC. Thus, in contrast to soluble CPY, the integral membrane protein, carboxypeptidase S, appears to be selectively concentrated in the internal vesicles (Odorizzi *et al.*, 1998).

12.2.2 A role for phosphatidylinositol (PI) 3-kinase in PLC/PVC structure and function

PI 3-kinase activity is usually associated with signal transduction, but there is a growing body of evidence for its participation in vesicle-mediated protein trafficking (reviewed by Corvera and Czech, 1998; Clague, 1998). Although there are some examples for the direct binding of a PI 3-kinase to the cytoplasmic domain of a receptor, without which further receptor-ligand transport cannot occur, for example platelet-derived growth factor (Joly *et al.*, 1994), the nature of various PI 3-kinases, their substrates, target molecules and effectors is manifold and it is difficult to obtain a clear picture. However, it has been established both for mammalian and yeast cells that PI 3-kinase binds to cysteine/histidine-rich RING-FYVE domains in proteins vital to vesicular transport (Gaullier *et al.*, 1998; Tall *et al.*, 1999).

Use of the PI 3-kinase inhibitor, wortmannin, provides ample evidence for the PLC as being a major site of PI 3-kinase action in mammalian cells. Wortmannin does not effect any morphological change on the TGN, but causes vacuolation of the PLC, a redistribution of MPR and lysosomal membrane proteins to different 'vacuolar' compartments, and a subsequent secretion of acid hydrolases (Brown *et al.*, 1995; Martys *et al.*, 1996; Reaves *et al.*, 1996). The dilation of the PLC is accompanied by a dramatic loss of internal vesicles (Fernandez-Borja *et al.*, 1999). Recent data suggest that wortmannin specifically blocks the export of MPR out of the TGN, by preventing the production of CCV, but does not prevent the recycling of MPR from the PLC to the TGN (Nakajima and Pfeffer, 1997; Karlsson and Carlsson, 1998).

In yeast, Vps34p, an essential protein for protein transport to the vacuole, is also a PI 3-kinase (Schu *et al.*, 1993), and is active in a complex together with Vps15p (Herman *et al.*, 1991). Another protein important for yeast vacuolar protein sorting is Fab1p, which has now been found to be a second PI kinase, responsible for the formation of phosphatidylinositol 3,5 bisphosphate (Odorizzi *et al.*, 1998). Phosphatidylinositol 3,5 phosphates are synthesized at the cytoplasmic leaflet of Golgi membranes (Herman *et al.*, 1991) but are degraded when they reach the vacuole. According to Wurmser and Emr (1998) this can only occur by a mechanism which delivers membrane into the lumen of the vacuole, where lipo- and proteolytic activities are present. They propose a situation whereby invagination and fission of the PVC membrane creates small internal vesicles, which upon fusion of the PVC with the vacuole are released into the interior of the vacuole. Support for this idea is provided by the *vma4Δ* mutant, which lacks the ability to acidify the vacuolar lumen. The vacuoles of these cells are filled with microvesicles. Furthermore, mutations in *VAM3* and *YPT7* genes whose products are required for PVC → vacuole transport show dramatically altered *in vivo* levels of phosphatidylinositol 3-phosphate.

Interestingly, yeast PI 3-kinase is very insensitive to wortmannin (Stack and Emr, 1994; Takegawa *et al.*, 1995). Instead, SST4, a PI 4-kinase appears to be the target for wortmannin action in yeast (Cutler *et al.*, 1997). However, the literature is confusing with regard to the *in vivo* effects of wortmannin on yeast cells: whereas Stack and Emr (1994) claimed that there is no effect on growth or CPY sorting, Cutler *et al.* (1997) have shown that wortmannin severely inhibits yeast growth at concentrations above 10 μg/ml. On the other hand, its toxicity decreases when the cells are grown on enriched medium, suggesting an interaction between wortmannin and substances in the growth medium.

12.2.3 Other important factors in TGN-PLC/PVC vesicle transport

A number of different types of protein at the surface of the membrane are involved in vacuolar vesicle targeting; they include: the rab GTPases (Novick and Zerial, 1997); various vesicle docking/fusion proteins, designated either as 'vesicle-associated membrane proteins' (VAMPs), syntaxins or SNAREs (soluble N-ethylmaleimide-sensitive factor attachment protein receptors) (Hay and Scheller, 1997); as well as vesicle coat proteins, for example adaptors and clathrin (Schmid, 1997). In addition, successful vesicle transport is also dependent on an interaction with components of the cytoskeleton (Itin *et al.*, 1999). The literature on these various factors in mammalian and yeast cells is huge and beset by confusing nomenclature. What follows is, therefore, a very brief summary of the information relevant to Golgi-lysosome/vacuole anterograde and retrograde transport. Importantly for plant researchers, homologs/orthologs of some of these factors can be used as potential diagnostica for the identification of PVCs in plant cells.

Specific Rab GTPases are localized to transport steps involving vesicles, and the 11 Rab/YPT genes in yeast may define the total number of vesicular pathways in that organism (Lazar *et al.*, 1997). A similar correlation appears to be true in mammalian cells. For example, in mammals, rab9p is required for TGN-PLC transport (Lombardi *et al.*, 1993), whereas rab7p is involved in the PLC-lysosome transfer step (Vitelli *et al.*, 1997). Their equivalents in yeast are, respectively, Vps21p and Ypt7 (Wichmann *et al.*, 1992; Horazdovsky *et al.*, 1994).

Vesicular traffic is also dependent upon the presence of specific v-(vesicle), and t-(target) SNAREs on appropriate membranes (Söllner *et al.*, 1993).

t-SNARES come in two forms: one associates with membranes via a lipid modification of the protein; while the other, termed a syntaxin, is inserted into the membrane via a C-terminal hydrophobic sequence. Organelle identity has been hypothesized to be dependent upon the presence of an organelle-specific syntaxin. Thus, the *Saccharomyces cerevisiae* genome contains genes for eight syntaxin homologs, with one specific for ER, one specific for traffic to *cis*-Golgi, two which correspond to yeast equivalents of the *trans*-Golgi network and early endosome, respectively, one for the prevacuolar compartment, one for the vacuole, and two on the plasma membrane (Holthuis *et al.*, 1998; Nichols and Pelham, 1998). In contrast, a certain v-SNARE may be found on vesicles in several different pathways, and the specificity of interactions in vesicle docking may be defined by combinations of several v-SNAREs acting together (Götte and Fischer von Mollard, 1998). Thus, an individual v-SNARE has less specificity as a marker for a pathway.

The major SNAREs for the TGN → PLC/PVC are Vti1p, a v-SNARE (yeast: Fischer von Mollard *et al.*, 1997; mammalian homolog: Fischer von Mollard and Stevens, 1999), and Pep12p/syntaxin 7 are the corresponding syntaxins in yeast and mammalian cells (Becherer *et al.*, 1996: Wang *et al.*, 1997). For the PLC/PVC → lysosome/vacuole the nature of the v-SNARE is unclear (see Section 12.7.3.5), although for homotypic vacuole-vacuole fusions in yeast it has been identified as Nyv1p (Nichols *et al.*, 1997). The major t-SNAREs in yeast for this step are Vam3p (syntaxin) and Vam7p (Darsow *et al.*, 1997; Sato *et al.*, 1998). A Vam7p homolog, mediating endosomal-lysosomal transport in mammalian cells, has recently been identified and localized (Advani *et al.*, 1999).

Retrograde transport from the lysosome/vacuole via the PLC/PVC to the TGN has been investigated in mammalian and yeast cells. Syntaxin 16 is localized to the TGN of mammalian cells (Simonsen *et al.*, 1998); the yeast homolog is Tlg2p (Holthuis *et al.*, 1998). Soluble proteins homologous to the yeast Sec1p, are also known to have important roles in vesicular traffic by participating in a poorly understood manner in the formation of SNARE complexes. The yeast Sec1p homolog, Vps45p, plays an important role in traffic between the TGN and PVC, and deletion of Vps45p severely perturbs protein transport to the yeast vacuole leading to the accumulation of 'post-Golgi transport vesicles' (Piper *et al.*, 1994). Interestingly, it has recently been found that Vps45p binds to Tlg2p (Nichols *et al.*, 1998), and that Vps45p also interacts with Pep12p (Burd *et al.*, 1997).

It is often implied in cell biology texts that CCVs are responsible for the recycling of the MPR to the TGN from the mammalian PLC, but in fact there is no evidence for the participation of either the AP-I adaptor complex or clathrin in this event (Draper *et al.*, 1990). It is known that recycling of MPR involves a diaromatic, FW sorting motif in the cytoplasmic tail of the receptor (Schweizer *et al.*, 1997), and that this is recognized by a cytosolic protein known as TIP47 ('tail interacting protein'); rab9p is also involved (Diaz and Pfeffer, 1998). The situation in yeast with respect to CCV and the recycling of the CPY receptor is similar. Genetic studies have identified five cytosolic proteins, constituting a so-called 'retromer' coat complex, which is vital to PVC-Golgi transport. Subcomplex I of the retromer consists of Vps35p, Vps29p and Vps27p, and subcomplex II contains Vps5p and Vps17p Seaman *et al.* (1997). Seaman and co-workers (1998) have proposed that subcomplex I interacts with the cytoplasmic domain of the Vps10p receptor, forming the inner shell of the coat. Subsequent assembly of the second subcomplex is considered to cause vesiculation. These authors have also pointed out that mammalian homologs of Vps5/26/29/35p exist, and that the Vps5 homolog, SNX1, is known to interact with the cytoplasmic tail of the epidermal growth factor receptor (Kurten *et al.*, 1996).

12.3 Evidence for post-Golgi/prevacuolar compartment(s) in plants

From the foregoing, there are good reasons to expect that plant cells will have a PVC, and that this will be multivesiculate in nature. So-called multivesicular bodies have frequently been described in the plant literature (reviewed in Robinson and Hinz, 1997; Robinson *et al.*, 1998c). Morphologically, they are very similar to their counterparts in mammalian cells, even possessing a filamentous plaque at the surface (see Figure 12.1; see also Figure 8a in Robinson and Hinz, 1997), but their unequivocal identification as a PVC is another matter. The identification of organelles that serve as PVCs in plant cells has been difficult, in part because a PVC must be defined functionally and the requisite assays have been difficult to establish; and, in part, because plant cells have two distinct vesicular pathways leading to two functionally distinct vacuolar compartments (Neuhaus and Rogers, 1998; Jauh *et al.*, 1999).

12.3.1 The effects of pH-perturbing agents on intracellular protein traffic

Although the pH of lytic vacuoles in plants is usually given as < 5, there are no data concerning the pH of the plant Golgi apparatus. However, as judged by the positive immunogold identification of V-H^+-ATPase and H^+-pyrophosphatase, it is to be expected that the *trans*-cisternae and *trans*-Golgi network are at least somewhat acidic (e.g. Zhang *et al.*, 1996; Robinson *et al.*, 1998e; Ratajczak *et al.*, 1999). This is supported by the frequently observed dilation of these *trans*-Golgi elements when plant cells are exposed to the ionophore monensin (e.g. Heupke and Robinson, 1985; Mollenhauer *et al.*, 1990; Zhang *et al.*, 1996).

Obviously, pH plays an important role in the targeting of vacuolar proteins in plants, since pH-perturbing agents, such as monensin and bafilomycin, lead to a

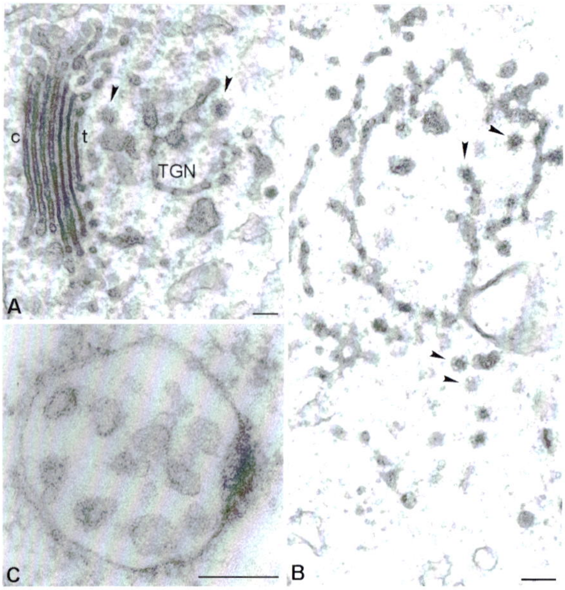

Figure 12.1 Post-Golgi structures in plants. A) Golgi stack and *trans*-Golgi network (TGN) from tobacco root cap cell. B) 'Partially-coated reticulum' from tobacco root cap cell. C) Multivesicular body from tobacco BY-2 suspension cultured cell. Arrowheads point to budding clathrin-coated vesicles. Abbreviations: c, *cis*; t, *trans*. (Internal scale bars = 100 nm).

missorting of these proteins (Craig and Goodchild, 1984; Matsuoka *et al.*, 1997). Working with tobacco cells, Matsuoka and co-workers (1997) have localized the site for the bafilomycin block to a post-Golgi, non-vacuolar organelle—a putative PLC. In the case of storage protein depositing seed tissue, monensin treatment causes the storage globulins to be directed instead to the cell surface, where they can be visualized under the electron microscope as small osmiophilic clumps on the plasma membrane (Robinson *et al.*, 1998a). Robinson and co-workers (1998a) have also shown that monensin treatment also causes the storage proteins in a putative, multivesicular body-like PVC in pea cotyledons to become less aggregated and more flocculent in appearance. In addition, the multivesicular bodies appeared to dilate. Since the major storage globulins in pea have a relatively low pI (Casey, 1979), an increase in pH would make them more soluble.

12.3.2 Wortmannin effects

The authors know of only four publications reporting the action of wortmannin on plant cells. Matsuoka and co-workers (Nakamura *et al.*, 1993; Matsuoka *et al.*, 1995; Koide *et al.*, 1997) applied wortmannin to suspension-cultured tobacco BY-2 cells expressing both barley lectin (normally containing an C-terminal targeting propeptide) and sweet potato sporamin (normally containing a N-terminal targeting propeptide). Under normal conditions, both proteins accumulated in a vacuole, even when their targeting propeptides were exchanged. However, at a concentration which had no effect on net protein synthesis (33 μM), wortmannin selectively redirected barley lectin with the C-terminal-targeting propeptide to the cell surface. *In vitro* assays showed that even at 1 μM, wortmannin completely inhibited PI 3-kinase activity. Based on these observations, Matsuoka and Neuhaus (1999) have divided soluble vacuolar proteins into two basic groups: those which have specific and conserved, but variably located, targeting motifs, and whose sorting is wortmannin insensitive; and a second group with C-terminal located, nonspecific targeting motifs, whose sorting is sensitive to wortmannin. However, it is important to point out that, in contrast to what is known about the site of wortmannin action in mammalian and yeast cells (see above), Matsuoka and Neuhaus (1999) consider the first group of proteins to exit the Golgi apparatus via CCV. If true, this would constitute a basic difference between plants and other eukaryotes in terms of TGN-based vesicle targeting.

Wortmannin (35 μM) has also been applied to tobacco mesophyll protoplasts expressing a vacuolar chitinase-green fluorescent protein (GFP) construct (Di Sebastiano *et al.*, 1998). Transport to the vacuole was effectively inhibited by wortmannin, but neither in this investigation nor in any of the others have ultrastructural data been presented that might provide a clue as to the exact site of wortmannin action in plant cells.

12.3.3 Localization of the vacuolar sorting receptor, BP-80, and homologs

Soluble plant vacuolar proteins do not share a common vacuolar sorting motif, as yeast or mammalian vacuolar/lysosomal proteins do. As Vitale and Raikhel (1999) have pointed out: '... there is no easily identifiable general consensus sequence for vacuolar sorting'. Nevertheless, plant vacuolar sorting signals can be roughly sorted into three groups: i) proteins possessing a conserved N-terminal NPIR sequence, such as barley aleurain or sweet potato sporamin; ii) proteins possessing C-terminal propeptides without any known consensus motif, such as barley lectin or tobacco chitinase; and finally iii) proteins with internal sequences or signal-patches again without any consensus motif, such as pea legumin or castor bean ricin (summarized in several recent reviews, Neuhaus and Rogers, 1998; Matsuoka and Neuhaus, 1999; Vitale and Raikhel, 1999).

Since most of these signals are both necessary and sufficient to sort proteins to the vacuole, receptors mediating this specific transport step must also exist in the plant secretory system. Kirsch *et al.* (1994) were the first to isolate such a receptor from a fraction enriched in clathrin-coated vesicles from developing pea cotyledons, using an affinity column prepared with the N-terminal vacuolar sorting sequence of the thiol protease barley aleurain, NPIR. They purified a type I integral membrane glycoprotein with a molecular mass of 80 kDa, called binding protein 80, BP-80. Molecular cloning and sequencing has revealed that this protein is a member of a novel family of membrane proteins, termed vacuolar sorting receptor (VSR) proteins, that are unique to plants (Paris and Rogers, 1996; Ahmed *et al.*, 1997a, 1997b; Paris *et al.*, 1997; Shimada *et al.*, 1997; Laval *et al.*, 1999).

The short C-terminal, cysteine-rich, cytoplasmic tail of BP-80 (and all homologs so far described) contains a Tyr-X-X-hydrophobic motif, typical for transmembrane proteins which are transported by CCV (Ahmed *et al.*, 1997a, 1997b; Paris *et al.*, 1997). Deletion of the transmembrane domain (TMD) and the cytoplasmic tail (CT) results in the protein being secreted (Paris *et al.*, 1997). It was, therefore, suggested that this Tyr-X-X-hydrophobic motif would probably interact with the clathrin coat of plant CCV. This interpretation was confirmed by the finding that BP-80 binds to mammalian adaptin proteins, subunits of the adaptor complex, which are essential for the formation of CCV (Butler *et al.*, 1997; Robinson *et al.*, 1998b; Sanderfoot *et al.*, 1998). Additionally, pea BP-80 and the *Arabidopsis* homolog, AtELP, co-purify during the isolation of CCV (Ahmed *et al.*, 1997a,b; Hinz *et al.*, 1999).

The N-terminal, lumenal domain of the protein contains an N-terminal region homologous to ReMembR-H2 (RMR) protein lumenal domains (Jiang *et al.*, 2000), a unique central region, and three C-terminal epidermal growth factor (EGF) repeats. The EGF repeat motif is found in numerous proteins, including growth factors, transmembrane proteins that function as receptors or as homeotic gene products, extracellular matrix proteins, and soluble secreted proteins that have highly-regulated protease functions (Davis, 1990). To the best of our knowledge, however, the VSR proteins are the only example of proteins carrying EGF repeats that are not directed to the cell exterior to perform their functions (Davis, 1990). VSR proteins contain two different types of EGF repeats, with the first two having a B.1 consensus and the third a B.2 consensus (Herz *et al.*, 1988). The functional difference in the two motifs is more easily appreciated by considering that B.2 contains conserved residues that are known to permit high affinity calcium binding; thus, the two types should be designated EGF and EGF-CB (for calcium binding) motifs (Davis, 1990). The third, EGF-CB motif is also a characteristic motif of β-integrins (Laval *et al.*, 1999). Indeed, Laval and co-workers (1999) have cloned an *Arabidopsis* homolog (AtELP1) using this conserved region of mammalian β-integrins as a probe. In other studies of receptors or receptor-like proteins, EGF repeats have variously been shown to comprise the ligand binding domain of thrombomodulin (Stearns *et al.*, 1989; Zushi *et al.*, 1989), to affect acid-dependent ligand dissociation from the low density lipoprotein receptor (Davis *et al.*, 1987), and to alter the conformation of a lectin domain in a peripheral lymph node homing receptor (Bowen *et al.*,

1990). Thus, they participate in protein-protein interactions, whether in *cis* or in *trans*.

Cargo specificity and binding constants have been determined for pea cotyledon BP-80 using heterologous *in vitro* assays with affinity columns loaded with immobilized synthetic peptides of the different sorting sequences. In these tests, pea BP-80 binds very specifically to the NPIR sorting sequence of barley proaleurain, with a binding constant of about 37 nM (Kirsch *et al.*, 1994). The pH dependency of this binding follows an optimum curve, with a sharp peak at ~pH 6.5 (Kirsch *et al.*, 1994), a pH value similar to that determined for the mammalian TGN and significantly higher than the pH in the mammalian endosome or the plant vacuole.

The lumenal domain of BP-80 has been expressed as a recombinant protein, secreted from insect cells and purified (Cao *et al.*, 2000). By protease digestion of the purified protein, and from ligand binding studies utilizing the intact form as well as a truncated form lacking the EGF repeats, three protease-resistant structural domains were defined: an N-terminal/RMR homology domain connected to a central domain together determine the NPIR-specific ligand binding site, and a C-terminal EGF repeat domain alters the conformation of the other two domains to enhance ligand binding. A fragment representing the central plus the C-terminal domains could bind ligand, but without NPIR specificity. These results indicate that two BP-80 ligand binding sites recognize two separate ligand determinants: a non-NPIR site defined by the central domain-EGF repeat domain structure, and an NPIR-specific site contributed by the interaction of the N-terminal/RMR homology domain and central domain (Cao *et al.*, 2000).

Although the native ligand(s) recognized by BP-80 in pea cotyledons remain(s) to be identified, it is reasonable to assume that pea BP-80 is a receptor for CCV-mediated transport of hydrolases. Indeed, in pea cotyledons BP-80 is absent from those vesicles responsible for the transport of globulins to the protein storage vacuole (Hinz *et al.*, 1999).

Pea BP-80 also binds to the N-terminal sorting sequence of sweet potato sporamin ($K_D > 100\,\mu M$) (Kirsch *et al.*, 1994), and to the C-terminal sorting sequence of brasil nut 2S albumin, but not the C-terminal sequence of barley lectin (Kirsch *et al.*, 1994, 1996). These results, however, have to be interpreted in the context that plant cells contain functionally different vacuoles. Barley aleurain is a cysteine protease present in the lumen of a lytic γ-tonoplast intrinsic protein (TIP) vacuole, sporamin is a protease inhibitor localized to vegetative storage γ/δ-TIP vacuoles, whereas barley lectin and brasil nut 2S albumin are proteins present in seed storage α- and δ-TIP vacuoles (Jauh *et al.*, 1998, 1999). Kirsch *et al.* (1996) have suggested that BP-80 has a broad specificity for different sorting signals. An alternative possibility is that different members of the BP-80 family of receptors are responsible for targeting different cargos to different vacuoles (Shimada *et al.*, 1997), although others have argued that the VSR proteins are likely to traffic only in the lytic vacuole pathway (Jiang and Rogers, 1998).

The interaction of BP-80 homologs with their native ligand has been investigated in two cases. Miller and co-workers (1999), working with tobacco stigmata, have been able to show that a tobacco BP-80 interacts with the precursor polypeptide of a proteinase inhibitor (Jiang and Rogers, 1999; Miller and Anderson, 1999; Miller *et al.*, 1999). Shimada *et al.* (1997) have shown that peptides derived from internal and C-terminal sequences of the 2S albumin precursor polypeptide from pumpkin seeds bind to two different BP-80 homologs. These results are puzzling because the vesicles carrying the 2S albumin in pumpkin are apparently formed at the ER and bypass the Golgi-apparatus on their way to the protein storage vacuole (Hara-Nishimura *et al.*, 1998). In the absence of measurements for markers of CCV- and Golgi-mediated transport, the results of Shimada *et al.* (1997) remain equivocal. In addition, the findings of Laval *et al.* (1999) indicate that one BP-80 homolog, AtELP1-located to the plasma membrane, may even be involved in the endocytotic pathway.

Like the mammalian and yeast lysosomal/vacuolar sorting receptors, members of the BP-80 family are not present in the vacuolar membrane, as has been shown by immunofluorescence studies (Paris *et al.*, 1997), immunoelectron microscopy (Paris *et al.*, 1997; Hinz *et al.*, 1999), and subcellular fractionation (Ahmed *et al.*, 1997a,b; Sanderfoot *et al.*, 1998; Hinz *et al.*, 1999; Miller *et al.*, 1999). BP-80 and its homolog AtELP, have been localized to the Golgi apparatus by immunoelectron microscopy (Paris *et al.*, 1997; Sanderfoot *et al.*, 1998; Hinz *et al.*, 1999). By subcellular fractionation, BP-80/AtELP partially cofractionate with Golgi membranes (Ahmed *et al.*, 1997a,b; Sanderfoot *et al.*, 1998; Hinz *et al.*, 1999). However, a significant portion of the receptor fractionates in a compartment which is neither Golgi nor vacuole (Ahmed *et al.*, 1997a,b; Hinz *et al.*, 1999).

12.3.4 Morphological and functional definition of a prevacuolar compartment in the lytic vacuole pathway

Immunoelectron microscopy studies have demonstrated BP-80 in 250 nm small vacuoles that appeared to be able to fuse with large vacuoles in pea root cells (Figure 12.2B and C); these structures were tentatively identified as PVCs for the lytic vacuole pathway (Paris *et al.*, 1997). Their functional characterization was carried out using a chimeric reporter protein system. This reporter contained a mutated form of the cysteine protease, proaleurain, that lacks vacuolar targeting determinants, attached through its C-terminus to the Ser/Thr-rich domain of BP-80 that is connected to its TMD and CT (Jiang and Rogers, 1998). Processing of the proaleurain moiety of the reporter protein into mature

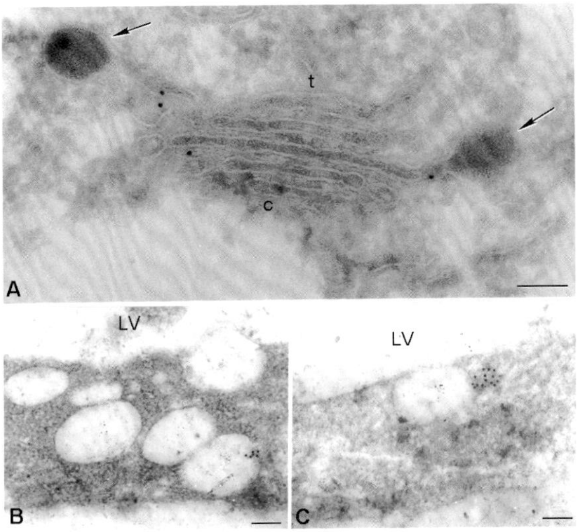

Figure 12.2 Immunogold localization of the vacuolar sorting receptor, BP-80. (A) Positive labeling of a Golgi stack in developing pea cotyledons. The cisternae are labeled, but the dense vesicles (arrows) are not (ultra-thin cryosections). (B/C) Labeling of putative prevacuolar compartments in pea root tip cells (Paris *et al.*, 1997; reprinted with permission from the American Society of Plant Physiologists) (plastic sections). Abbreviations: c, *cis*; t, *trans*; LV, lytic vacuole; BP, binding protein. (Internal scale bars = 100 nm).

aleurain would only occur in a post-Golgi, acidified compartment (Holwerda *et al.*, 1990), and would, therefore, provide a functional definition for correct targeting of the reporter protein to a PVC or to the lytic vacuole itself. When expressed in tobacco protoplasts, this reporter reached a destination where the mutated proaleurain was proteolytically processed into mature aleurain and separated from membrane association (Jiang and Rogers, 1998). The reporter protein trafficked through the Golgi because brefeldin A (BFA) prevented its processing, and because it acquired complex modifications to Asn-linked glycans, an alteration specific to the Golgi in plant cells. In addition, confocal immunofluorescence studies showed that the reporter protein co-localized with tobacco VSR proteins, markers for Golgi and PVC, but did not co-localize with tobacco binding protein (BiP) or calnexin, markers for ER. Additionally, the reporter protein was not associated with the central vacuole tonoplast (Jiang and Rogers, 1998). These results, together with the knowledge that BP-80 was not present in vacuolar membranes, strongly indicated that processing of the reporter proaleurain moiety occurred in the lytic PVC.

Transit through the Golgi by the reporter was also supported by experiments utilizing a modified reporter, where three tandemly repeated Kex2 cleavage sites were introduced in front of the TMD. Kex2p is the prototype of a Golgi-resident protease responsible for the processing of prohormones at pairs of basic residues in yeast and mammalian cells, and the presence of an appropriate Kex2p-like protease activity in tobacco cells has been documented (Kinal *et al.*, 1995). When expressed in tobacco cells, the Kex2 cleavage sites of the modified reporter were specifically cleaved by a Golgi-localized plant Kex2p-like protease. This cleavage resulted in secretion of the mutated proaleurain moiety of the reporter (Jiang and Rogers, 1999). Thus, the BP-80 TMD and CT sequences were shown to contain the information necessary to direct the reporter to the PVC via the Golgi complex.

To determine the role of the CT in PVC targeting of the reporter protein, a series of deletions were made in the CT sequences, with the interesting result that complete CT deletion still resulted in PVC targeting and Golgi trafficking of the reporter, with an efficiency similar to that of the unmutated reporter (Jiang and Rogers, 1998). This result indicates that the BP-80 TMD alone is sufficient for targeting to the lytic PVC. This result also raises an interesting question, since a reporter missing its CT should not be packaged in CCVs. How is a reporter protein containing only TMD sequences selected in the Golgi/TGN for traffic in a specific pathway? This observation is not unique for the plant system; similar observations were previously made in yeast (Roberts *et al.*, 1992). As these authors suggested, the lytic vacuole may be the default compartment for integral membrane proteins lacking a CT, but it is not clear how such proteins would reach that compartment. It is possible that the N-terminal portions of these proteins interact with other proteins to form dimers or multimers in order to be recruited into the same CCVs. However, the mutated form of proaleurain used in the chimeric reporter protein studies would not seem likely to interact with another protein in this manner. Alternatively, there may be a non-CCV pathway for transport of the reporter protein lacking a CT to the lytic vacuole. This hypothesis is attractive because it would explain how net flow of membrane for lytic vacuole biogenesis and maintenance could occur; as CCVs are thought to recycle from the TGN to the prevacuole to the TGN, it is difficult to see how their traffic could result in any net delivery of membrane.

The reporter protein assay only measured forward traffic, and it was possible that the CT would be important in retrieval from the prevacuole to Golgi via CCVs. Studies on Vps10p, a type I transmembrane receptor protein that sorts carboxypeptidase Y to the yeast vacuole, indicated that deletion of its CT resulted in shortened receptor half-life and degradation of the receptor in the vacuole (Cereghino *et al.*, 1995). Shortened survival of the CT-deleted receptor was due to its inability to be recycled from the PVC to the TGN (Cereghino *et al.*, 1995; Bryant *et al.*, 1998). To test a similar role of the BP-80 CT, the stability of intact BP-80 was compared to that of a truncated BP-80 missing its CT; the half-life of the truncated BP-80 was substantially reduced compared to that of the intact receptor (Jiang and Rogers, 1998). These results are consistent with those obtained in the yeast system, and indicate that the BP-80 CT may be important for recycling the receptor back to the TGN from the PVC.

12.3.5 Golgi-vacuole SNAREs

The last 4 yrs have seen some progress on the identification and localization of plant SNARE proteins (reviewed by Sanderfoot and Raikhel, 1999). Four homologs/orthologs of SNAREs belonging to the late endomembrane system have been described from *Arabidopsis*: AtPep12p, AtVti1p, AtVps45p and AtVam3p. *AtPEP12* can functionally complement for Pep12p in the yeast mutant, *pep12Δ* (Bassham *et al.*, 1995), is membrane-anchored at its C-terminus, and does not localize to the ER, Golgi or tonoplast (Conceicao *et al.*, 1997). In continuous sucrose density gradients of *Arabidopsis* root membranes, AtPep12p antibodies recognize proteins in three different fractions, two of which showed cross-reactivity towards antibodies generated against the vacuole protein sorting receptor, AtELP (Sanderfoot *et al.*, 1998). Similarly, AtELP and a tobacco BP-80 homolog co-fractionate in sucrose gradients prepared from tobacco stigmata (Miller *et al.*, 1999).

AtVti1a, an integral membrane protein, is 30% identical to its yeast homolog, and can functionally complement for its yeast counterpart. AtVtip and AtPep12p can be co-immunoprecipitated, and the two proteins co-fractionate in sucrose density gradients of *Arabidopsis* root membranes (Zheng *et al.*, 1999). AtVps45p, an extrinsic membrane protein, is 45% identical to its yeast counterpart, and can also functionally complement yeast *vps45Δ* mutants; it is also 47% identical to mammalian Vsp45p. In sucrose density gradients, AtVsp45p co-fractionated with AtELP but not AtPep12p (Bassham and Raikhel, 1998). AtVAM3p also functionally complements for its yeast counterpart, and was initially localized to the tonoplast (Sato *et al.*, 1997). However, in a more recent paper, Sanderfoot *et al.* (1999) have pointed to the very high sequence similarity between AtPep12p and AtVam3p, and have suggested that AtVam3p is more likely to be a t-SNARE of the PVC than a true homolog of the yeast vacuolar t-SNARE, ScVamp3.

With the appropriate SNARE antisera now available, one would have expected immunogold electron microscopy (e.g. Figure 12.3) to have provided clear-cut identification of the plant PVC and TGN → PVC transport vesicles. Unfortunately, this is not the case. Whilst AtPep12p is definitely not localized to either Golgi or tonoplast membranes, it must be clearly stated that the small 100 nm diameter immunogold labeled structures bear no resemblance to any immediately recognizable organelle in plant cells, in particular to a multivesicular body (Conceicao *et al.*, 1997). The same structures also label positively for AtELP (Sanderfoot *et al.*, 1998). Since AtELP and the closely related BP-80 are definitely enriched in plant CCVs, which are roughly 100 nm in diameter, it is conceivable, considering the possible exchangeability of SNAREs and their lack of specificity (for discussion, see Sanderfoot and Raikhel, 1999), that the structures depicted are derivatives of the TGN rather than the PVC. The same applies for the recent immunolocalizations of AtVti1p and AtVamp3, both of

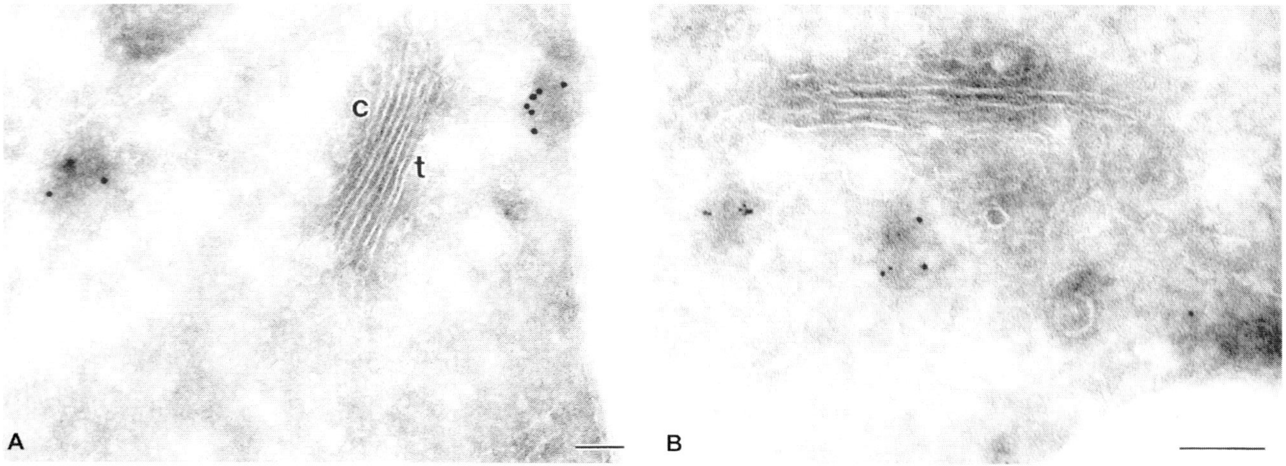

Figure 12.3 Immunogold localization of plant prevacuolar t-SNAREs in cryosectioned *Arabidopsis* root cells. (Micrographs courtesy of N. Raikhel). A) AtVAM3p (10 nm gold); B) AtPEP12p (5 nm gold). Abbreviations: c, *cis*; t, *trans*; t-SNARE, target soluble N-ethylmaleimide-sensitive factor attachment protein receptor. (Internal scale bars = 100 nm).

which are localized to the same AtPEP12p-positive structures and lie very close to the Golgi stack (Zheng *et al.*, 1999).

12.3.6 Possible PVCs for storage vacuoles

Protein storage vacuoles may develop by transformation of a pre-existing lytic vacuole, for example, in the case of vegetative storage proteins in the paraveinal mesophyll (Klauer and Franceschi, 1997; Jauh *et al.*, 1998). They also appear to originate *de novo* during the development of seed tissues (Hoh *et al.*, 1995; Robinson *et al.*, 1995), and are present in the meristematic zone of the primary root of germinating seedlings (Paris *et al.*, 1996). Studies on the intracellular transport of the proteins deposited in the storage vacuoles of legume cotyledons indicate that they are sorted via a mechanism involving physical aggregation, which may or may not involve receptors (recently reviewed by Robinson and Hinz, 1999). In contrast to soluble acid hydrolases, the sorting of storage globulins begins at the *cis*-Golgi (Robinson *et al.*, 1997), and they are released in special 150–200 nm diameter 'dense vesicles' (Hinz *et al.*, 1999). *In situ* immunogold electron microscopy has shown that dense vesicles do not contain BP-80, but carry α-TIP, the aquaporin typical of the storage vacuole tonoplast (Hinz *et al.*, 1999). This observation contradicts the proposal of Gomez and Chrispeels (1993), that storage vacuole lumenal and membrane proteins are transported via different vesicles, a situation which might occur in a transgenic system (for a further discussion on this point see Robinson and Hinz, 1999).

Robinson and co-workers (Robinson *et al.*, 1998a; Robinson and Hinz, 1999) have described the presence of large ($\geq 1\,\mu m$), storage globulin-filled multivesicular bodies in developing pea cotyledons. The limiting membrane of these structures has α-TIP, but BP-80 is absent (Hinz *et al.*, in preparation), an observation in keeping with transport to a non-lytic vacuole. Observations on cotyledon cells made at a developmental stage when vicilin deposition is already underway but legumin expression is just starting, have shown that whilst the storage protein deposits located at the periphery of the protein storage vacuole contained predominantly vicilin, the multivesicular bodies were highly concentrated in legumin (see Figure 12.4; see also Figure 4a in Robinson *et al.*, 1998 and Figure 4a in Robinson and Hinz, 1999). At later stages in the development of pea cotyledon, the storage protein deposits in the vacuole are seen to be stratified, with vicilin towards the lumen of the vacuole and legumin nearest the tonoplast, Robinson and co-workers suggested that this stratification was a consequence of the fusion of a legumin-filled multivesicular body onto a vicilin-containing vacuole protein deposit. A double-label immunogold localization micrograph has been presented to support this hypothesis (see Figure 4b in Robinson *et al.*, 1998).

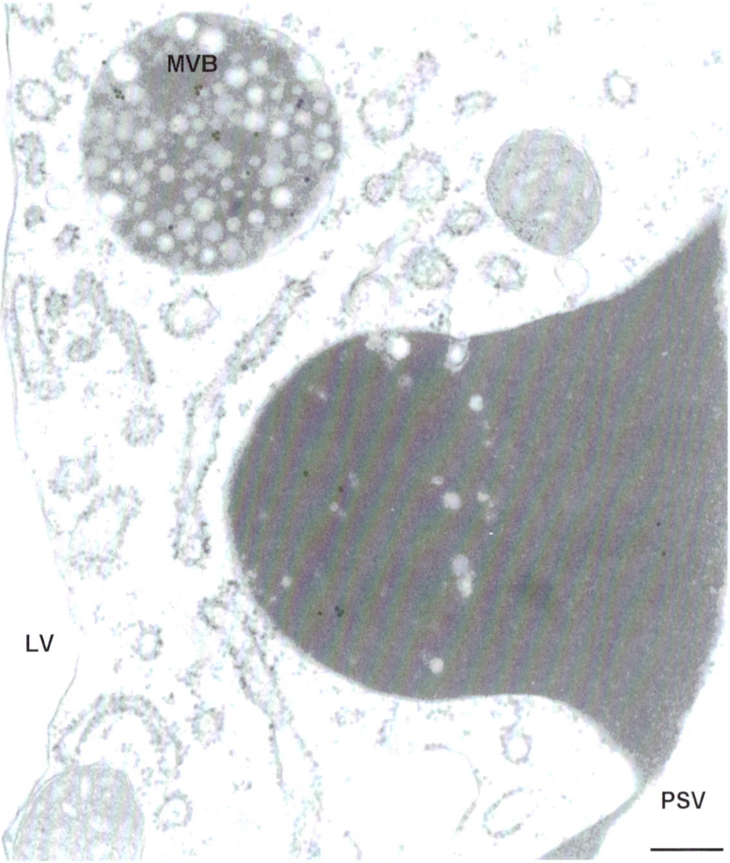

Figure 12.4 Multivesicular body (MVB) and dome-shaped globulin deposit in the protein storage vacuole (PSV) of a developing pea cotyledon. Immunogold (10 nm gold) labeling with legumin antibodies. Abbreviation: LV, lytic vacuole. (Internal scale bar = 250 nm).

12.3.7 A unique PVC in the protein storage vacuole pathway of non-leguminous dicotyledonous plants

Consideration of the plant PSV system is complicated by the fact that PSVs in most seeds are complex organelles with three distinct components: a matrix, which contains most of the soluble storage proteins; a crystalloid composed of unknown proteins organized in a lattice structure; and globoid cavities that may contain phytic acid or oxalate crystals (Lott, 1980; Weber and Neumann, 1980). In contrast, PSVs in legume seeds (including peas) lack crystalloid (Lott, 1980; Weber and Neumann, 1980). The matrix and crystalloid appear to be destinations for different types of proteins. When expressed in transgenic tobacco seeds, the bean seed storage proteins, phaseolin and phytohemagglutinin, were localized to the PSV matrix (Greenwood and Chrispeels, 1985; Sturm and Chrispeels, 1988), while the maize 15 kDa zein was localized to the PSV crystalloid (Hoffman *et al.*, 1987).

Results from studies of traffic of chimeric integral membrane proteins to vacuoles in tobacco suspension culture protoplasts have indicated that some membrane components of organelles in the PSV pathway may derive directly from ER (Jiang and Rogers, 1998), a conclusion that would be consistent with those from precursor-accumulating vesicle studies (Hara-Nishimura *et al.*, 1998). As discussed in Section 12.3.4, a chimeric protein comprised of a lumenal reporter domain attached to only the BP-80 TMD trafficked from ER through the Golgi to the lytic PVC. In contrast, attachment of the 15 amino acid C-terminal CT of the PSV-specific α-TIP caused that reporter (now designated the α-TIP CT reporter) to move directly from ER, bypassing the Golgi, to a second type of vacuole. This vacuole was also the destination of intact α-TIP when it and the chimeric reporter were expressed together, and was therefore considered to be a PSV equivalent in the tobacco protoplasts (Jiang and Rogers, 1998).

To determine further details of this pathway, transgenic tobacco plants expressing the α-TIP CT reporter were studied (Jiang *et al.*, 2000). The reporter protein localized to 1–2 μm diameter organelles in root tip cells and developing seeds that were also marked by the presence of a unique tonoplast intrinsic protein, termed 'dark intrinsic protein' (DIP) (Culianez-Macia and Martin, 1993) and were predominantly located in the cytoplasm. As seed development proceeded, these organelles were taken up by PSVs and aggregated, such that they formed the PSV crystalloid in mature seeds (Jiang *et al.*, 2000). These organelles, termed 'DIP organelles' to emphasize the specific localization of this tonoplast intrinsic protein in their membranes, represent an intermediate for vesicular traffic from the Golgi because they also abundantly accumulate a unique receptor-like protein of the ReMembR-H2 (RMR) protein family that acquires Golgi-specific Asn-linked oligosaccharide modifications. Thus, DIP organelles serve as PVCs for proteins destined for PSVs that traffic from the Golgi as well as directly from the ER (Jiang *et al.*, 2000).

As little or no DIP labeling was detected in the tonoplast of mature PSVs, and little or no labeling for δ-TIP or α-TIP (PSV tonoplast markers) was detected around or in the crystalloid, it was hypothesized that the DIP organelles in the cytoplasm are surrounded by a membrane that fuses with the PSV tonoplast to deliver the internal, DIP-containing membranes to the vacuole interior (Jiang *et al.*, 2000). The validity of this hypothesis remains to be tested.

The morphological similarity of cytoplasmic DIP organelles (Jiang *et al.*, 2000) with precursor-accumulating (PAC) vesicles (Hara-Nishimura *et al.*, 1998; Hayashi *et al.*, 1999), the fact that PAC vesicles were identified in species where PSVs in mature seeds contain crystalloid, and the fact that PAC vesicles contain proteins derived both directly from the ER as well as from Golgi traffic

(Hara-Nishimura *et al.*, 1998) all indicate that DIP organelles and PAC vesicles may be functionally and structurally related. As pea PSVs lack crystalloid, it is much more difficult to relate PAC vesicles/DIP organelles to the multivesicular bodies that appear to serve as PVCs in developing pea cotyledons (see Section 12.3.6). In this respect, it may be helpful to prepare antibodies that can recognize the pea DIP and RMR proteins, so that their localization in developing pea cotyledons can be determined.

12.4 Interaction of PLC/PVC with endocytotic pathways

12.4.1 Mammals and yeasts

Receptor-mediated endocytosis is a well-documented feature of mammalian cells (reviewed by Trowbridge *et al.*, 1994; Mukherjee *et al.*, 1997), and has also been established for yeast cells (Geli and Riezman, 1998). The endocytotic pathway in both cases ends in the lytic compartment of the cell, i.e. either in the lysosome or vacuole. In both cases, the biosynthetic (TGN → PLC/PVC → lysosome/vacuole) and endocytotic pathways to the vacuole converge at the PLC/PVC. However, receptor-ligand complexes which are internalized at the plasma membrane via CCV are not delivered directly to the PLC/PVC. They are first transported to a low pH compartment, where the ligand dissociates from its receptor and from which the receptor is recycled to the plasma membrane. This compartment constitutes the 'early endosome', in contrast to the PLC/PVC, which is termed the 'late endosome'. The early endosome in mammalian cells tends to adopt a tubular, fenestrating morphology, sometimes with internal vesicles (e.g. Schmid *et al.*, 1988; Tooze and Hollinshead, 1991; Escola *et al.*, 1998). In yeast, the early endosome lies in the cortical cytoplasm and is also pleiomorphic in appearance (Hicke *et al.*, 1997).

Interestingly, mammalian cells employ both direct (biosynthetic) and indirect (endocytotic) pathways to bring proteins to the lysosome. In addition to the MPR-sorted soluble acid hydrolases (see Section 12.7.2.2) and LAMP 1 (Höning *et al.*, 1996), which exit the TGN via CCV, at least one lysosomal membrane protein, lysosomal acid phosphatase (LAP), follows the secretory pathway to the cell surface and is then internalized and sent along the endocytotic pathway to the lysosome (Braun *et al.*, 1989; Sosa *et al.*, 1993). We are not aware of a similar situation in yeast, although in dynamin (*vps1-ts*) and clathrin (*chc1-ts*) temperature-sensitive mutants, Golgi and vacuolar membrane proteins reach the vacuole via the plasma membrane and are not sent to the PVC (Seeger and Payne, 1992; Nothwehr *et al.*, 1995).

12.4.2 Plants

Although unequivocal evidence for receptor-mediated endocytosis in plants via plasma membrane derived CCV still has to be published, studies on the uptake of nonspecific electron dense tracers indicate that an endocytotic pathway does exist in plant cells (last reviewed by Battey *et al.*, 1999). Time course experiments show that, after internalization in CCV, tracer molecules are first detected in a structure termed the 'partially-coated reticulum', before entering multivesicular bodies, and finally the vacuole (e.g. Tanchak *et al.*, 1988). Robinson and Hillmer (1990) have discussed the possibility that the partially-coated reticulum may be the early endosome of plant cells, and have drawn attention to its morphological similarity to, and possible origin from, the TGN of the plant Golgi apparatus. One might speculate that, analogous to the mammalian early endosome (see Stoorvogel *et al.*, 1996), the CCVs seen budding from the partially-coated reticulum are involved in recycling to the plasma membrane, but otherwise this structure remains enigmatic. So far, the only direct evidence for a convergence of endocytotic and biosynthetic pathways is the demonstration of (vacuolar destined?) acid phosphatase and endocytosed ferritin in multivesicular bodies from soybean protoplasts (Record and Griffing, 1988).

12.5 The final step: delivery into the lysosome/vacuole

Three possibilities exist for the arrival of the contents and the membrane of the PLC/PVC into the lysosome/vacuole: maturation and gradual transformation of the PLC/PVC into lysosome/vacuole (Murphy, 1991); fusion of the PLC/PVC with a pre-existing lysosome/vacuole (Griffiths and Gruenberg, 1991); and by vesicle transport. Although the latter mechanism would provide the most obvious way to prevent sorting receptors from reaching the lysosome/vacuole, the great majority of evidence obtained on mammalian cells now seems to be in support of a fusion process. This has been demonstrated by *in vitro* fusion experiments between isolated endosomes and lysosomes (Mullock *et al.*, 1998; Schmid *et al.*, 1998), and also by the depiction of putative fusion profiles in thin sections (e.g. van Deurs *et al.*, 1995; Futter *et al.*, 1996). The fusion model is not necessarily incompatible with the need to keep sorting receptors and other components of the PLC/PVC membrane out of the lysosome/vacuole, provided the fusion is only temporary, thereby allowing the transfer of lumenal molecules and microvesicles without a mixing of the limiting membranes ('kiss and run' model; Storrie and Desjardins, 1996). Data in support of such a scenario have, in fact, been presented (Bright *et al.*, 1997).

The situation in yeast cells is not so clear as it is in mammals, but the delivery of PVC microvesicles into the lumen of vacuoles unable to acidify (see Section 12.2.2, Wurmser and Emr, 1998) is highly suggestive of a fusion between the PVC and the vacuole. In plants, the fusion of multivesicular bodies with lytic vacuoles has been reported (Tanchak and Fowke, 1987; Coulomb and Coulomb, 1996). In order to explain the stratification of the globulins, legumin and vicilin, in the storage protein deposits at the tonoplast of storage vacuoles in developing pea cotyledons, Robinson *et al.* (1998) also invoked a fusion process between a multivesicular body and the tonoplast. Indeed, microvesicles similar to those in the multivesicular bodies are seen in the storage protein deposits within the vacuole (see Figure 12.4 and also Figure 4b, Robinson *et al.*, 1998).

The lysosome and yeast vacuole are not terminal compartments. Much evidence has accumulated showing that LAMP-1(Akazaki *et al.*, 1993), major histocompatibility complex class II molecules (Brachet *et al.*, 1997), and some soluble lysosomal constituents (Jahrous *et al.*, 1994) participate in retrograde transport out of the lysosome. It has also been shown that the AP-2 adaptor complex of CCV can assemble at the surface of lysosomes, suggesting perhaps that CCV may bud from lysosomes (Traub *et al.*, 1996). In yeast, data have recently been presented which indicate that the v-SNARE, Vti1p, is also recycled out of the vacuolar membrane, a process in which the gene, *VAC7*, is involved (Bryant *et al.*, 1998). It has been known for some time that hydrolases usually localized to the vacuole can be released into the medium of suspension cultured cells (Wink, 1984). This has often been interpreted as being a consequence of saturation of the vacuolar targeting machinery resulting in secretion. However, the results by Kunze *et al.* (1998), which demonstrate that vacuole-processed forms of chitinase I and β1–3 glucanase are secreted by suspension-cultured tobacco cells, provide the first indication that transport out of the vacuole of plant cells may also occur.

References

Advani, R.J., Yang, B., Prekeris, R., Lee, K.C., Klumpermann, J. and Scheller, R.H. (1999) VAMP-7 mediates vesicular transport from endosomes to lysosomes. *J. Cell Biol.*, **146** 765-75.

Ahmed, S.U., Bar-Peled, M. and Raikhel, N.V. (1997a) Cloning and subcellular location of an *Arabidopsis* receptor-like protein that shares common features with protein sorting receptors of eukaryotic cells. *Plant Physiol.*, **114** 325-36.

Ahmed, S.U., Bar-Peled, M. and Raikhel, N.V. (1997b) Cloning and subcellular location of an *Arabidopsis* receptor-like protein that shares common features with protein-sorting receptors of eukaryotic cells. (Addendum). *Plant Physiol.*, **115** 311-12.

Akasaki, K., Fukuzawa, M., Kinoshita, H., Furuno, K. and Tsuji, H. (1993) Cycling of two endogenous lysosomal membrane proteins, lamp-2 and acid phosphatase, between the cell surface and lysosomes in cultured rat hepatocytes. *J. Biochem.*, **114** 598-604.

Anderson, R.G.W. and Orci, L. (1988) A view of acidic intracellular compartments. *J. Cell Biol.*, **106** 539-43.

Bassham, D.C. and Raikhel, N.V. (1998) An *Arabidopsis* VPS45p homolog implicated in protein transport to the vacuole. *Plant Physiol.*, **117** 407-15.

Bassham, D.C., Gal, S., Conceicao, A.S. and Raikhel, N.V. (1995) An *Arabidopsis* syntaxin homolog isolated by functional complementation of a yeast *pep12* mutant. *Proc. Natl. Acad. Sci. USA*, **92** 7262-66.

Battey, N.H., James, N.C., Greenland, A.J. and Brownlee, C. (1999) Exocytosis and endocytosis. *Plant Cell*, **11** 643-59.

Becherer, K.A., Rieder, S.E., Emr, S.D. and Jones, E.W. (1996) Novel syntaxin homolog, Pep12p, required for the sorting of lumenal hydrolases to the lysosome-like vacuole in yeast. *Mol. Biol. Cell*, **7** 579-94.

Bowen, B.R., Fennie, C. and Lasky, L.A. (1990) The Mel 14 antibody binds to the lectin domain of the murine peripheral lymph node homing receptor. *J. Cell Biol.*, **110** 147-53.

Brachet, V., Raposo, G., Amigorena, S. and Mellman, I. (1997) II chain controls the transport of major histocompatibility complex class II molecules to and from lysosomes. *J. Cell Biol.*, **137** 51-65.

Braulke, T. (1996) Origin of lysosomal proteins, in *Subcellular Biochemistry*, Vol. 27 (eds. J.B. Lloyd and R.W. Mason) Plenum Press, New York, pp. 15-49.

Braun, M., Waheed, A. and von Figura, K. (1989) Lysosomal acid phosphatase is transported to lysosomes via the cell surface. *EMBO J.*, **8** 3633-40.

Bright, N.A., Reaves, B.J., Mullock, B.M. and Luzio, J.P. (1997) Dense core lysosomes can fuse with late endosomes and are re-formed from the resultant hybrid organelles. *J. Cell Sci.*, **110** 2027-40.

Brown, W.J., DeWald, D.B., Emr, S.D., Plutner, H. and Balch, W.E. (1995) Role for phosphatidylinositol 3-kinase in the sorting and transport of newly synthesized lysosomal enzymes in mammalian cells. *J. Cell Biol.*, **130** 781-96.

Bryant, N.J., Piper, R.C., Weisman, L.S. and Stevens, T.H. (1998) Retrograde traffic out of the yeast vacuole occurs via the prevacuolar/endosomal compartment. *J. Cell Biol.*, **142** 651-63.

Burd, C.G., Peterson, M., Cowles, C.R. and Emr, S.D. (1997) A novel Sec18p/NSF-dependent complex required for Golgi-to-endosome transport in yeast. *Mol. Biol. Cell*, **8** 1089-104.

Butler, J.M., Kirsch, T., Watson, B., Paris, N., Rogers, J.C. and Bevvers, L. (1997) Interaction of the vacuolar targeting receptor, BP-80, with clathrin-adaptors. *Plant Physiol.*, **114** (Suppl.) 1210.

Cao, X., Rogers, S.W., Butler, J., Beevers, L. and Rogers, J.C. (2000) Structural requirements for ligand binding by a plant vacuolar sorting receptor. *Plant Cell*, **12** 1493-506.

Casey, R. (1979) Genetic variability in the structure of the α-subunits of legumin from *Pisum sativum*: a two-dimensional gel electrophoresis study. *Heredity*, **43** 265-72.

Cereghino, J.L., Marcusson, E.G. and Emr, S.D. (1995) The cytoplasmic tail domain of the vacuolar protein sorting receptor, Vps10p, and a subset of VPS gene products regulate receptor stability, function and localization. *Mol. Biol. Cell*, **6** 1089-102.

Clague, M.J. (1998) Molecular aspects of the endocytic pathway. *Biochem. J.*, **336** 271-82.

Conceicao, A.S., Marty-Mazars, D., Bassham, D.C., Sanderfoot, A.A., Marty, F. and Raikhel, N.V. (1997) The syntaxin homolog, AtPEP12p, resides on a late post-Golgi compartment in plants. *Plant Cell*, **9** 571-82.

Conibear, E. and Stevens, T.H. (1998) Multiple sorting pathways between the late-Golgi and the vacuole in yeast. *Biochim. Biophys. Acta*, **1404** 211-30.

Cooper, A.A. and Stevens, T.H. (1996) Vps10p cycles between the late-Golgi and prevacuolar compartments in its function as the sorting receptor for multiple yeast vacuolar hydrolases. *J. Cell Biol.*, **133** 529-41.

Corvera, S. and Czech, M.P. (1998) Direct targets of phosphoinositide 3-kinase products in membrane traffic and signal transduction. *Trends Cell Biol.*, **8** 442-46.

Coulomb, S. and Coulomb, S. (1996) Endocytosis in *Cucurbita pepo* root meristems: coated vesicles, multivesicular bodies and vacuole relationship. *Comptes Rendus Acad. Sci. Paris*, **III 319** 377-83.

Craig, S. and Goodchild, D.J. (1984) Golgi-mediated vicilin accumulation in pea cotyledon cells is redirected by monensin and nigericin. *Protoplasma*, **122** 91-97.

Culianez-Macia, F.A. and Martin, C. (1993) DIP: a member of the MIP family of membrane proteins that is expressed in mature seeds and dark-grown seedlings of *Antirrhinum majus*. *Plant J.*, **4** 717-25.

Cutler, N.S., Heitman, J. and Cardenas, M.E. (1997) SST4 is an essential phosphatidylinositol 4-kinase that is a target of wortmannin in *Saccharomyces cerevisiae*. *J. Biol. Chem.*, **272** 27671-77.

Darsow, T., Rieder, S.E. and Emr, S.D. (1997) A multi-specificity syntaxin homolog, Vam3p, essential for autophagy and biosynthetic protein transport to the vacuole. *J. Cell Biol.*, **138** 517-29.

Davis, C.G. (1990) The many faces of epidermal growth factor repeats. *New Biol.*, **2** 410-19.

Davis, C.G., Goldstein, J.L., Südhof, T.C., Anderson, R.G.W., Russell, D.W. and Brown, M.S. (1987) Acid-dependent ligand dissociation and recycling of LDL receptor mediated by growth factor homology region. *Nature*, **326** 760-65.

Dautry-Varsat, A., Ciechanover, A. and Lodish, H.F. (1983) pH and the recycling of transferrin during receptor-mediated endocytosis. *Proc. Natl. Acad. Sci. USA*, **80** 2258-62.

Di Sebastiano, G.-P., Paris, N., Marc-Martin, S. and Neuhaus, J.-M. (1998) Specific accumulation of GFP in a non-acidic vacuolar compartment via a C-terminal propeptide-mediated sorting pathway. *Plant J.*, **15** 449-57.

Diaz, E. and Pfeffer, S.R. (1998) TIP47: a cargo selective device for mannose 6-phosphate receptor trafficking. *Cell*, **93** 433-43.

Draper, R.K., Goda, Y., Brodsky, E.M. and Pfeffer, S.R. (1990) Antibodies to clathrin inhibit endocytosis but not recycling to the *trans*-Golgi network *in vitro*. *Science*, **248** 1539-41.

Duncan, J.R. and Kornfeld, S. (1988) Intracellular movement of the two mannose 6-phosphate receptors: return to the Golgi apparatus. *J. Cell Biol.*, **106** 617-28.

Dunn, W.A., Conolly, T.P. and Hubbard, A. (1986) Receptor-mediated endocytosis of the epidermal growth factor by rat hepatocytes: receptor pathway. *J. Cell Biol.*, **102** 24-36.

Escola, J.-M., Kleijmeer, M.J., Stoorvogel, W., Griffith, J.M., Yoshie, O. and Geuze, H.J. (1998) Selective enrichment of tetraspan proteins on the internal vesicles of multivesicular endosomes and on exosomes secreted by human B-lymphocytes. *J. Biol. Chem.*, **273** 20121-27.

Felder, S., Miller, K., Moehren, G., Ullrich, A., Schlesinger, J. and Hopkins, C.R. (1990) Kinase activity controls the sorting of the epidermal growth factor receptor within the multivesicular body. *Cell*, **61** 623-34.

Fernandez-Borja, M., Wubbolts, R., Calafat, J., Janssen, H., Divecha, N., Dusseljee, S. and Neefjes, J. (1999) Multivesicular body morphogenesis requires phosphatidylinositol 3-kinase activity. *Curr. Biol.*, **9** 55-58.

Fischer von Mollard, G. and Stevens, T.H. (1999) The *Saccharomyces cerevisiae* v-SNARE, Vti1p, is required for multiple membrane transport pathways to the vacuole. *Mol. Biol. Cell*, **10** 1719-32.

Fischer von Mollard, G., Nothwehr, S.F. and Stevens, T.H. (1997) The yeast v-SNARE, Vti1p, mediates two vesicle transport pathways through interactions with the t-SNAREs, Sed5p and Pep12p. *J. Cell Biol.*, **137** 1511-24.

Friend, D.S. and Farquhar, M.S. (1967) Functions of coated vesicles during protein absorption in the rat vas deferens. *J. Cell Biol.*, **35** 357-76.

Futter, C.E., Pearse, A., Hewlett, L.J. and Hopkins, C.R. (1996) Multivesicular endosomes containing internalized EGF-EGF receptor complexes mature and then fuse directly with lysosomes. *J. Cell Biol.*, **132** 1011-23.

Gaullier, J.M., Simonsen, A., D'Arrigo, A., Bremnes, B., Stenmark, H. and Aasland, R. (1998) FYVE fingers bind Ptdins (3) P. *Nature*, **394** 432-33.

Geli, M.I. and Riezman, H. (1998) Endocytic internalization in yeast and animal cells: similar and different. *J. Cell Sci.*, **111** 1031-37.

Gomez, L. and Chrispeels, M.J. (1993) Tonoplast and soluble vacuolar proteins are targeted by different mechanisms. *Plant Cell*, **5** 1113-24.

Götte, M. and Fischer von Mollard, G. (1998) A new beat for the SNARE drum. *Trends Cell Biol.*, **8** 215-18.

Greenwood, J.S. and Chrispeels, M.J. (1985) Correct targeting of the bean storage protein, phaseolin, in the seeds of transformed tobacco. *Plant Physiol.*, **79** 65-71.

Griffiths, G. and Gruenberg, J. (1991) The arguments for pre-existing early and late endosomes. *Trends Cell Biol.*, **1** 5-9.

Griffiths, G., Hoflack, B., Simons, K., Mellman, I. and Kornfeld, S. (1988) The mannose 6-phosphate receptor and the biogenesis of lysosomes. *Cell*, **52** 329-41.

Hara-Nishimura, I., Shimada, T., Hatano, K., Takeuchi, Y. and Nishimura, M. (1998) Transport of storage proteins to protein storage vacuoles is mediated by large precursor-accumulating vesicles. *Plant Cell*, **10** 825-36.

Hay, J.C. and Scheller, R.H. (1997) SNAREs and NSF in targeted membrane fusion. *Curr. Opin. Cell Biol.*, **9** 505-12.

Hayashi, M., Toriyama, K., Kondo, M., Hara-Nishimura, I. and Nishimura, M. (1999) Accumulation of a fusion protein containing 2S albumin induces novel vesicles in vegetative cells of *Arabidopsis*. *Plant Cell Physiol.*, **40** 263-72.

Herman, P.K., Stack, J.H., DeModena, J.A. and Emr, S.D. (1991) A novel protein kinase homolog essential for protein sorting to the yeast lysosome-like vacuole. *Cell*, **64** 425-37.

Herz, J., Hamann, U., Rogne, S., Myklebost, O., Gausepohl, H. and Stanley, K.K. (1988) Surface location and high affinity for calcium of a 500 kD liver membrane protein closely related to the LDL-receptor suggest a physiological role as lipoprotein receptor. *EMBO J.*, **7** 4119-27.

Heupke, H.-J. and Robinson, D.G. (1985) Intracellular transport of α-amylase in barley aleurone cells: evidence for the participation of the Golgi apparatus. *Eur. J. Cell Biol.*, **39** 265-72.

Hicke, L., Zanolari, B., Pypaert, M., Rohrer, J. and Riezman, H. (1997) Transport through the yeast endocytic pathway occurs through morphologically distinct compartments and requires an active secretory pathway and Sec18p/N-ethylmaleimide-sensitive fusion protein. *Mol. Biol. Cell*, **8** 13-31.

Hinz, G., Hillmer, S., Bäumer, M. and Hohl, I. (1999) Vacuolar storage proteins and the putative vacuolar sorting receptor, BP-80, exit the Golgi apparatus of developing pea cotyledons in different transport vesicles. *Plant Cell*, **11** 1509-24.

Hoffman, L.M., Donaldson, D.D., Bookland, R., Rashka, K. and Herman, E.M. (1987) Synthesis and protein deposition of maize 15 kD zein in transgenic tobacco seeds. *EMBO J.*, **6** 3213-21.

Hoh, B., Hinz, G., Jeong, B.-K. and Robinson, D.G. (1995) Protein storage vacuoles form *de novo* during pea cotyledon development. *J. Cell Sci.*, **108** 299-310.

Holthuis, J.C.M., Nichols, B.J., Dhruvakumar, S. and Pelham, H.R.B. (1998) Two syntaxin homologs in the TGN/endosomal system of yeast. *EMBO J.*, **17** 113-26.

Holwerda, B.C., Galvin, N.J., Baranski, T.J. and Rogers, J.C. (1990) *In vitro* processing of aleurain, a barley vacuolar thiol protease. *Plant Cell*, **2** 1091-106.

Höning, S., Griffith, J., Geuze, H.J. and Hunziker, W. (1996) The tyrosine-based lysosomal targeting signal in Lamp1 mediates sorting into Golgi-derived clathrin-coated vesicles. *EMBO J.*, **15** 5230-39.

Höning, S., Sosa, M., Hille-Rehfeld, A. and von Figura, K. (1997) The 46 kDa mannose 6-phosphate receptor contains multiple binding sites for clathrin adaptors. *J. Biol. Chem.*, **272** 19884-90.

Horazdovsky, B.F., Busch, G.R. and Emr, S.D. (1994) *VPS21* encodes a rab5-like GTP-binding protein that is required for the sorting of yeast proteins. *EMBO J.*, **13** 1297-1309.

Itin, C., Ulitzur, N., Mühlbauer, B. and Pfeffer, S.R. (1999) Mapmodulin, cytoplasmic dynein, and microtubules enhance the transport of mannose 6-phosphate receptors from endosomes to the *trans*-Golgi network. *Mol. Biol. Cell*, **10** 2191-97.

Jahrous, A., Storrie, B., Griffiths, G. and Desjardins, M. (1994) Evidence for retrograde traffic between terminal lysosomes and the prelysosomal/late endosome compartment. *J. Cell Sci.*, **107** 145-57.

Jauh, G.Y., Fischer, A.M., Grimes, H.D., Ryan, C.A. and Rogers, J.C.(1998). δ-Tonoplast intrinsic protein defines unique plant vacuole functions. *Proc. Natl. Acad. Sci. USA*, **95** 12995-99.

Jauh, G.-Y., Phillips, T.E. and Rogers, J.C. (1999) Tonoplast intrinsic protein isoforms as markers for vacuolar functions. *Plant Cell*, **11** 1867-82.

Jiang, L. and Rogers, J.C. (1998) Integral membrane protein sorting to vacuoles in plant cells: evidence for two pathways. *J. Cell Biol.*, **143** 1183-99.

Jiang, L. and Rogers, J.C. (1999) The role of BP-80 and homologs in sorting proteins to vacuoles. *Plant Cell*, **11** 2069-71.

Jiang, L., Phillips, T.E., Rogers, S.W. and Rogers, J.C. (2000) Biogenesis of the protein storage vacuole crystalloid. *J. Cell Biol.* (in press).

Joly, M., Kazlaukas, A., Fay, F.S. and Corvera, S. (1994) Disruption of PDGF receptor trafficking by mutation of its PI 3-kinase binding sites. *Science*, **263** 684-87.

Jørgensen, M.U., Emr, S.D. and Winther, J.R. (1999) Ligand recognition and domain structure of Vps10p, a vacuolar protein sorting receptor in *Saccharomyces cerevisiae*. *Eur. J. Biochem.*, **260** 461-69.

Karlsson, K. and Carlsson, S.R. (1998) Sorting of lysosomal membrane glycoproteins lamp-1 and lamp-2, into vesicles distinct from mannose 6-phosphate receptor/γ-adaptin vesicles at the *trans*-Golgi network. *J. Biol. Chem.*, **273** 18966-73.

Kinal, H., Park, C.-M., Berry, J.O., Koltin, Y. and Bruenn, J.A. (1995) Processing and secretion of a virally encoded antifungal toxin in transgenic tobacco plants: evidence for a Kex2p pathway in plants. *Plant Cell*, **7** 677-88.

Kirsch, T., Paris, N., Butler, J.M., Beevers, L. and Rogers, J.C. (1994) Purification and initial characterization of a potential plant vacuolar targeting receptor. *Proc. Natl. Acad. Sci. USA*, **91** 3403-407.

Kirsch, T., Saalbach, G., Raikhel, N.V. and Beevers, L. (1996) Interaction of a potential vacuolar targeting receptor with amino- and carboxyl-terminal targeting determinants. *Plant Physiol.*, **111** 469-74.

Klauer, S.F. and Franceschi, V.R. (1997) Mechanism of transport of vegetative storage proteins to the vacuole of the paraveinal mesophyll of soybean leaves. *Protoplasma*, **200** 174-85.

Kobayashi, T., Stang, E., Fang, K.S., de Moerloose, P., Parton, R.G. and Gruenberg, J. (1998) A lipid associated with the antiphospholipid syndrome regulates endosome structure and function. *Nature*, **392** 193-97.

Koide, Y., Hirano, H., Matsuoka, K. and Nakamura, K. (1997) The N-terminal propeptide of the precursor to sporamin acts as a vacuole-targeting signal even at the C-terminus of the mature part in tobacco cells. *Plant Physiol.*, **114** 863-70.

Kornfeld, S. and Mellman, I. (1989) The biogenesis of lysosomes. *Annu. Rev. Cell Biol.*, **5** 483-525.

Kunze, I., Kunze, G., Bröker, M., Manteuffel, R., Meins, F. and Müntz, K. (1998) Evidence for secretion of vacuolar α-mannosidase, class I chitinase and class I β-1,3-glucanase in suspension cultures of tobacco cells. *Planta*, **205** 92-99.

Kurten, R.C., Cadena, D.L. and Gill, G.N. (1996) Enhanced degradation of EGF receptors by a sorting nexin, SSNX1. *Science*, **272** 1008-10.

Laval, V., Chabannes, M., Carrière, M., Canut, H., Barre, A., Rougé, R., Pont-Lezica, R. and Galaud, J.-P. (1999) A family of *Arabidopsis* plasma membrane receptors presenting animal β-integrin domains. *Biochem. Biophys. Acta*, **1435** 61-70.

Lazar, T., Götte, M. and Gallwitz, D. (1997) Vesicular transport: how many Ypt/Rab-GTPases make a eukaryotic cell? *Trends Cell Biol.*, **22** 468-72.

Le Borgne, R. and Hoflack, B. (1997) Mannose 6-phosphate receptors regulate the formation of clathrin-coated vesicles in the TGN. *J. Cell Biol.*, **137** 335-45.

Le Borgne, R. and Hoflack, B. (1998) Protein transport from the secretory to the endocytic pathway in mammalian cells. *Biochim. Biophys. Acta*, **1404** 195-209.

Lombardi, D., Soldati, T., Riederer, M.A., Goda, Y., Zerial, M. and Pfeffer, S.R. (1993) Rab9 functions in transport between late endosomes and the *trans*-Golgi network. *EMBO J.*, **12** 677-82.

Lott, J.N.A. (1980) Protein bodies, in *The Biochemistry of Plants* (Ed. N.E. Tolbert), Academic Press, New York, pp. 589-623.

Martys, J.L., Wjasow, C., Gangi, D.M., Kielian, M.C., McGraw, T.E. and Backer, J.M. (1996) Wortmannin-sensitive trafficking pathways in Chinese hamster ovary cells: differential effects on endocytosis and lysosomal sorting. *J. Biol. Chem.*, **271** 10953-62.

Matsuoka, K. and Neuhaus, J.-M. (1999) *Cis*-elements of protein transport to the plant vacuoles. *J. Exp. Bot.*, **50** 165-74.

Matsuoka, K., Bassham, D.C., Raikhel, N.V. and Nakamura, K. (1995) Different sensitivity to wortmannin of two vacuolar sorting signals indicates the presence of distinct sorting machineries in tobacco cells. *J. Cell Biol.*, **130** 1307-18.

Matsuoka, K., Higuchi, T., Maeshima, M. and Nakamura, K. (1997) A vacuolar-type H^+-ATPase in a non-vacuolar organelle is required for the sorting of soluble vacuolar protein precursors in tobacco cells. *Plant Cell*, **9** 533-46.

Mellman, I. (1996) Endocytosis and molecular sorting. *Annu. Rev. Cell Dev. Biol.*, **12** 575-625.

Mellman, I., Fuchs, R. and Helenius, A. (1986) Acidification of the endocytic and exocytic pathways. *Annu. Rev. Biochem.*, **55** 663-700.

Miesenböck, G., De Angelis, D. and Rothman, J.E. (1998) Visualizing secretion and synaptic transmission with pH-sensitive green fluorescent proteins. *Nature*, **394** 192-95.

Miller, E. and Anderson, M. (1999) Reply: the role of BP-80 in sorting to the vacuole in stigmas. *Plant Cell*, **11** 2071-73.

Miller, E.A., Lee, M.C.S. and Anderson, M.A. (1999) Identification and characterization of a prevacuolar compartment in stigmas of *Nicotiana alata*. *Plant Cell*, **11** 1499-508.

Mollenhauer, H.H., Morré, D.J. and Rowe, L.D. (1990) Alteration of intracellular traffic by monensin: mechanism, specificity and relationship to toxicity. *Biochim. Biophys. Acta*, **1031** 225-46.

Mukherjee, S.M., Ghosh, R.N. and Maxfield, F.R. (1997) Endocytosis. *Physiol. Rev.*, **77** 759-803.

Mulholland, J., Konopka, J., Singer-Kruger, B., Zerial, M. and Botstein, D. (1999) Visualization of receptor-mediated endocytosis in yeast. *Mol. Biol. Cell*, **10** 799-817.

Mullock, B.M., Bright, N.A., Frearon, C.W., Gray, S.R. and Luzio, J.P. (1998) Fusion of lysosomes with late endosomes produces a hybrid organelle of intermediate density and is NSF dependent. *J. Cell Biol.*, **140** 591-601.

Murphy, R.F. (1991) Maturation models for endosome and lysosome biogenesis. *Trends Cell Biol.*, **1** 77-82.

Nakajima, Y. and Pfeffer, S.R. (1997) Phosphatidylinositol 3-kinase is not required for recycling of mannose 6-phosphate receptors from late endosomes to the *trans*-Golgi network. *Mol. Biol. Cell*, **8** 577-82.

Nakamura, K., Matsuoka, K., Mukumoto, F. and Watanabe, N. (1993) Processing and transport to the vacuole of a precursor to sweet potato sporamin in transformed tobacco cell line, BY-2. *J. Exp. Bot.*, **44** 331-38.

Nelson, N. and Harvey, W.R. (1999) Vacuolar and plasma membrane proton-adenosine triphosphatases. *Physiol. Rev.*, **79** 361-85.

Neuhaus, J.M. and Rogers, J.C. (1998) Sorting of proteins to vacuoles in plants. *Plant Mol. Biol.*, **38** 127-44.

Nichols, B.J. and Pelham, H.R.B. (1998) SNAREs and membrane fusion in the Golgi apparatus. *Biochim. Biophys. Acta*, **1404** 9-31.

Nichols, B.J., Ungermann, C., Pelham, H.R.B., Wickner, W.T. and Haas, A. (1997) Homotypic fusion mediated by t- and v-SNAREs. *Nature*, **387** 199-202.

Nichols, B.J., Holthuis, J.C.M. and Pelham, H.R.B. (1998) The Sec1p homolog, Vps45p, binds to the syntaxin, Tlg2p. *Eur. J. Cell Biol.*, **77** 263-68.

Nothwehr, S.F., Conibear, E. and Stevens, T.H. (1995) Golgi and vacuolar membrane proteins reach the vacuole in *vps1* mutant yeast cells via the plasma membrane. *J. Cell Biol.*, **129** 35-46.

Novick, P. and Zerial, M. (1997) The diversity of Rab proteins in vesicle transport. *Curr. Opin. Cell Biol.*, **9** 496-504.

Odorizzi, G., Babst, M. and Emr, S.D. (1998) Fab1p PtdIns(3)P 5-kinase function essential for protein sorting in the multivesicular body. *Cell*, **95** 847-58.

Paris, N. and Rogers, J.C. (1996) The role of receptors in targeting soluble proteins from the secretory pathway to the vacuole. *Plant Physiol. Biochem.*, **34** 223-27.

Paris, N., Stanley, M.C., Jones, R.L. and Rogers, J.C (1996) Plant cells contain two functionally distinct vacuolar compartments. *Cell*, **85** 563-72.

Paris, N., Rogers, S.W., Jiang, L., Kirsch, T., Beevers, L., Phillips, T.E. and Rogers, J.C. (1997) Molecular cloning and further characterization of a probable plant vacuolar sorting receptor. *Plant Physiol.*, **115** 29-39.

Piper, R.C., Whitters, E.A. and Stevens, T.H. (1994) Yeast Vps45p is a Sec1p-like protein required for the consumption of vacuole-targeted, post-Golgi transport vesicles. *Eur. J. Cell Biol.*, **65** 305-18.

Ratajczak, R., Hinz, G. and Robinson, D.G. (1999) Localization of pyrophosphatase in membranes of cauliflower inflorescence cells. *Planta*, **208** 205-11.

Raymond, C.K., Howald-Stevenson, I., Vater, C.A. and Stevens, T.H. (1992) Morphological classification of the yeast vacuolar protein sorting mutants: evidence for a prevacuolar compartment in class E mutants. *Mol. Biol. Cell*, **3** 1389-402.

Reaves, B.J., Bright, N.A., Mullock, B.M. and Luzio, J.P. (1996) The effect of wortmannin on the localisation of lysosomal type I integral membrane glycoproteins suggests a role for phosphoinositide 3-kinase activity in regulating membrane traffic late in the endocytic pathway. *J. Cell Sci.*, **109** 749-62.

Record, R.D. and Griffing, L.R. (1988) Convergence of the endocytic and lysosomal pathways in soybean protoplasts. *Planta*, **176** 425-32.

Rieder, S.E., Banta, L.M., Köhrer, K., McCaffery, J.M. and Emr, S.D. (1996) Multilamellar endosome-like compartment accumulates in the yeast *vps28* vacuolar protein sorting mutant. *Mol. Biol. Cell*, **7** 985-99.

Roberts, C.J., Nothwehr, S.F. and Stevens, T.H. (1992) Membrane protein sorting in the yeast secretory pathway: evidence that the vacuole may be the default compartment. *J. Cell Biol.*, **119** 63-83.

Robinson, D.G. and Hillmer, S. (1990) Endocytosis in plants. *Physiol. Plant.*, **79** 96-104.

Robinson, D.G. and Hinz, G. (1997) Vacuole biogenesis and protein transport to the plant vacuole: a comparison with the yeast vacuole and the mammalian lysosome. *Protoplasma*, **197** 1-25.

Robinson, D.G. and Hinz, G. (1999) Golgi-mediated transport of seed storage proteins. *Seed Sci. Res.*, **9** 263-83.

Robinson, D.G., Hoh, B., Hinz, G. and Hohl, I. (1995) One vacuole or two vacuoles: do protein storage vacuoles arise *de novo* during pea cotyledon development? *J. Plant Physiol.*, **45** 654-64.

Robinson, D.G., Bäumer, A., Hinz, G. and Hohl, I. (1997) Ultrastructure of the pea cotyledon Golgi apparatus: origin of dense vesicles and action of brefeldine A. *Protoplasma*, **200** 198-209.

Robinson, D.G., Bäumer, M., Hinz, G. and Hohl, I. (1998a) Vesicle transfer of storage proteins to the vacuole: the role of the Golgi apparatus and multivesicular bodies. *J. Plant Physiol.*, **152** 659-67.

Robinson, D.G., Braulke, T., Denecke, J., Happel, N., Movafeghi, A. and Pimpl, P. (1998b) New developments in plant coated-vesicle research. *Proceedings of the First Balkan Botanical Congress*.

Robinson, D.G., Galili, G., Herman, E. and Hillmer, S. (1998c) Topical aspects of vacuolar protein transport: autophagy and prevacuolar compartments. *J. Exp. Bot.*, **49** 1263-70.

Robinson, D.G., Hinz, G. and Holstein, S.E.H. (1998d) The molecular biology of transport vesicles. *Plant Mol. Biol.*, **38** 49-76.

Robinson, D.G., Hoppenrath, M., Oberbeck, K., Luykx, P. and Ratajczak, R. (1998e) Localization of pyrophosphatase and V-ATPase in *Chlamidymonas reinhardii*. *Bot. Acta*, **111** 108-122.

Sahagian, G.G. and Steer, C.J. (1985) Transmembrane orientation of the mannose-6-phosphate receptor in isolated clathrin-coated vesicles. *J. Biol. Chem.*, **260** 9838-42.

Sanderfoot, A. and Raikhel, N.V. (1999) The specificity of vesicle trafficking: coat proteins and SNAREs. *Plant Cell*, **11** 629-41.

Sanderfoot, A.A., Ahmed, S.U., Marty-Mazars, D., Rapaport, I., Kirchhausen, T., Marty, F. and Raikhel, N.V. (1998) A putative vacuolar cargo receptor partially co-localizes with AtPEP12p on a prevacuolar compartment in *Arabidopsis* roots. *Proc. Natl. Acad. Sci. USA*, **95** 9920-25.

Sanderfoot, A.A., Kovaleva, V., Zheng, H. and Raikhel, N.V. (1999) The t-SNARE, AtVAM3p, resides on the prevacuolar compartment in *Arabidopsis* root cells. *Plant Physiol.*, **121** 929-38.

Sato, M.H., Nakamura, N., Ohsumi, Y., Kouchi, H., Kondo, M., Hara-Nishimura, I., Nishimura, I. and Wada, Y. (1997) The AtVAM3 encodes a syntaxin-related molecule implicated in the vacuolar assembly in *Arabidopsis thaliana*. *J. Biol. Chem.*, **272** 24530-35.

Sato, T.K., Darsow, T. and Emr, S.D. (1998) Vam7p, a SNAP-25-like molecule, and Vam3p, a syntaxin homolog, function together in yeast vacuolar protein trafficking. *Mol. Biol. Cell*, **18** 5308-19.

Schmid, J.A., Ellinger, I. and Kosma, P. (1998) *In vitro* fusion of tissue-derived endosomes and lysosomes. *Eur. J. Cell Biol.*, **77** 166-74.

Schmid, S.L. (1997) Clathrin-coated vesicle formation and protein sorting: an integrated process. *Ann. Rev. Biochem.*, **66** 511-48.

Schmid, S., Fuchs, R., Male, P. and Mellman, I. (1988) Two distinct populations of endosomes involved in membrane recycling and transport to lysosomes. *Cell*, **52** 73-83.

Schu, P.V., Takegawa, K., Fry, M.J., Stack, J.H., Waterfield, M.D. and Emr, S.D. (1993) Phosphatidylinositol 3-kinase encoded by yeast VPS34 gene essential for protein sorting. *Science*, **260** 88-91.

Schweizer, A., Kornfeld, S. and Rohrer, J. (1997) Proper sorting of the cation-dependent mannose 6-phosphate receptor in endosomes depends on a pair of aromatic amino acids in its cytoplasmic tail. *Proc. Natl. Acad. Sci. (USA)*, **94** 14471-76.

Seaman, M.N.J., Marcusson, E.G., Cereghino, J.L. and Emr, S.D. (1997) Endosome-to-Golgi retrieval of the vacuolar protein sorting receptor, Vps10p, requires the function of the *VPS29*, *VPS30* and *VPS35* gene products. *J. Cell Biol.*, **137** 79-92.

Seaman, M.N.J., McCaffery, J.M. and Emr, S.D. (1998) A membrane coat complex essential for endosome-to-Golgi retrograde transport in yeast. *J. Cell Biol.*, **142** 665-81.

Seeger, M. and Payne, G.S. (1992) Selective and immediate effects of clathrin heavy chain mutations on Golgi membrane protein retention in *Saccharomyces cerevisiae*. *J. Cell Biol.*, **118** 531-40.

Shimada, T., Kuroyanagi, M., Nishimura, M. and Hara-Nishimura, I. (1997) A pumpkin 72 kDa membrane protein of precursor-accumulating vesicles has characteristics of a vacuolar sorting receptor. *Plant Cell Physiol.*, **38** 1414-20.

Simonsen, A., Bremnes, B., Ronning, E., Aasland, R. and Stenmark, H. (1998) Syntaxin, a putative Golgi t-SNARE. *Eur. J. Cell Biol.*, **75** 223-31.

Söllner, T., Whiteheart, S.W., Brunner, M., Erdjument-Bromage, H., Geromanos, S., Tempst, P. and Rothman, J.E. (1993) SNAP receptors implicated in vesicle targeting and fusion. *Nature*, **362** 318-24.

Sosa, M.A., Schmidt, B., Von Figura, K. and Hille-Rehfeld, A. (1993) *In vitro* binding of plasma membrane-coated vesicle adaptors to the cytoplasmic domain of lysosomal acid phosphatase. *J. Biol. Chem.*, **268** 12537-43.

Stack, J.H. and Emr, S.D. (1994) Vps34p required for yeast vacuolar protein sorting is a multiple specificity kinase that exhibits both protein kinase and phosphatidylinositol-specific PI 3-kinase activities. *J. Biol. Chem.*, **269** 31552-62.

Stearns, D.J., Kurosawa, S. and Esmon, C.T. (1989) Microthrombomodulin: residues 310–486 from the epidermal growth factor precursor homology domain of thrombomodulin will accelerate protein C activation. *J. Biol. Chem.*, **264** 3352-56.

Stoorvogel, W., Oorschot, V. and Geuze, H.J. (1996) A novel class of clathrin-coated vesicles budding from endosomes. *J.Cell Biol.*, **132** 21-33.

Storrie, B. and Desjardins, M. (1996) The biogenesis of lysosomes: is it a kiss and run continuous fusion and fission process? *Bioessays*, **18** 895-903.

Sturm, A. and Chrispeels, M.J. (1988) Correct glycosylation, Golgi-processing, and targeting to protein bodies of the vacuolar protein phytohemagglutinin in transgenic tobacco. *Planta*, **175** 170-83.

Takegawa, K., DeWald, D.B. and Emr, S.D. (1995) *Schizosaccharomyces pombe* Vps34p, a phosphatidylinositol-specific PI 3-kinase essential for normal cell growth and vacuole morphology. *J. Cell Sci.*, **108** 3745-56.

Tall, G.G., Hama, H., DeWald, D.B. and Horazdovsky, B.F. (1999) The phosphatidylinositol 3-phosphate binding protein, Vac1p, interacts with a Rab GTPase and a Sec1p homolog to facilitate vesicle-mediated vacuolar protein sorting. *Mol. Biol. Cell.*, **10** 1873-89.

Tanchak, M.A. and Fowke, L.C. (1987) The morphology of multivesicular bodies in soybean protoplasts and their role in endocytosis. *Protoplasma*, **138** 173-82.

Tanchak, M.A., Rennie, P.J. and Fowke, L.C. (1988) Ultrastructure of the partially coated reticulum and dicytosomes during endocytosis by soybean protoplasts. *Planta*, **175** 433-41.

Tooze, J. and Hollinshead, M. (1991) Tubular early endosomal network in AtT20 and other cells. *J. Cell Biol.*, **115** 635-53.

Traub, L.M., Bannykh, S.I., Rodel, J.E., Aridor, M., Balch, W.E. and Kornfeld, S. (1996) AP-2-containing clathrin coats assemble on mature lysosomes. *J. Cell Biol.*, **135** 1801-14.

Trowbridge, I.S., Collawn, J.F. and Hopkins, C.R. (1994) Signal-dependent membrane protein targeting in the endocytic pathway. *Annu. Rev. Cell Biol.*, **9** 129-61.

van Deurs, B., Holm, P.K., Kayser, L., Sandvig, K. and Hansen, S.H. (1993) Multivesicular bodies in Hep-2 cells are maturing endosomes. *Eur. J. Cell Biol.*, **61** 208-24.

van Deurs, B., Holm, P.K., Kayser, L. and Sandvig, K. (1995) Delivery to lysosomes in the human carcinoma cell line, Hep-2, involves an actin filament-facilitated fusion between mature endosomes and pre-existing lysosomes. *Eur. J. Cell Biol.*, **66** 309-23.

van Deurs, B., Holm, P.K. and Sandvig, K. (1996) Inhibition of the vacuolar H^+-ATPase with bafilomycin reduces delivery of internalized molecules from mature multivesicular endosomes to lysosomes in Hep-2 cells. *Eur. J. Cell Biol.*, **69** 343-50.

Vitale, A. and Raikhel, N.V. (1999) What do proteins need to reach different vacuoles? *Trends Plant Sci.*, **4** 149-55.

Vitelli, R., Santello, M., Lattero, D., Chiarello, M., Bifulco, M., Bruni, C.B. and Bucci, C. (1997) *J. Biol. Chem.*, **272** 4391-97.

Wang, H., Frelin, L. and Pevsner, J. (1997) Human syntaxin 7: a Pep12p/Vps6p homolog implicated in vesicle trafficking to lysosomes. *Gene*, **199** 39-48.

Weber, E. and Neumann, D. (1980) Protein bodies, storage organelles in plant seeds. *Biochem. Physiol. Pflanzen*, **175** 279-306.

Wink, M. (1984) Evidence for an extracellular compartment of plant cell suspension cultures: the cell culture medium. *Naturwissensch.*, **71** 635-37.

Wichmann, H., Hengst, L. and Gallwitz, D. (1992) Endocytosis in yeast: evidence for the involvement of a small GTP-binding protein (Ypt7p). *Cell*, **71** 1131-42.

Wurmser, A.E. and Emr, S.D. (1998) Phosphoinositide signaling and turnover: PtdIns(3)p, a regulator of membrane traffic, is transported to the vacuole and degraded by a process that requires lumenal vacuolar hydrolase activities. *EMBO J.*, **17** 4930-42.

Zhang, G.-F., Driouich, A. and Staehelin, L.A. (1996) Monensin-induced redistribution of enzymes and products from the Golgi stacks to swollen vesicles in plant cells. *Eur. J. Cell Biol.*, **71** 332-40.

Zheng, H., Fischer von Mollard, G., Kovaleva, V., Stevens, T.H. and Raikhel, N.V. (1999) The plant vesicle-associated SNARE, AtVTI1a, likely mediates vesicle transport from the *trans*-Golgi network to the prevacuolar compartment. *Mol. Biol. Cell*, **10** 2251-2264.

Zushi, M., Gomi, K., Yamamoto, S., Maruyama, I., Hayashi, T. and Suzuki, K. (1989) The last three consecutive epidermal growth factor-like structures of human thrombomodulin comprise the minimum functional domain for protein C-activating co-factor activity and anticoagulant activity. *J. Biol. Chem.*, **264** 10351-53.

Index